Risk, Resilience and Reliability Analysis for Sustainable Management

Risk, Resilience and Reliability Analysis for Sustainable Management

Editors

Esmaeil Zarei
Samuel Yousefi
Mohsen Omidvar

MDPI

Basel • Beijing • Wuhan • Barcelona • Belgrade • Novi Sad • Cluj • Manchester

Editors

Esmaeil Zarei
Embry-Riddle Aeronautical University
Prescott, AZ, USA

Samuel Yousefi
University of British Columbia
Kelowna, BC, Canada

Mohsen Omidvar
Bushehr University of Medical Sciences
Bushehr, Iran

Editorial Office
MDPI
St. Alban-Anlage 66
4052 Basel, Switzerland

This is a reprint of articles from the Special Issue published online in the open access journal *Sustainability* (ISSN 2071-1050) (available at: https://www.mdpi.com/journal/sustainability/special_issues/Risk_Resilience_and_Reliability_Analysis).

For citation purposes, cite each article independently as indicated on the article page online and as indicated below:

Lastname, A.A.; Lastname, B.B. Article Title. *Journal Name* **Year**, *Volume Number*, Page Range.

ISBN 978-3-0365-8758-5 (Hbk)
ISBN 978-3-0365-8759-2 (PDF)
doi.org/10.3390/books978-3-0365-8759-2

Contents

About the Editors

Esmaeil Zarei

Esmaeil Zarei has been an Associate Professor of Safety Science at Embry-Riddle Aeronautical University (ERAU) since October 2022. He has also been affiliated with the Centre for Risk, Integrity, and Safety Engineering (C-RISE) at Memorial University, Canada, since 2020. His academic journey began as a faculty instructor from 2012 to 2014 and an Assistant Professor from 2017 to 2020 at the Department of Occupational Health and Safety Engineering in Iran. Before academia, he worked as a Safety Engineer for three years at the Iranian Petroleum Health Organization. Throughout his career, Esmaeil has held various leadership roles, including Founder and Director of Safety, Occupational Health Labs, and Student Research Center. He also served as an Executive Manager in the Industry Innovations and Partnerships Office. Notably, he was recognized for establishing undergraduate and graduate programs. Esmaeil's research focuses on Safety, Risk, and Human Factors/Reliability analysis, with a particular emphasis on emerging sectors such as the Hydrogen Economy and critical industries such as Aviation and Process Industries. At ERAU's Robertson Safety Institute (RSI), he generates foundational knowledge and applies it to complex sociotechnical systems.

Samuel Yousefi

Samuel Yousefi is currently a Graduate Research Assistant and Doctoral Student at the School of Engineering, University of British Columbia (UBC), Kelowna, Canada. He has focused on developing analytical approaches and decision support frameworks using system modelling and multi-criteria decision-making techniques to address various problems, such as the adoption of new technologies (e.g., blockchain) in sustainable supply chains and logistics systems. He received his B.Sc. and M.Sc. degrees in Industrial Engineering from the Urmia University of Technology, Iran, in August 2012 and September 2014, respectively. After graduation, he worked as a researcher and research assistant for about five years until December 2019. He has working experience in the industry, mostly as a quality assurance specialist and consultant for more than two years as well. He has published more than 35 articles in leading international journals in engineering and management (e.g., the *International Journal of Production Economics, Computers & Industrial Engineering, Energy Policy, Journal of Cleaner Production, and Safety Science*). In addition to organizing several Special Issues in the field of data and decision analytics, he has collaborated with more than 80 international journals and conferences as a reviewer. His main research interests include Operations Management, Supply Chain Management, Risk Analysis, Decision Science, and business intelligence.

Mohsen Omidvar

Mohsen Omidvar is currently working as Department Chair and Assistant Professor of Health, Safety, and Environment Engineering (HSE) at Bushehr University of Medical Sciences, Bushehr, Iran. His advanced research is primarily concerned with process safety; risk management; resilience engineering; and human error analysis, and he has published more than 30 peer-reviewed papers.

Preface

Due to the increasing importance of considering socio-environmental issues in the modern day, modern critical systems should pursue sustainability-related objectives while meeting operational goals. Furthermore, sociotechnical systems working with complex operations represent dynamic complexity, relative ignorance, and intractability, which entail interactive and dependent social elements and organizational and human activities. Considering the influence of and the relationship between operational concerns such as risk, reliability, and resilience, and strategic concerns such as sustainability, helps managers and policy-makers make more reliable and efficient decisions in a wide range of engineering and management systems. This book aims to extend the available knowledge on the extent and quality of such interactions and discusses how one can ensure that reliability and resilience are maintained in dynamic conditions to achieve sustainable operation. Under these conditions, most existing engineering and management systems in various industries (e.g., food, mining, and construction) should be required to undergo adaptive improvements to become more resilient to potential future typical or extraordinary circumstances. This book also sheds light on the challenges and future directions which the research community should focus on and introduces various approaches and applications to develop more sustainable and resilient solutions in both engineering and management systems.

This book is a collection of 11 articles demonstrating the recent developments in risk, resilience, and reliability analysis for sustainable management. Several novel analytical approaches and fascinating applications of risk, resilience, and reliability analysis related to supply chain management, project and construction management, health, safety, and environmental management, sustainable food production, and safety engineering are introduced in this book.

Esmaeil Zarei, Samuel Yousefi, and Mohsen Omidvar

Editors

sustainability MDPI

Review

Systems Thinking Accident Analysis Models: A Systematic Review for Sustainable Safety Management

Mahdieh Delikhoon [1], Esmaeil Zarei [2], Osiris Valdez Banda [3], Mohammad Faridan [4] and Ehsanollah Habibi [1,*]

1 Department of Occupational Health and Safety Engineering, Faculty of Health, Isfahan University of Medical Sciences, Isfahan 81746-73461, Iran; mdelikhon@yahoo.com
2 Centre for Risk, Integrity and Safety Engineering (C-RISE), Faculty of Engineering and Applied Science, Memorial University of Newfoundland, St. John's, NL A1B 3X5, Canada; ezarei@mun.ca
3 Research Group on Maritime Risk and Safety, Department of Applied Mechanics, Aalto University, 00076 Espoo, Finland; osiris.valdez.banda@aalto.fi
4 Environmental Health Research Center, Department of Occupational Health and Safety at Work Engineering, Lorestan University of Medical Sciences, Khorramabad 68138-33946, Iran; mfereidan@yahoo.com
* Correspondence: habibi@hlth.mui.ac.ir; Tel.: +98-913-113-1623

Abstract: Accident models are mental models that make it possible to understand the causality of adverse events. This research was conducted based on five major objectives: (i) to systematically review the relevant literature about AcciMap, STAMP, and FRAM models and synthesize the theoretical and experimental findings, as well as the main research flows; (ii) to examine the standalone and hybrid applications for modeling the leading factors of the accident and the behavior of sociotechnical systems; (iii) to highlight the strengths and weaknesses of exploring the research opportunities; (iv) to describe the safety and accident models in terms of safety-I-II-III; and finally, to investigate the impact of the systemic models' applications in enhancing the system's sustainability. The systematic models can identify contributory factors, functions, and relationships in different system levels which helps to increase the awareness of systems and enhance the sustainability of safety management. Furthermore, their hybrid extensions can significantly overcome the limitations of these models and provide more reliable information. Applying the safety II and III concepts and their approaches in the system can also progress their safety levels. Finally, the ethical control of sophisticated systems suggests that further research utilizing these methodologies should be conducted to enhance system analysis and safety evaluations.

Keywords: accident analyses; AcciMap; STAMP; FRAM; safety-III; sustainable system

Citation: Delikhoon, M.; Zarei, E.; Banda, O.V.; Faridan, M.; Habibi, E. Systems Thinking Accident Analysis Models: A Systematic Review for Sustainable Safety Management. *Sustainability* **2022**, *14*, 5869. https://doi.org/10.3390/su14105869

Academic Editor: Victoria Gitelman

Received: 23 March 2022
Accepted: 9 May 2022
Published: 12 May 2022

1. Introduction

The protection of human resources and environments along with reducing the risk of losses are the major concerns of system managers all over the world. Safety management has also shown to have a vital role in establishing the sustainable progress of a system [1]. The concept of sustainability refers to the effective management of the environment in short and long-term procurement in order to ensure that resources and social provisions meet the needs of future generations. It also takes into account the potential for long-term risk reduction [2].

In that regard, establishing a sustainable organization requires proactively managing risk in an integrated way to decrease unplanned chains of events and losses—particularly, in order to promote the quality of performance and productivity. On the other hand, one of the key elements for achieving sustainability, improving safety, and maintaining low incident rates is to perform a comprehensive, accurate and detailed analysis of an organization's incidents and accidents [3].

An accident is defined as an unplanned chain of events resulting from inadequate risk control or the application of safety constraints that causes injury, illness or damage to people, property, the environment, or credit [4]. The ILO states that occupational accidents or illnesses cause the death of one worker every 15 s. It also declares that 153 accidents occur due to work practices at the same time, and 6300 workers die every day from work-related illnesses or accidents at work.

The ILO also declares that shortcomings in taking appropriate health and safety measures at work lead to an economic burden equal to 4% of global GDP per year [5–7]. Illnesses and accidents induced by work activities have also proved to affect economic growth much more than several other common illnesses and disorders, such as cancer, cardiovascular disorders, Alzheimer, and HIV/AIDS [6]. It is worth noting that the socio-economic costs of accidents are significantly higher than their financial ones and such costs cannot be easily estimated. This highlights the importance of risk assessment, reliability analysis and modeling of the causation of the accidents [8,9].

Occupational accidents usually occur due to several factors, such as human factors, job design, environmental and economic conditions, lack of experience, long working hours, fatigue, sleep disorders, noise, physical pressures, workload, role ambiguity and conflicts, and demographic characteristics and lifestyle [10–18].

Some studies suggest that the human factors contribute to approximately 80% of occupational accidents and that human error is a main contributing factor for workplace accidents [13,19].

Most industrial facilities are complex engineered sociotechnical systems where the social, human, organizational, and technical factors are considered in their design and structure. Internal and external interactions between physical equipment and people also exist in such facilities [20]. In other words, with the increasing advancement of technology and complex engineering systems, accidents are not simply the result of a minor failure. Although they emerge from complex interactions between system components, they are usually related to latent factors such as human error, technical failures, external factors and abnormal process situations [21]. Due to the complexity of modern industrial technological systems, the risk of accidents involving such systems has become more concerning [22,23]. The continual recurrence of catastrophic events such as Bhopal, Piper Alpha, BP Texas City, Bunce field, and Gulf of Mexico, as complex technological systems, has contributed to serious losses and raised social and legislative stakeholders' concerns over the last decades [24,25]. The accident in the Gulf of Mexico highlighted some critical issues in system safety and common thinking about defining the causality of accidents. It also revealed that the linear models are incapable of determining the interaction between the leading factors, and, despite their wide use in accident analysis techniques, do not enable systems to reach the zero-accident target [24]. Therefore, as highlighted by Hollnagel et al. (2006), in order to control the adverse consequences of these accidents, it is essential to know the background, future complications, control measures, and resources that can be achieved through using accident modeling strategies [26]. In other words, accident models are mental models upon which it is possible to understand how and why accidents occur in terms of causality. They are also used as a means of risk assessment to determine appropriate safety measures for enhancing the stability of systems [27,28]. Therefore, these concepts have been promoted in recent years as effective tools in enhancing safety and preventing accidents through applying proactive rather than reactive methods. The most important step necessary to achieve this goal is to enhance awareness about the technological, organizational and human factors affecting the system [3].

Various classifications of the accident models have therefore been introduced and evaluated in the literature [29–38]. Accident models are usually divided into sequential, epidemiological and systemic models [39]. While the focus of the first two models is on the linear investigation of accident causality, the systemic models mainly consider the interaction among the major system components (technical, human, organizational, and managerial). In other words, the interrelations among the causes of the accident according to the systematic model are non-linear and include multiple feedback loops [40,41].

Nonetheless, the application of these advanced models and their associated methods have already been expanded and criticized at a number of different levels [42,43]. Therefore, it is timely to systematically subject the studies of accident analysis models to a thorough review. Furthermore, much of the research in this field, up to now has focused on the review of the specific methodologies (e.g., AcciMap) or distinctive accident models [44]. Hence, we believe that broad review on systemic analysis methods should be conducted to fully provide ample indications about how they can be more applicable to conduct practical analysis as well as preventing the accidents.

Therefore, the principal objectives of this systematic review were defined as follows: First, an overview of the papers that had applied the methodologies of AcciMap, STAMP, FRAM in their analyses to synthesize the theoretical and experimental findings—particularly for recognizing the main research flows. Second, to examine the application of the mentioned approaches combined with other methods for modeling causal factors of the accidents and the behavior of sociotechnical systems. Third, highlighting the advantages and disadvantages of these approaches to explore the opportunities for research and practice. Fourth, to describe the safety and accident models in terms of safety-I ("as few things as possible go wrong") and safety-II ("as many things as possible go right"), as well as safety-III ("freedom from unacceptable losses"). To describe these three paradigms of safety in detail: In the safety-I paradigm, accidents occur due to system failures and performance malfunctions, according to which safety management is reactive because the response is to the time that events occurred and any contributory factors were identified. In the safety-II paradigm, the system is adjusted to respond to events and to eliminate the problems before they occurred and its effort is to make functions "go right". Based on this concept, safety management is proactive. The safety-III concept represents that inadequate hazards control is the main cause of accidents. In this paradigm, safety management does not regard the identification of the root cause. Instead, it investigates the reason for control malfunctions, preventing accidents, and system performance auditing [4,45].

The final objective of this work was to investigate the impact of employing the systemic models for enhancing the systems sustainability.

Accordingly, the following research questions were defined for this review:

What research flows in sociotechnical systems have been examined from the perspective of these three systemic accident models?
How has previous research contributed to the three systemic accident models and what are the needs and shortcomings in these studies?
How are the current problems best dealt with and what challenges do accident analysts face?
What is the role of systemic accident models in improving system sustainability?

1.1. *Evaluation of Accident Models*

Generally, there are three categories of accident models: sequential, epidemiological and systemic models [46]. The classification of these models and their subset methods are illustrated in Figure 1.

1.2. *Sequential Accident Models*

According to these models, the leading cause of an accident is a linear sequence of events. In other words, the causes of these accidents stem from a series of separate events that occur in a specific chronological order. Most of the traditional accident models such as Domino theory, CCA, FTA, ETA, and FMEA are classified within this type. Domino theory is different from domino effect as the second involves extensive resonance creating events in the process and chemical industries [39,47]. This category, however, suffers from some limitations in determining the contributing factors of the accidents in the complex sociotechnical systems that were developed in the second half of the twentieth century [48]. Accidents have always proved to have more than just one single cause. Thus, the need for more robust methods of overcoming the limitation of sequential models that explain the underlying causes of accidents lead to the development of epidemiological models in the 1980s [49].

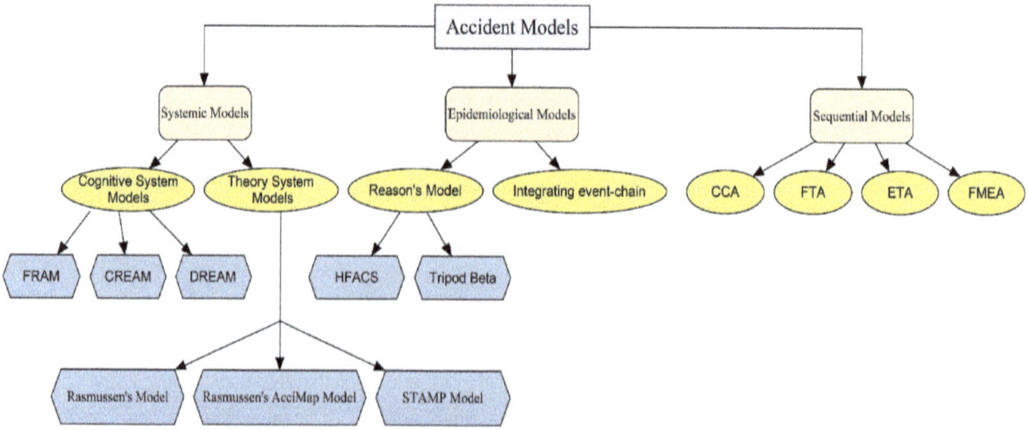

Figure 1. Accident model classifications and subset methods [29,36,37].

1.3. Epidemiological Accident Models

In these models, accidents are considered to be caused as a combination of "latent" factors such as management functions and organizational culture, as well as "active failures" [50]. Reason's Swiss cheese model is one of the subsets of this category which regards the critical role of organizational safety and the contributory factors of failures of the relevant protective barriers. In this model, the human errors that directly interact with the regulation of the process or technology are the first leading factors for inducing the accidents [51]. In the Reason's model, the dynamics of the accident causation states that failures are transient between barriers, and holes (latent errors) are moving continuously [52]. Bow-tie [53], Threat and Error Management [54], and Tripod [55] are other examples of the models in this category where the use of protective barriers compatible with probable failures is common. The epidemiological models are static and follow the causal pattern in sequential models. Therefore, it may be difficult to also find the explicit factors or critical causes [48,56]. In contrast, the interactions among organizational factors which lead to accidents in the sociotechnical system are more complex and dynamic than the sequential and epidemiological models [57].

1.4. Systemic Accident Models

The causes of new accidents in complex sociotechnical systems do not necessarily result from simple defects, and leading factors for accidents occurring in such systems are relevant to the interactions among the system components [21,58,59]. According to the sociotechnical theory, since human and social identities are integral parts of the technical systems, an organization can fulfil its objectives by optimizing the technical as well as the social aspects of the system rather than by merely optimizing the technical aspects of the system [60–62]. Therefore, in order to investigate the causes of accidents in sociotechnical systems, it is necessary to understand the interactions among the principal aspects (e.g., social, technical, human, and organizational) of the system.

Modern sociotechnical systems have drastically modified human activities over the past decades. One of the most noticeable examples of such a shift is the transition from predominantly manual tasks to more cognitive and knowledge-based ones. In fact, various failures and safety issues have already emerged and most of the accidents in such systems cannot be analyzed sufficiently using traditional accident models. Therefore, a new model for risk and safety management with the basis of systems theory was also introduced as a systemic accident model [48].

In systemic models, the study of accidents is based on the uncommon interrelationship and unusual conditions related to accidents. This indicates that there is variability in the system and in order to prevent uncontrollable variability, which is intolerable for the system and leads to an accident, the system performance should be monitored continuously [63]. Some notable systems-modeling approaches of this type include STAMP [39], AcciMap, the hierarchical sociotechnical framework [64] and FRAM [48]. Theoretically, these models are similar; however, their development, methodology, and outputs might differ considerably. These models are described further down.

1.4.1. Rasmussen's Sociotechnical Framework and AcciMap Accident Analysis Technique Overview

The concept of Rasmussen's framework for risk management is based on the control theory, in which the control of system processes is a main concern of safety. In other words, in this framework view, accidents in the sociotechnical systems result from a control problem. Rasmussen's structure of risk management in the sociotechnical systems consists of several levels, from the legislator to the operator (top-down) of the system, respectively (Figure 2). This framework is the basis for the AcciMap accident analysis model [64,65]. Accordingly, the main approach in the AcciMap is the analysis of causal chains of events in the selected accident scenarios using a cause-consequence chart with the aim of analyzing the control layers of the sociotechnical system at the lowest level. On the other hand, in order to extend the cause-consequence chart, a vertical analysis of the mapped accident contributing factors at the hierarchical levels must be conducted [66].

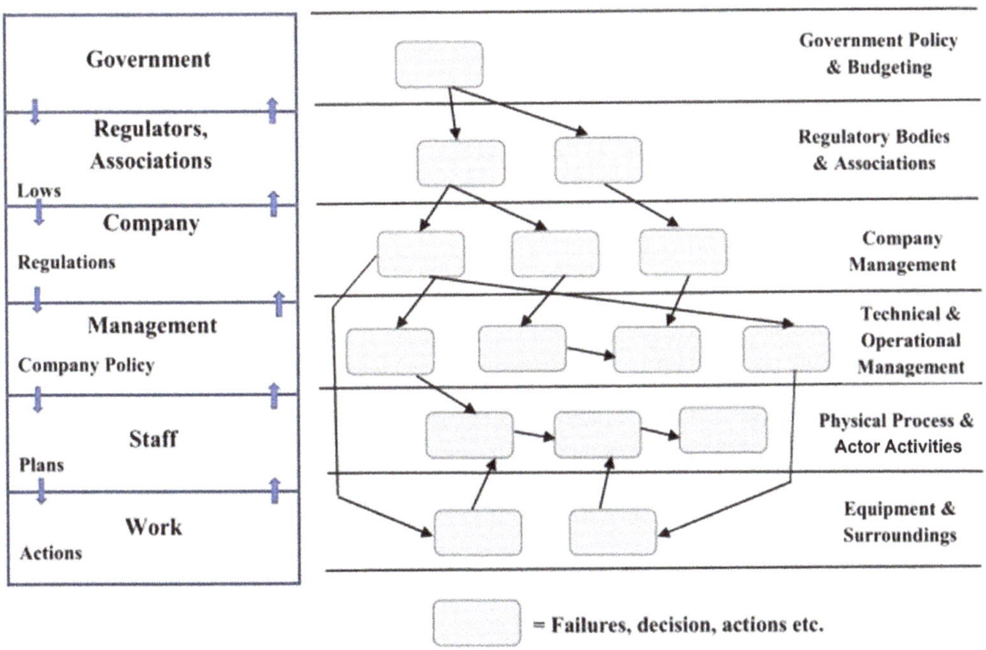

Figure 2. Rasmussen's Framework and AcciMap technique [65].

1.4.2. STAMP Analysis Approach Overview

STAMP is a new non-linear system-based accident theory established by Leveson (2011). According to this model, system components are interrelated and enforced by the specific safety constraints [42]. This theory allows for the determination of the dynamics of the interrelationships between system components, as well as a better description of the systems' degree of complexity and technical originality [42].

From the perspective of STAMP, the system is described as a control structure that includes control and feedback loops, and the superior level controls the lower level by applying safety restrictions. Controls and feedbacks are transmitted through every control loop via a collection of relative channels (Figure 3). In the view of organization, controls can be over economic practices and priorities, as well as feedback on reportages and requisitions [42]. Accidents, according to STAMP, are caused by inadequate system components controls which contribute to unsafe component interactions and failures [28].

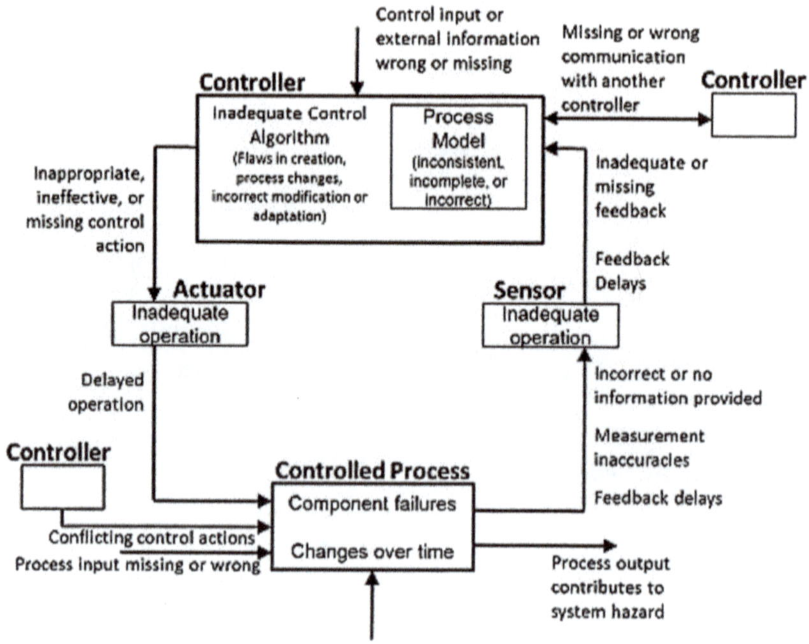

Figure 3. General factors in unsafe control used to create STAMP model [42].

STPA and CAST are the two methodologies to be extended and developed from the general STAMP theory. These techniques are usually employed in the analysis of hazards and accidents, respectively [42].

1.4.3. FRAM Analysis Approach Overview

FRAM was first presented as a tool for analyzing accidents in complex systems—particularly, with the aim of evaluating how the functions of a system can interact and trigger accidents. The term "function" refers to the tasks, activities, or components that a system performs or employs in order to achieve a goal [67]. FRAM enables the analysis of the complicated non-linear relationships among functional activities. It also allows for evaluation of the way that functions interact to induce an accident [48]. FRAM can also be utilized for accident analysis and risk assessment based on the operational perspective and the unpredictability of functions [68].

This model has been applied in accident analysis to determine the cause of the accident by documenting typical system performances and their variability in order to manage them. When the method is used with the aim of risk assessment, it examines how variability in one function can affect the performance of other related functions, detects the disruptive variability and finally, controls and minimizes risk levels [69]. Figure 4 demonstrates a schematic of FRAM [48,67] in which each system function is represented by a hexagonal shape with six aspects, representing I as an input, O as an output, P as a precondition, R as a resource, T as a time, and C as a control. Analyzing system performance to develop models and conceptualize the variability and resonance according to FRAM approach can also be performed using the computer-based tool 'FMV'; http://functionalresonance.com/ 14 June 2021) [70].

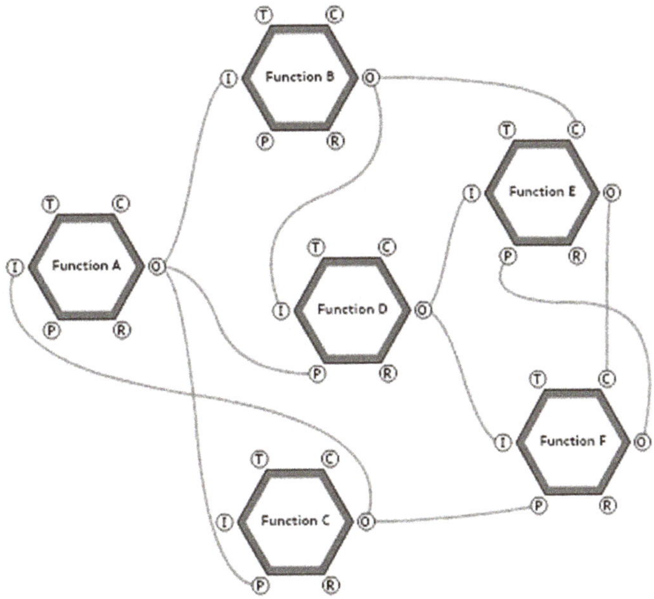

Figure 4. General model of FRAM [71].

2. Materials and Methods

2.1. Search Strategy

We began this investigation by formulating the title as a query in order to locate all papers published in this context. The following question, 'how many articles have been published describing the application of systemic accident analysis models (AcciMap, STAMP, and FRAM)?' was then taken into account and according to the lines of our search, several selected keywords and limiters were used as well: ("STAMP" OR "CAST" OR "STPA" OR "FRAM" OR "AcciMap" OR "Rasmussen's risk management framework" OR "Rasmussen's framework" OR "systemic accident models") AND ("accident analysis" OR "risk assessment" OR "hazard analysis"). Published studies from five international databases (Scopus, Medline/PubMed, Web of science, Science Direct and Google Scholar) were searched. When scanning databases, our search was limited to articles published in the English language with publication dates from 1 January 1990 to 1 October 2021.

2.2. Research Screening and Eligibility Criteria

In order to select the studies for inclusion in the current systematic review, we used the Preferred Reporting Items for Systematic Reviews and Meta-analysis (PRISMA) methodology. In the identification phase of this method, after downloading the relevant studies, the duplicates, non-English language research, review articles, letters, and conference proceedings were excluded from our list. Following that, the titles and abstracts of the papers were examined in order to identify those that were particularly relevant. For more screening, full text articles were then retrieved.

The eligibility of the selected papers was then assessed according to predefined inclusion and exclusion criteria.

The inclusion criteria were as follows: original articles that used the AcciMap, STAMP, and FRAM methodologies in their analyses, studies conducting a systemic analysis with the goals of improving the system safety and resilience through system redesign, and articles that combined other accident analysis methods with systemic methods.

Studies were excluded if they had different data sources, study dates and used additional analyses with either incomplete or insufficient coverage of the systemic models in their methodologies.

In cases where it was not possible to select suitable papers according to the defined criteria, we studied the full text of the paper and if appropriate, it was selected. Finally, we reviewed the full text of the selected articles and extracted information and included them in the tables with the relevant titles.

3. Results

3.1. Descriptive Results

According to the study plan, 527 records were collected, as shown in Figure 1. Prior to performing screening, 125 duplicates and non-English papers, along with four letters and conference proceedings were excluded from the first list. The anthology of results was then reduced to 398. It should also be noted that this study focused on the research literature that were consistent with our methodology, study goals, and method of application. Additionally, papers that combined alternative methodologies with systemic models to improve their findings were considered. We excluded 167 studies after an examination of the remaining abstracts in terms of relevance. A more thorough analysis of the selected publications' methods and results sections resulted in the elimination of a further 64 papers. Eventually, 63 papers were selected for conducting the analyses in the current study. The results of the search are depicted in the PRISMA flow diagram (Figure 5). Furthermore, as shown in Figure 6, the frequency of 63 systemic methods studies were presented. Accordingly, among 25 AcciMap studies, seven papers were published from the years 2003 to 2010 and 18 works were published from 2011 to 2021. This frequency for 16 STAMP studies in similar ranges was 1 and 16 with a higher frequency in 2018. For 22 FRAM studies, the frequency was 1 and 21, with a higher frequency in 2021. Overall, considering the trend of using these methods, the number of articles increased from 2016, which indicated their capability to understand the behaviors of complex sociotechnical systems.

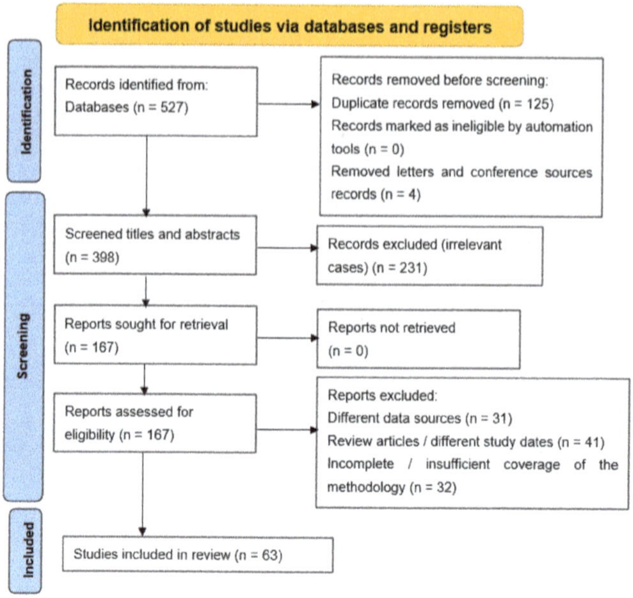

Figure 5. PRISMA 2020 flow diagram of the structured literature review.

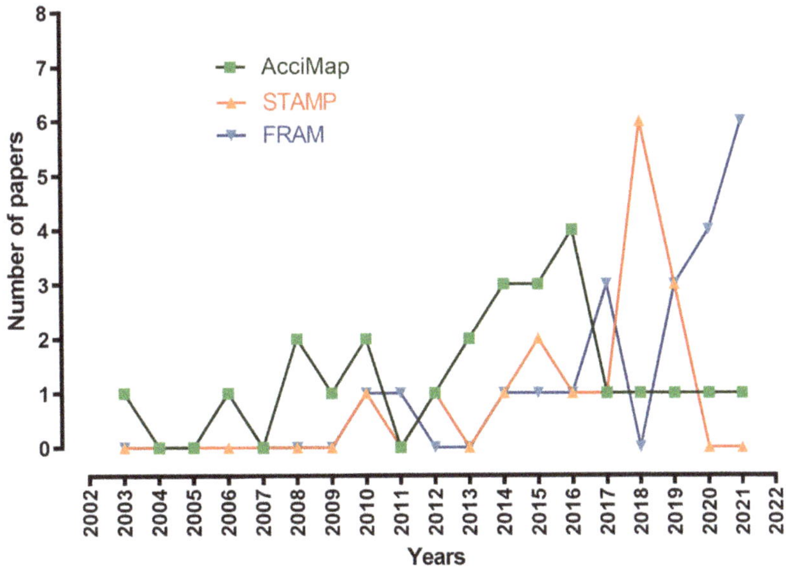

Figure 6. The frequency of use of per techniques in recent 20 years by researchers.

3.2. *Key Findings of AcciMap Studies*

As a result of searching the aforementioned databases, we found 25 publications that employed the AcciMap approach to analyze an incident or accident and conduct a safety or risk assessment. Of the AcciMap investigations, 44% (11 studies) and 24% (6 studies) were, respectively, undertaken in the transportation and public health sectors.

Two of the six studies found in the context of healthcare systems had considered the complex interactions among all levels of a complex sociotechnical system using the logic gates or decision trees incorporated with AcciMap. This was to particularly demonstrate the priority and sequence of determined causality for designing public policies by reducing the risk levels in complex systems and investigating the disasters and outbreaks related to the water distribution systems in Canada.

They found a distinction between low-level physical and individual variables, and similar causes of events at the governmental and regulatory factors level [72,73].

Two additional studies conducted in the United Kingdom assessed the level of safety and examined the major events and factors contributing to outbreaks in the food production industry in order to proactively prevent accidents and improve the safety management system [74,75].

One study within the scope of public health examined the factors that contribute to infection outbreaks and provided strategies and interventions for limiting and preventing their occurrence [76]. Additionally, four studies employed AcciMap to connect risk management, accident analysis, and learning from accidents in the context of outdoor recreation. Two studies, on the other hand, utilized a hybrid method to better support the implementation of the AcciMap technique. One of these studies used AcciMap in combination with the CWA to identify accident-related variables and describe conditions within which the accidents occurred. CWA also specified constraints that affect system behavior [77]. Another study used AcciMap in conjunction with the fuzzy ISM and Matrix of Cross Impact Multiplications in which fuzzy ISM was used with the aim of determining the interactions and the hierarchical representation of contributing factors, and Matrix of Cross Impact Multiplications was implied for categorizing and determining the most important factors [78]. Furthermore, two other studies utilized a coding template for the AcciMap technique to quantitatively assess the relationships among accident causes based on the reported frequency of incidents [43,75]. In this regard, another paper was allocated the codes to accident contributing factors to create a contextual view of the event. They demonstrated the time and place in which decisions and responses were performed [79]. Akyuz et al. applied ANP methods to determine the priority of accident related factors via weighting factors [80]. Other publications were performed in the contexts such as marine, disaster response, navigation, civil engineering, systems thinking principles, and healthcare-related incidents. Overall, AcciMap was used in studies with six hierarchical levels developed based on Rasmussen's (1997) framework. A few works used the five levels of AcciMap and one depicted the contributing factors in the outcome level [81–83]. Table A1 outlines the details of these works (Appendix A).

3.3. Key Findings of STAMP Studies

STAMP was found to be the subject of 16 studies, which are listed in Table A2.

These studies were carried out in a variety of contexts and with multiple objectives. Three of the reviewed studies employed this methodology for the risk assessment and identified abnormal system behaviors and potentially unsafe situations in terms of STAMP-STPA. The results from the risk analysis were also utilized to improve and update situational awareness and to prevent accidents through the introduction of safety limitations [84–87]. Moreover, with the aim of accident analysis, some studies used another form of STAMP (CAST methodology) to model and investigate the control deficiency, flaws or missings in a similar way, based on Leveson's (2004) taxonomy, and suggesting corresponding adjustments to increase system sustainability [84–86,88–93].

Additionally, some studies also utilized STAMP in conjunction with other approaches to extend their research beyond the control flaws to fundamental patterns of failures and their implications for the organization's compliance and direction of functions [88,89,91,94]. For example, Lower et al. used HFACS combined with STAMP to improve the accident analysis. This framework incorporated the HFACS levels into a controlling structure of STAMP which can depict the interrelationship between human, technical, and the environmental factors and can be used for hazard, safety and accident analysis [95]. Another study used STAMP in conjunction with SD to provide an integrated framework for analyzing and elaborating on the dynamics and interconnections of human error [86]. Generally, it is clear from reviewing studies that they analyzed and investigated the existing components of system structure(s) and did not elaborate on designing systems by relying on safety properties and system resilience. Table A2 provides a summary of the studies (Appendix B).

3.4. Key Findings of FRAM Studies

The FRAM model was utilized in 22 studies and in terms of the contexts, aviation accounted for 28% (6 studies) of the total reviewed papers. The model was also used in other contexts such as the construction and transportation industries, hydrocarbon release accidents, public health and chemical process, and hazard and resilience analysis for complex sociotechnical systems and emergency response systems. Risk analysis, accident analysis, comparison with other approaches, and hybrid usage of FRAM combined with other methods were among the main objectives of the papers that employed FRAM methodology. According to Table A3, 16 studies were conducted with the objective of conducting prospective analyses of risk, hazard, safety, and system behavior as a result of complex interactions between sociotechnical system components. Additionally, they provided controlling strategies for minimizing the risk of function variability or functional resonance in order to improve system operation resilience and sustainability.

A group of researchers considered integrating FRAM with other methodologies such as MCs, GMTA, fuzzy logic, and BN to conduct quantitative and more accurate analyses for increasing the methods' applicability [94,96–100]. For instance, MCs was applied for the quantification of performance variability and the determination of critical couplings through allocating score and probability distribution to each variability [96]. In addition, the hybrid framework including TASM and the combination of FRAM and GMTA was applied in aviation settings to provide the concept maps [101]. In another research, Q-FRAM provided quantitative concepts in which key indicators of performances were excluded from FRAM and allocated to four concepts of resilience, including anticipate, response, monitor and learn via an MSDM hierarchical approach [97]. Fuzzy logic was also used by Slim et al., in which the performance couplings were weighted and variability of the performances was evaluated with the aim of an aircraft de-icing simulation [98].

Furthermore, two retrospective studies employed FRAM-AHP to evaluate the accidents by determining the main and important criteria to identify the essential functions and relationships between them. These papers would ultimately offer recommendations for enhancing the system operation sustainability [102,103]. Table A3 summarizes the findings of these investigations (Appendix C).

4. Discussion

The primary goals of this work were to provide an overview of the papers that had employed AcciMap, STAMP, and FRAM methodologies in their analyses—particularly, in order to: identify the major research flows in terms of the accident analysis, risk assessment and safety analysis of sociotechnical systems; to examine the applicability of hybrid methods for modeling the behavior of accidents and sociotechnical systems; to highlight the advantages and disadvantages of these approaches; to describe safety and accident models in terms of safety-I and safety-II as well as safety-III; and to investigate the impact of using system models for enhancing the systems' sustainability.

4.1. The Main Research Flows on Three Systemic Approaches

4.1.1. AcciMap Approach

According to the findings of the related studies, the advantages of the AcciMap application for accident analysis are its ease of use, capability of recognizing factors related to sociotechnical systems, and time-saving nature. Additionally, the most common accident factors at the system's lower levels were "physical practice and operator's function" as well as "instrument and environment". Therefore, it can be concluded that the AcciMap approach in almost all studies can effectively identify the leading factors of the accident, especially at higher levels.

This would also highlight the role of regulatory and governmental bodies in creating a safe environment, demonstrate the interaction of factors at different levels of the system and recommend methods by which the system might be used to prevent accidents proactively [79].

4.1.2. STAMP Approach

The results of related studies showed a similar pattern in which control deficiencies such as "management and the operational process" and the "company" were identified at lower levels of the system.

These contributory factors may be due to the information available to analysts instead of a fixed feature of the accident's leading factors. However, the detected factors at higher levels of the system indicate that controllers at these levels employ strategies to design and provide interventions on human and technical factors which highlight the need for accident prevention.

4.1.3. FRAM Approach

A search of the literature revealed that this method has been used for analysis in construction, transportation, hydrocarbon release accidents, public health, and chemical process sectors. In the FRAM approach, the variability of depicting normal functions is used to determine the emergent behavior of hazards and there is no need for an accident occurrence [103,104]. The model's outputs showed that FRAM has a complicated methodology and procedure and is a challenging model to interpret. As a result, researchers employed novel and innovative techniques to circumvent this problem [99,101,105]. All reviewed studies which used each of the three mentioned methods also identified multiple contributory factors, functions, and relationships.

4.2. Hybrid Use of the Systemic Methods

In this section, we discuss the utilization of systemic techniques integrated with other methods or expanding to the larger methodology as the qualitative and (semi) quantitative approaches. According to the theory of systemic analysis approach, these methods describe and analyze the sociotechnical systems qualitatively. However, a shortcoming is that these methods are only qualitative in nature, particularly due to focusing on constructing a perception model [60]. QRA has shown to have a significant role in effective risk control, as well as addressing the issue of a qualitative structure of systemic analysis methods, mainly in complex sociotechnical systems. Several studies have already proposed quantifying these methods using fuzzy AHP, SME and the MCs and MCMCs methods as the compliment [96,99,106]. The proposed method represents the system more realistically with a quantitative value [100].

MCs allows for reliability indicators to be estimated using real processes and random system behavior simulation in order to make a reality-based scenario by employing a computer-based model. One of the most important applications of MCs is in risk and reliability analysis in the engineering systems. The outputs from MCs simplifies the estimation of the PoFs [107]. According to our literature review, some studies have utilized FRAM and MCs for the enhancement of the traditional safety assessment techniques. For example, Patriarca et al. (2017) used MCs for the first time in their work for quantifying the performance variability in a FRAM model. Their main objective was highlighting the critical functions and links among these functions as well as facilitating the process of safety analysis [96]. Similarly, Kaya et al. integrated MCs as well as a criticality matrix with the FRAM to study how they may be used to enhance the quantification of a system-based risk analysis and critical condition evaluation [94]. Kim et al. proposed a layout to apply the FRAM quantitatively in order to perform the risk assessment. Such layout regarded regulations for variability's aggregation and allocated values for functions and their interactions and therefore showed that the system was more realistic [100]. A FRAM-based tool was also developed utilizing AHP to support in decision-making by quantifying the resilience of urban planning systems [97,99,106].

Contrastingly, Slim et al. engaged predictive FRAM combined with Fuzzy logic to generate numerical indicators for a more comprehensible representation of potential performance variability with the aim of an aircraft system simulation [98]. Moreover, the N-K model was recently introduced by Huang et al. (2021) with the aim of quantitative evaluation of the FRAM model. This model uses the theory of risk pulse according to which the severity of functional coupling can be calculated. According to the model, each coupling with a higher frequency of operation is more likely to have an accident and poses a greater risk. It is worth noting that, unlike earlier studies, this model is constructed on historical data and was not affected by subject matter experts [105]. Furthermore, among AcciMap studies, other authors utilized a coding template for the AcciMap technique to quantitatively assess the accident related factors for assessing the level of safety, proactively preventing the accident and improving the safety management system [43,75]. In order to better support the implementation of this method, AcciMap was also used together with the fuzzy ISM and Matrix of Cross Impact Multiplications to determine and classify the interactions and hierarchical structure of the contributory factors of the accident [78]. Moreover, Wang et al. reported that the simultaneous use of the BN method and systemic methods can provide a quantitative correlation between numerical calculation values and the probability of occurrence [108]. Using the SD method, which explicitly highlights the interrelated time processes, integrated with a BN modeling framework (Dynamic Bayesian Network) for assessing and modeling accidents can overcome the limitations [109]. In this regard, Rong et al. used SD modeling in conjunction with STAMP to demonstrate the dynamic processes which lead to the system changes and to generate safety control structures with STAMP [86]. Banda et al. also applied the STAMP and BN for the operational use and design of the safety management system [110]. FRAM was also used along with DBN in another study to quantitatively assess and model the system resilience that helps systems to better adjust to unwanted events and restore from major losses. [99].

In the qualitative manner of developing a wider methodology, AcciMap was employed in conjunction with the CWA that enhanced the identification of the causes of accidents and their relationship with the management and system rules in term of the cultural, economic, and social aspects. CWA also specified constraints that affect the system behavior [77]. Kontogiannis et al. investigated the patterns of organizational breakdowns in accidents using the VSM along with STAMP—particularly, with the aim of creating a link between control flaws and organizational breakdowns [85].

However, another study applied Rasmussen's AH combined with FRAM and provided a new structure of FRAM by functional analysis at the hierarchical layers of the system [104]. Additionally, Studic et al. used a hybrid approach including the TASM, the combination of FRAM and GMTA to conduct a system-based modelling of the safety and to provide concept maps in aviation settings [101].

Hence, using the mentioned methods together with systemic accident analysis models as a compliment can improve the process of analysis by providing more reliable information to decision makers. Therefore, future research should consider the dynamic aspects of complex sociotechnical systems in their analysis and more studies should be performed in the context of the resilience analysis of safety management and system behavior using a systemic approach in a dynamic manner.

4.3. *Advantages and Drawbacks of Systemic Methods*

The field of systemic events and analytical modeling describes the system performance and variation control by establishing connections between functions and components of organizational accidents with multiple causes in line with the human factor at different levels of the company in complex modern technologies [111]. They also highlight the influences and possible effects of an unforeseeable occurrence of complex combinations of events and the study of the interactions which exist among system elements. In the present study, we carefully examined the various literature to present the most reasonable and fair presentation of each method and to remain completely neutral in reviewing each method. Moreover, we indicated that each method can be adapted (the mentioned drawbacks will be addressed). According to the peer reviewed studies [39,69,111–113], the main advantages and drawbacks of the three investigated accidents models are shown in Table 1.

It is worth noting that, in accordance with the control characteristics of systemic accident analysis approaches, the application of social, organizational, and managerial controls, collectively referred to as non-technical controls, should be considered in addition to technical controls. As a result, the issue of accident analysis became even more crucial [113] and the primary concern is how inadequate non-technical controls, in addition to the failures of physical controls, can contribute to the occurrence of an accident.

4.4. *Safety and Accidents Methods in Terms of Safety-I, Safety-II and Safety-III*

"Safety" is commonly defined as the absence of an accident, or a system's ability to ensure that the number of harmful events is kept to a minimum and acceptable level [114]. In other words, the purpose of applying safety is to protect, maintain, and gain access to significant and valuable objectives. As a result, safety and sustainability are inextricably linked or even synonymous, as when a system is unsafe, it cannot be sustainable, and vice versa [3].

Table 1. The main advantages (Yes), and drawbacks (No) of systemic approaches.

Descriptions	AcciMap	STAMP	FRAM
Description of accidents with a single diagram	Yes	No	Yes
Proximal sequence of events and influences	Yes	Yes	Yes
Simplicity of identifying the causes of accident	Yes	No	Yes
Identification of contributing factors close to or far from the accident	Yes	Yes	Yes
Provision of recommendations for the control structure	Yes	Yes	Yes
Description of events and actions	Yes	Yes	No
Description of components of system	No	Yes	Yes
Providing enough information about system structure	No	No	No
Focus on operators and functions	No	Yes	Yes
Considering the environmental conditions (equipment and surroundings)	Yes	Yes	Yes
Identifying singular root causes for accidents	No	No	No
Definition of system boundaries	Yes	Yes	No
Providing a context to identify system safety improvements	Yes	Yes	Yes
Identification of the control and feedback inadequacies	No	Yes	No
Empirical data are not required	Yes	Yes	Yes
Minimized level of system information is required for analysis	No	No	No
Easier to be implemented	Yes	No	No
Providing adequate guidance regarding the methodology	Yes	No	Yes
Appropriate for use in a variety of contexts	Yes	Yes	Yes
Ability to quantify the accident occurrence and yield probabilities	No	No	No
Is not affected by analyst bias	No	No	No
Easy to disseminate results to non-experts	No	No	No

From this perspective, the three concepts of safety (i.e., safety-I, II and III) in relation to accident analysis models are discussed in the following. In the traditional safety-engineering paradigm, safety-I implies that as few things as possible should go wrong during the design process [115,116]. As systems become more advanced and sophisticated, it becomes increasingly vital to focus on enhancing safety while also maintaining the performance modifications to an acceptable level [4].

Complex systems, however, present a different set of safety challenges due to their inherent complexities, ambiguities, and potential for conflicts. Contrary to the apparent significance of these challenges, the traditional management of safety has relatively overlooked this issue [116–120]. According to a safety-I perspective, performance variability should be prevented as it is harmful. In the safety-II approach, is it inevitable, but it may also be useful, so it should be monitored and managed. Therefore, safety-I should progress to a safety-II perspective, in which considerable improvements are established, and we can rely on the system's capacity to react to daily performance variations under varied conditions and maintenance of safety [121]. Therefore, the effort is made for systems to respond to or prevent the hazards by providing suitable controls and interfaces.

In addition, the perspective of the risk assessment "to identify causes and contributory factors" in safety-I should become "understanding the conditions in which performance variability occur" in safety-II [122]. Hence, companies were looking for techniques to implement in varied circumstances according to a safety-II perspective. From a safety-II perspective, since the focus is on monitoring and controlling the determined performance variability, traditional methods are not considered to be sufficient. In that regard, approaches such as the FRAM model [123] were established to explain the system's necessary activities, their connectivity, variability, and resonance, as well as to offer strategies for monitoring and dampening the variability that contributes to accidents [124].

More recently, Hollnagel advanced the concept of safety-III, while its properties remained unspecified beyond those of safety-II. According to this system theory, Leveson defines safety-III as "freedom from intolerable losses" [124,125]. Safety-III defines the concept of accident casualty differently by shifting its focus on the inadequacy of hazard controls as well as relying on the system theory. Considering the concept of sustainability, it also refers to the maintenance of the safety constraints and prevention of losses upon exposure to the control inadequacy, hazards and unexpected events. Safety-III is primarily concerned with engagement in the design of complex systems' safety management structures in which an appropriate safety culture is created, effective information is available, and the structure of safety management is extensively and carefully constructed. Thus, it is critical to design a sustainable system that is achievable using STAMP, or other tools based on the principle of STAMP (e.g., by using STPA and CAST). System theory approaches identify and analyze controls, hazards, unplanned changes, and associated adaptations in order to mitigate the risk and identify emerging hazards [126].

Nevertheless, it is worth noting that safety-III needs to be extended and improved. It would be preferable if a comprehensive method were developed to analyze sociotechnical systems holistically and to improve integration and communication between human factors and technical aspects for engineers during the early stages of the complex design process, as well as to be capable of being used for highly automated system analysis [126].

4.5. System Thinking and Improvement in Sustainability of Safety Management

A system is defined as a collection of interrelated elements that are structured to accomplish a specific purpose. Understanding how system components interact and are organized is critical at the system thinking level. Systems thinking was defined as the science of gathering information about the systems' behavior by creating a rising deep awareness of their components [2]. Moreover, in the systems thinking concept, system components and their environmental interactions have the same importance for the system components behavior. This concept also attends to emergent features, regards complexity, and determines feedback loops, hierarchy, and self-organization, as well as discovering the dynamics and their outcomes [127]. Complex systems have dynamic behavior that needs to be sustained in normal operations. They must also deal with the disturbances and variability of their behavior in order to prevent accidents [26]. Depending on the level of existing risk at work, each company has its own unique health and safety management system. In order to prevent degradation of the system, despite proper design and policy, it is necessary to manage and monitor the system continuously [1].

Therefore, the major element for establishing a sustainable safety management system and ensuring the longevity of safe and healthy organizations is planning and engaging a systemic approach to manage and control the risks. However, in order to execute this, the application of effective methodologies, tools and principles is required. Systems thinking concepts and approaches are able to provide awareness about systems and solve complex issues and for this reason it has been used in a numerous type of fields and disciplines 6. To present a thorough overview of scientists' growing awareness of the notion of safety, and to determine how safety has progressed over time, it is essential to approach these concepts via a system thinking perspective. In order to develop an in depth understanding and awareness of the various layers of the system, this perspective recommends opportunities to act in accordance with one's own human level of awareness. Basically, risk and safety management sought to construct socio-technical systems capable of generating events in the desired locations and preventing or omitting undesirable ones. Nowadays, safety science is concerned with increasing the generation of sustainable systems through using proactive rather than reactive approaches to system safety enhancement. Thus, through increasing system and subsystem awareness, systems thinking approaches can create proactiveness. This approach recommended intervening at the root-cause level rather than focusing on observed symptoms and occurrences. Proposed approaches for this purpose are systemic models that can be used for the analysis of a system's resilience. In that regard, STAMP methodology has already been employed to analyze and assess an organization's sustainable performance or the integration of sustainability in an organization—particularly, by incorporating high-hazard and high-functional-requirement scenarios with predictive objectives [26]. Some studies have used this method in different contexts. They identified abnormal system behaviors and potentially unsafe situations that led to the improvement and updating of system awareness, and the prevention of accidents through the introduction of safety limitations [84,85,88]. It was also employed in accidents analysis in a variety of contexts for identifying insufficient system control limitations and suggesting corresponding adjustments to increase system sustainability [88–93,95,128,129].

Accordingly, sustainable safety management can also be assessed and analyzed through FRAM which is a performance-based risk identification method [48]. This model was employed to evaluate the accidents as well as identify the essential functions and relationships between them and ultimately, offered recommendations for increasing the sustainability of system operations [102,103].

5. Conclusions

Our research provided a comprehensive review of systemic approaches of accident analysis utilized in the field of safety investigations. According to the inclusion criteria of this study, a total of 63 research publications employed the three systemic analysis methodologies. AcciMap, STAMP and FRAM were included.

Considering our key findings, all the reviewed research that employed one of these three methods discovered multiple contributing elements, functions, and interactions at various system levels. For instance, for the AcciMap and STAMP methods, the majority of contributing elements and controlling flaws were discovered at the system's lower levels.

Furthermore, the FRAM framework demonstrates the normal functions of the sociotechnical system, defines their variability and identifies the out-of-range variability as the leading indicators of the accident. Due to the relative complexity and difficulty in the interpretation of this model, various novel modifications need to be considered. In addition to an investigation of the advantages and drawbacks associated with the systemic methods, the static and qualitative nature of systemic models and the dynamic structure and ethical control of sophisticated systems were investigated. Safety and accidents analysis methods were also described in terms of safety-I, safety-II and safety-III. Furthermore, this research introduced certain approaches that may be employed in conjunction with the three examined models—particularly, to optimize their applications.

Nonetheless, further research is required to elucidate the critical variables underlying selected systems thinking methodologies for accident causation.

Author Contributions: Supervision, E.H; project administration, M.D.; conceptualization, E.Z. and O.V.B.; methodology, M.D.; writing—original draft preparation, M.D.; writing—review and editing, E.Z., M.F., and O.V.B.; funding acquisition, E.H. All authors have read and agreed to the published version of the manuscript.

Funding: This research was funded by the Isfahan University of Medical Sciences, Isfahan, Iran (Grant number 340011 and ethical number IR.MUI.REC. 1400-018).

Institutional Review Board Statement: Not applicable.

Informed Consent Statement: Not applicable.

Data Availability Statement: Not applicable.

Acknowledgments: This article was extracted from the thesis written by Mahdieh Delikhoon, a student of Occupational Health and Safety Engineering.

Conflicts of Interest: The authors declare no conflict of interest.

Abbreviations

The following abbreviations are used in this manuscript:

ILO	International Labor Organization
GDP	global gross domestic product
STAMP	Systems-Theoretic Accident Model and Processes
FRAM	Functional Resonance Accident Model
CCA	Cause-Consequence Analysis
FTA	Fault Tree Analysis
ETA	Event Tree Analysis
FMEA	Failure Modes and Effect Analysis
STPA	System Theoretic Process Analysis
CAST	Causal Analysis based on STAMP
FMV	FRAM Model Visualizer
CWA	Cognitive Work Analysis
ISM	Interpretive Structural Modeling
VSM	Viable Systems Model
HEMS	Helicopter Emergency Medical Service
SD	System Dynamics
SMD	Soma Mine Disaster
SMS	Safety Management System
MCs	Monte Carlo simulations
GMTA	Goals-Means Task Analysis
BN	Bayesian Networks
AH	Abstraction Hierarchy
TASM	Total Apron Safety Management
DBN	Dynamic Bayesian Network
QRA	Quantitative Risk Analysis
AHP	Analytical Hierarchy Process
SME	Subject Matter Experts
MCMCs	Markov Chain Monte Carlo simulation
PoFs	Probability of Failures
MCDM	Multi Criteria Decision Making

Appendix A

Table A1. General information and findings from 25 AcciMap studies.

Objective	Scope of the Study	Main Findings	Location	Reference
To find the causes of the disasters related to drinking water distribution systems.	Public health	• Implies complex interactions among all levels of a complex sociotechnical system for designing the public policies to reduce risk in complex systems. • There was a distinction between low-level physical and individual variables, as well as a parallelism between high-level governmental and regulatory factors.	Saskatchewan, Canada	[72]
Investigation of leading factors of the water transportation system outbreaks.	Public health	• Describes the causes of accidents. • Specifies how to prevent an accident.	Walkerton, Ontario, Canada	[73]
Investigation of the incidents/accidents causality of space programme's launch vehicle.	Aerospace	• Provides a broad framework of leading events, particularly at higher levels, indicating the involvement of regulatory and political authorities in accident formation.	São Paulo, Brazil	[128]

Table A1. *Cont.*

Objective	Scope of the Study	Main Findings	Location	Reference
Assessing the food system safety accidents.	Public health	• Identifies methods for preventing accidents caused by similar sources of hazards.	UK	[74]
Analysis of the contributory factors for the infection outbreaks.	Public health	• Demonstrates the strategies and interventions that can be taken to limit and prevent the occurrence of the outbreaks.	Maidstone and Tunbridge Wells, UK	[76]
Modeling the events leading up to the Stockwell Underground station accident in July 2005	Public health	• Proposes a dynamic structure for organization in response to the type of operations and obvious events.	London, UK	[79]
Evaluating the led outdoor activity domain.	Led outdoor recreation	• AcciMap is a comprehensive approach to the risk management and accidents analysis developed based on the concept of 'learning from the accident'.	Dorset, UK	[129]
Comparing the AcciMap, the HFACS and the STAMP methods to analyze the Mangatepopo gorge tragedy.	Led outdoor recreation	• Describes the failures through the six levels of the studied system.	New Zealand	[130]
Assessment of organizational factors in aircraft accidents.	Transport (aircraft)	• The causal remoteness that interlinked to the fatal accident increases as we move up the vertical axis from the accident.	Australia	[131]
Examining the incident of rail level crossing system.	Transport (rail)	• In addition to the primary cause of the incident, various system-wide factors contribute to the occurrence of an incident.	Victoria, Australia	[132]
Assessment of applicability of systemic frameworks for incident data analysis.	Led outdoor recreation	• Capability of framework to classify contributory factors at various levels of the led outdoor activity was confirmed.	New Zealand	[133]
Testing applicability of the method for the analysis the risks associated to the studied case.	Disaster response	• Provides more extensive comprehension of the performance of the case.	Victoria, Australia	[134]
Accident analysis using AcciMap, STAMP and SCM methods.	Transport (rail)	• Levels 4 and 5 had the most effective factors in accident and Level 1 of the system, i.e., national government did not include any factors.	Cumbria, UK	[135]
Using AcciMap and Analytical Network Process for the assessment of the contributory factors of the marine accidents.	Navigation	• Reveals the main leading factors of accident. • Essential precautionary measures have already been proposed.	Turkey	[80]
Identifying the factors that contribute to the collapse of a bridge.	Civil engineering	• Several levels of failure modes were detected. • Demonstrated that human error is a leading contributor element in the occurrence of accidents.	China	[136]
Developing a coding template to quantitatively analyze the causes of road freight crashes.	Transport; (road accidents)	• Highlighted the role of systemic approach in enhancement of the safety knowledge. • Recommended preventive measures in the critical domain.	Australia	[75]
Identifying the human and systemic causes of outbreaks in the food production domain.	Public health	• The contributory macro and micro factors and their interactions were identified.	South Wales, UK	[81]
Using AcciMap and CWA approaches to systemic analysis of a case.	Transport (off-road)	• Hybrid method enhanced the identifying the causes of accidents and their relationship with the management and system rules in term of the cultural, economic, and social aspects.	Queensland; Australia	[77]

Table A1. *Cont.*

Objective	Scope of the Study	Main Findings	Location	Reference
Systemic analysis of South Korea Sewol ferry accident.	Maritime	• Highlighted the importance of allocating resources to safety management in a proactive manner, ongoing monitoring, and having independent and well-informed personnel in charge of continuously monitoring risk to prevent safety migration.	South Korea	[82]
Investigating the tragic Sewol Ferry accident.	Maritime; Ferry accidents	• Emphasized the significance of organizational and human variables in the occurrence of accidents.	South Korea	[83]
Developing the incidents reporting system as well as emphasizing the importance of learning from the accidents.	Led outdoor recreation	• Indicate the ability of Rasmussen's method of expansion through the safety critical domains.	Australia	[43]
Assessing the factors for systemic accidents causation.	Ship grounding accidents	• Used the fuzzy Interpretive Structural Modeling, and Matrix of Cross Impact Multiplications to overcome the limitations of the present AcciMap technique.	China	[78]
Performing the risk management proactively.	Road accidents	• Demonstrated that the effectiveness of good management and concern for safety at various levels of the sociotechnical system is a key issue for managing the risks proactively.	Bangladesh	[137]
Recognizing the principles of systems thinking in a range of varied systems and events.	Systems thinking tenets	• Declared that the systems thinking tenets can be related to accident causation.	Australia	[138]
Evaluating the formalized AcciMap for assessing the causation of accidents.	Healthcare accidents	• Applied leading factors for formulation of safety recommendations.	Scotland, UK	[139]

Appendix B

Table A2. General information and findings from 16 STAMP studies.

Objective	Scope of Study	Main Findings	Location	Reference
Analyzing the railway accidents and providing improvement measures	Transport (accident in railway)	• Spread accidents analysis in wide sense. • Made impressive urgent actions for case of the study.	China	[84]
Using joint STAMP–VSM framework to systemic accidents analysis.	Aviation (HEMS)	• Analyzed the control flaws. • Reviewed the infrastructure of safety. • Models loops and constraints information. • Regarded the conformity and direction of organizational activities. • Developed vast strength interventions	Greece	[85]
Demonstration of practicality and validity of the STAMP model.	Industry (a case study in the oil and gas)	• Violations against existing safety constraints that lead to accidents at any level of the organization were identified.	USA	[88]
Development of human error causal analysis framework through the STAMP-SD based analysis.	Military	• In whole, 41 leading items related to a broad view of sociotechnical systems were identified and categorized into four types of human errors.	USA	[86]

Table A2. *Cont.*

Objective	Scope of Study	Main Findings	Location	Reference
Demonstration of adaptive and integrated safety management based on STAMP concept.	Maritime Transport System	• The authors recommended using the control loop of STAMP as a basis to develop and implement the integrated safety management.	Finland	[87]
Analysis of Korean Sewol ferry accident based on STAMP.	Maritime	• The study developed some continuous improvements and corrective actions to prevent occurrences of catastrophic accidents.	South Korea	[89]
Evaluation of hazard control measures effectiveness using STAMP.	Maritime, safety management of traffic	• Determined the level of system hazards. • Identified unsafe situations. • Established control measures of maneuvers. • Updated the situational awareness. • Implemented the real-time safety restrictions.	Finland	[126]
Investigated the patient safety incident practices.	Public health	• Offered insights to integration of Human factors and Ergonomics into current practice.	UK	[90]
The STAMP was used for the SMD analyzing.	Mine accident	• Identified the inadequate system control constraints. • Suggested the related improvements. • Demonstrated the robustness of method for the cases with high degree of uncertainty.	USA	[91]
Analyzing the contributing factors of pipeline leakage and explosion accident.	Process industries accident	• Expanded the causal analysis from a systematic perspective. • Illustrated the utility of model to this case.	China	[92]
Analyzing the human factors and taxonomy of system.	Accident analysis	• Analyzed the accidents that occurred due to a major mismatch among components.	Poland	[95]
Designing maritime safety management systems.	Safety management systems	• A descriptive process of analysis and key performance indicators was provided for designing maritime safety management systems.	Finland	[116]
Hazard analysis of Software-Controlled Systems based on STPA.	Software-Controlled Systems	• A new method HCAT-STPA was proposed for analyzing the software control systems hazards.	China	[140]
Using of the STAMP and Bayesian Networks to operational use and design of the safety SMS.	Maritime	• Developed maritime SMS auditing processes.	Finland	[110]
Application of systemic methods for the analysis of coal mines accidents.	Coal mines accident	• STAMP model was shown to be a comprehensive and systematic technique. • The model characteristics and analysis processes were complex.	China	[127]
Identifying the contributing factors of abnormal behaviors of system that cause process malfunctions using STAMP.	Indoor environment safety	• STAMP effectively identified causes of physical process anomalies.	Japan	[93]

Appendix C

Table A3. General information and findings of 22 FRAM studies.

Objective	Scope of Study	Main Findings	Location	References
Analyzing aircraft accidents induced by automation autopilots.	Aviation	• Predicted the possible hazard occurrence which may result from complex interactions among human, technological and organizational factors.	Japan	[141]
Comparing the two methods: STEP and FRAM	Aviation	• FRAM demonstrated the dynamic interactions of sociotechnical systems. • Described non-linear interrelations among the functions. • Determined the conditions, variability and performance resonance of the functions.	Norway	[142]
Analyzing an accident related to the ATM system.	Aviation	• Proposed some recommendation on the system operation resilience. • Indicated that a more profound understanding on the system function is need.	Brazil	[102]
Hazard analysis of software system using FRAM and System Hazard Analysis.	Airline	• Established a requirements-based methodology.	Australia	[143]
Assessing risk in sustainable construction via FRAM methodology.	Construction	• Control strategies were developed to reduce the risk for function variability or functional resonance.	Brazil	[103]
Analysis of the hazards attributed to the sociotechnical system.	Maritime	• Determined the occurrence and aggregation of functions variability. • Illustrated the interactions of functions of system. • Determined how safety constraints are violated.	China	[144]
Investigating the compatibility of FRAM model and Rasmussen's AH	Transport (railway)	• Provided a new structure of FRAM by functional analysis at hierarchical layers of the system.	UK	[104]
Enhancement of the traditional safety assessment based on semi quantitative FRAM and MCs.	Aviation (ATM system)	• Highlighted the critical functions and critical links among these functions. • Facilitated the safety analysis by considering the system response to different operating conditions and different risk conditions.	Los Angeles	[96]
Using a hybrid approach as combining FRAM and TASM to system-based modelling of the safety	Ground handling services	• Advocated the benefits of systemic approaches. • Demonstrated the suitability of the TASM framework for hazard and accident analysis.	UK	[101]
Risk assessment and modeling the performance interactions for the maintenance of system.	Hydrocarbon Release Accidents	• The event investigated by connecting various activities and risk influencing factors from a functional perspective.	Norway	[145]

Table A3. *Cont.*

Objective	Scope of Study	Main Findings	Location	References
Quantifying the FRAM.	Resilience Quantification	• The model excluded the main leading indexes. • Resilience bases of the FRAM (anticipate, respond, monitor, learn) were demonstrated. • Overall system variability was demonstrated.	Italy	[97]
Predictive performance assessment and improvement of a framework through the integration of FRAM and fuzzy logic.	Complex Sociotechnical Systems	• Generated numerical indicators for a more comprehensible representation of potential performance variability.	Canada	[98]
Developing a theory of change to support intervention development.	Public health; care safety	• Supported the theory of change to develop a guide for future safety interventions.	UK	[146]
To explore how tensions and contradictions are managed by people.	Public health; patient safety	• Highlighted the main areas of performance variability.	UK	[147]
Qualitative risk analysis of shipping operations.	Maritime accident	• Determined the variability of events underlying the accident. • Provided suggestions to examine these events.	Turkey	[121]
Risk assessment of highlyautomated vehicles using FRAM.	Automated driving	• The risk and safety assessment were performed. • Proposed recommendations for system design. • Required perspectives on work validation were represented. • Suitability of model was evaluated in detail.	Germany	[148]
Analyzing human factors and non-technical skills by modeling the performed activities.	Offshore drilling operations	• Underlined the role of human factors and non-technical skills for the productivity and safety of the work in both normal and critical operation situations.	Brazil	[149]
Quantitative assessment of resilience through FRAM and DBN	Chemical process systems	• An effective tool for the purpose of the study was provided.	Kazakhstan	[99]
Identifying the challenges within the case of the study	Transition process	• It revealed some challenges affecting the transition process.	Canada	[150]
Investigating the applicability of quantified systemic method for risk analysis of the case of study using FRAM and MCs.	Tram operating system	• Systemic method determined functional interactions of the system. • Aggregation of variability was determined. • Comprehensive risk analysis of the case of study was performed.	Turkey	[94]

Table A3. *Cont.*

Objective	Scope of Study	Main Findings	Location	References
Use of quantitative FRAM for risk assessment.	System of COVID-19 pandemic emergency response	• Potential risks and critical conditions were assessed • Highlighted the role of emergency response strategies at the governance scale.	Republic of Korea	[100]
To survey the role of resilience engineering in identifying the system requirements.	Software	• New strategies for meeting the requirements of software for complex systems were represented.	Brazil	[151]

References

1. Lee, D. The Effect of Safety Management and Sustainable Activities on Sustainable Performance: Focusing on Suppliers. *Sustainability* **2018**, *10*, 4796. [CrossRef]
2. Lee, D.; Schniederjans, M.J. How corporate social responsibility commitment influences sustainable supply chain management performance within the social capital framework: A propositional framework. *Int. J. Corp. Strategy Soc. Responsib.* **2017**, *1*, 208–233.
3. Blokland, P.; Reniers, G. Safety Science, a Systems Thinking Perspective: From Events to Mental Models and Sustainable Safety. *Sustainability* **2020**, *12*, 5164. [CrossRef]
4. Aven, T. A risk science perspective on the discussion concerning Safety I, Safety II and Safety III. *Reliab. Eng. Syst. Saf.* **2021**, *217*, 108077. [CrossRef]
5. Habibi, E. *A Safety Analysis of Industrial Accidents. Accident Records of Major Coal Producing Countries Are Analysed to Obtain Fatal and Non-Fatal Accident Rates. Significant Factors Influencing These Rates Are Identified with Efficacy of Preventive Measures*; University of Bradford: Bradford, UK, 2010.
6. Habibi, E.; Karimi, A.; Shahreza, H.D.; Mahaki, B.; Nouri, A. A study of the relationship between the components of the five-factor model of personality and the occurrence of occupational accidents in industry workers. *Iran. J. Health Saf. Environ.* **2016**, *3*, 499–505.
7. ILO. Work Hazards Kill Millions, Cost Billions. World of Work Magazine. 2003. Available online: https://www.ilo.org/global/about-the-ilo/mission-and-objectives/features/WCMS_075615/lang--en/index.htm (accessed on 12 October 2021).
8. Zarei, E.; Gholamizadeh, K.; Khan, F.; Khakzad, N. A dynamic domino effect risk analysis model for rail transport of hazardous material. *J. Loss Prev. Process Ind.* **2022**, *74*, 104666. [CrossRef]
9. Zarei, E.; Karimi, A.; Habibi, E.; Barkhordari, A.; Reniers, G. Dynamic occupational accidents modeling using dynamic hybrid Bayesian confirmatory factor analysis: An in-depth psychometrics study. *Saf. Sci.* **2021**, *136*, 105146. [CrossRef]
10. García-Herrero, S.; Mariscal, M.; García-Rodríguez, J.; Ritzel, D.O. Working conditions, psychological/physical symptoms and occupational accidents. Bayesian network models. *Saf. Sci.* **2012**, *50*, 1760–1774. [CrossRef]
11. Ghamari, F.; Mohammadfam, I.; Mohammadbeigi, A.; Ebrahimi, H.; Khodayari, M. Determination of effective risk factors in incidence of occupational accidents in one of the large metal industries, Arak (2005–2007). *Iran Occup. Health* **2012**, *9*, 89–96.
12. Moghaddam, A.M.; Tabibi, Z.; Sadeghi, A.; Ayati, E.; Ravandi, A.G. Screening out accident-prone Iranian drivers: Are their at-fault accidents related to driving behavior? *Transp. Res. Part F Traffic Psychol. Behav.* **2017**, *46*, 451–461. [CrossRef]
13. Omidi, L.; Zakerian, S.A.; Saraji, J.N.; Hadavandi, E.; Yekaninejad, M.S. Prioritization of Human Factors Variables in the Management of Major Accident Hazards in Process Industries Using Fuzzy AHP Approach. *Health Scope* **2018**, *7*, e61649. [CrossRef]
14. Pordanjani, T.R.; Ebrahimi, A.M. Safety Motivation and Work Pressure as Predictors of Occupational Accidents in the Petrochemical Industry. *Health Scope* **2015**, *4*, 33. [CrossRef]
15. Swaen, G.; van Amelsvoort, L.; Bültmann, U.; Slangen, J.; Kant, I. Psychosocial Work Characteristics as Risk Factors for Being Injured in an Occupational Accident. *J. Occup. Environ. Med.* **2004**, *46*, 521–527. [CrossRef]
16. Fabiano, B.; Curro', F.; Reverberi, A.P.; Pastorino, R. Port safety and the container revolution: A statistical study on human factor and occupational accidents over the long period. *Saf. Sci.* **2010**, *48*, 980–990. [CrossRef]
17. Habibi, E.; Gharib, S.; Mohammadfam, I.; Rismanchian, M. Human error assessment in Isfahan oil refinery's work station operators using systematic human error reduction prediction approach technique. *Int. J. Environ. Health Eng.* **2013**, *2*, 25. [CrossRef]
18. Pietilä, J.; Räsänen, T.; Reiman, A.; Ratilainen, H.; Helander, E. Characteristics and determinants of recurrent occupational accidents. *Saf. Sci.* **2018**, *108*, 269–277. [CrossRef]
19. Islam, R.; Khan, F.; Abbassi, R.; Garaniya, V. Human error probability assessment during maintenance activities of marine systems. *Saf. Health Work* **2018**, *9*, 42–52. [CrossRef]

20. Baxter, G.; Sommerville, I. Socio-technical systems: From design methods to systems engineering. *Interact. Comput.* **2011**, *23*, 4–17. [CrossRef]
21. Zarei, E.; Khakzad, N.; Cozzani, V.; Reniers, G. Safety analysis of process systems using Fuzzy Bayesian Network (FBN). *J. Loss Prev. Process Ind.* **2018**, *57*, 7–16. [CrossRef]
22. Rathnayaka, S.; Khan, F.; Amyotte, P. SHIPP methodology: Predictive accident modeling approach. Part I: Methodology and model description. *Process Saf. Environ. Prot.* **2011**, *89*, 151–164. [CrossRef]
23. Tan, Q.; Chen, G.; Zhang, L.; Fu, J.; Li, Z. Dynamic accident modeling for high-sulfur natural gas gathering station. *Process Saf. Environ. Prot.* **2014**, *92*, 565–576. [CrossRef]
24. Ale, B.; van Gulijk, C.; Hanea, A.; Hanea, D.; Hudson, P.; Lin, P.-H.; Sillem, S. Towards BBN based risk modelling of process plants. *Saf. Sci.* **2014**, *69*, 48–56. [CrossRef]
25. Zarei, E.; Khan, F.; Abbassi, R. Importance of human reliability in process operation: A critical analysis. *Reliab. Eng. Syst. Saf.* **2021**, *211*, 107607. [CrossRef]
26. Hollnagel, E.; Woods, D.D.; Leveson, N. *Resilience Engineering: Concepts and Precepts*; Ashgate: Aldershot, UK, 2006.
27. Wagenaar, W.; Hudson, P. *The Analysis of Accidents with a View to Prevention*; Report for Shell International SIPM; Department of Experimental Psychology, Leiden University: The Hague, The Netherlands, 1987.
28. Leveson, N. A systems approach to risk management through leading safety indicators. *Reliab. Eng. Syst. Saf.* **2015**, *136*, 17–34. [CrossRef]
29. Al-Shanini, A.; Ahmad, A.; Khan, F. Accident modelling and analysis in process industries. *J. Loss Prev. Process Ind.* **2014**, *32*, 319–334. [CrossRef]
30. Attwood, D.; Khan, F.; Veitch, B. Occupational accident models—Where have we been and where are we going? *J. Loss Prev. Process Ind.* **2006**, *19*, 664–682. [CrossRef]
31. HaSPA. *The Core Body of Knowledge for Generalist OHS Professionals*; Safety Institute of Australia Tullamarine: Tullamarine, VIC, Australia, 2012.
32. Hermitte, T.; Phan, V. *Review of Accident Causation Models Used in Road Accident Research*; DaCoTA: Bryn Mawr, PA, USA, 2012.
33. Hollnagel, E.; Speziali, J. *Study on Developments in Accident Investigation Methods: A Survey of the 'State-of-the-Art'*; Swedish Nuclear Power Inspectorate: Stockholm, Sweden, 2008.
34. Katsakiori, P.; Sakellaropoulos, G.; Manatakis, E. Towards an evaluation of accident investigation methods in terms of their alignment with accident causation models. *Saf. Sci.* **2009**, *47*, 1007–1015. [CrossRef]
35. Leveson, N.G. *Evaluating Accident Models Using Recent Aerospace Accidents, Part 1: Event-Based Model*; MIT Libraries: Cambridge, MA, USA, 2001.
36. Qureshi, Z.H. A review of accident modelling approaches for complex critical sociotechnical systems. In Proceedings of the 12th Australian Workshop on Safety Related Programmable Systems (SCS'07), Adelaide, Australia, 30–31 August 2007.
37. Underwood, P.; Waterson, P. *Accident Analysis Models and Methods: Guidance for Safety Professionals*; Loughborough University: Loughborough, UK, 2013.
38. Wienen, H.C.; Bukhsh, F.A.; Vriezekolk, E.; Wieringa, R.J. *Accident Analysis Methods and Models—A Systematic Literature Review*; Centre for Telematics and Information Technology (CTIT): Enschede, The Netherlands, 2017.
39. Leveson, N. A new accident model for engineering safer systems. *Saf. Sci.* **2003**, *42*, 237–270. [CrossRef]
40. Simms, D.L.; Perrow, C. Normal Accidents: Living with High-Risk Technologies. *Technol. Cult.* **1986**, *27*, 903. [CrossRef]
41. Valdez Banda, O.A.; Kannos, S.; Goerlandt, F.; van Gelder, P.H.A.J.M.; Bergström, M.; Kujala, P. A systemic hazard analysis and management process for the concept design phase of an autonomous vessel. *Reliab. Eng. Syst. Saf.* **2019**, *191*, 106584. [CrossRef]
42. Leveson, N. *Engineering a Safer World: Applying Systems Thinking to Safety*; MIT Press: Cambridge, MA, USA, 2012.
43. Salmon, P.M.; Goode, N.; Taylor, N.; Lenné, M.G.; Dallat, C.E.; Finch, C.F. Rasmussen's legacy in the great outdoors: A new incident reporting and learning system for led outdoor activities. *Appl. Ergon.* **2017**, *59*, 637–648. [CrossRef]
44. Waterson, P.; Jenkins, D.P.; Salmon, P.M.; Underwood, P. 'Remixing Rasmussen': The evolution of Accimaps within systemic accident analysis. *Appl. Ergon.* **2017**, *59*, 483–503. [CrossRef]
45. Pouyakian, M.; Jafari, M.J.; Laal, F.; Nourai, F.; Zarei, E. A comprehensive approach to analyze the risk of floating roof storage tanks. *Process Saf. Environ. Prot.* **2020**, *146*, 811–836. [CrossRef]
46. Underwood, P.; Waterson, P. A critical review of the STAMP, FRAM and Accimap systemic accident analysis models. In *Advances in Human Aspects of Road and Rail Transportation*; CRC Press: Boca Raton, FL, USA, 2012; pp. 385–394.
47. Chen, C.; Reniers, G.; Khakzad, N. A thorough classification and discussion of approaches for modeling and managing domino effects in the process industries. *Saf. Sci.* **2020**, *125*, 104618. [CrossRef]
48. Hollnagel, E. *Barriers and Accident Prevention*; Ashgate: Hampshire, UK, 2004.
49. Hollnagel, E. *Anticipating Failures: What Should Predictions Be About?* Linkoeping University Graduate School for Human-Machine Interaction: Linköping, Sweden, 2001.
50. Gordon, J.E. The epidemiology of accidents. *Am. J. Public Health Nations Health* **1949**, *39*, 504–515. [CrossRef]
51. Woods, D.D.; Johannesen, L.J.; Cook, R.I.; Sarter, N.B. *Behind Human Error: Cognitive Systems, Computers and Hindsight*; Dayton University Research Institute (Urdi) OH: Beavercreek, OH, USA, 1994.
52. Reason, J. *Managing the Risks of Organizational Accidents*; Routledge: London, UK, 2016.
53. Boishu, Y. *SMS and Risk Assessment Automation*; SM ICG Industry Day: Bern, Switzerland, 2014.

54. Maurino, D.; Seminar (CASS). Threat and error management (TEM). In Proceedings of the Canadian Aviation Safety Seminar (CASS), Vancouver, BC, Canada, 18–20 April 2005.
55. Kjellen, U. *Prevention of Accidents Through Experience Feedback*; CRC Press: Boca Raton, FL, USA, 2000. [CrossRef]
56. de Carvalho PV, R.; Gomes, J.O.; Huber, G.J.; Vidal, M.C. Normal people working in normal organizations with normal equipment: System safety and cognition in a mid-air collision. *Appl. Ergon.* **2009**, *40*, 325–340. [CrossRef]
57. Young, M.; Shorrock, S.; Faulkner, J.; Braithwaite, G. *Who Moved My (Swiss) Cheese*; ISASI: Sterling, VA, USA, 2005; pp. 31–33.
58. Yousefi, A.; Hernandez, M.R.; Peña, V.L. Systemic accident analysis models: A comparison study between AcciMap, FRAM, and STAMP. *Process Saf. Prog.* **2018**, *38*, e12002. [CrossRef]
59. Zarei, E.; Yazdi, M.; Abbassi, R.; Khan, F. A hybrid model for human factor analysis in process accidents: FBN-HFACS. *J. Loss Prev. Process Ind.* **2018**, *57*, 142–155. [CrossRef]
60. Rostamabadi, A.; Jahangiri, M.; Zarei, E.; Kamalinia, M.; Banaee, S.; Samaei, M.R. A Novel Fuzzy Bayesian Network-HFACS (FBN-HFACS) model for analyzing Human and Organization Factors (HOFs) in process accidents. *Process Saf. Environ. Prot.* **2019**, *132*, 59–72. [CrossRef]
61. Rostamabadi, A.; Jahangiri, M.; Zarei, E.; Kamalinia, M.; Alimohammadlou, M. A novel Fuzzy Bayesian Network approach for safety analysis of process systems; An application of HFACS and SHIPP methodology. *J. Clean. Prod.* **2020**, *244*, 118761. [CrossRef]
62. Zarei, E.; Ramavandi, B.; Darabi, A.H.; Omidvar, M. A framework for resilience assessment in process systems using a fuzzy hybrid MCDM model. *J. Loss Prev. Process Ind.* **2020**, *69*, 104375. [CrossRef]
63. Hollnagel, E. Understanding accidents-from root causes to performance variability. In Proceedings of the IEEE 7th Conference on Human Factors and Power Plants, New York, NY, USA, 15–19 September 2002.
64. Rasmussen, J. Risk management in a dynamic society: A modelling problem. *Saf. Sci.* **1997**, *27*, 183–213. [CrossRef]
65. Rasmussen, J.; Suedung, I. *Proactive Risk Management in a Dynamic Society*; Swedish Rescue Services Agency: Karlstad, Sweden, 2000.
66. Clarkson, J.; Hopkins, A.; Taylor, K. *Report of the Board of Inquiry into F-111 (Fuel Tank) Deseal/Reseal and Spray Seal Programs*; Royal Australian Air Force: Canberra, ACT, Australia, 2001; Volume I.
67. Hollnagel, E. *FRAM, the Functional Resonance Analysis Method: Modelling Complex SOCIO-technical Systems*; Ashgate Publishing, Ltd.: Farnham, UK, 2012.
68. Riccardo, P.; Gianluca, D.P.; Giulio, D.G.; Francesco, C. FRAM for Systemic Accident Analysis: A Matrix Representation of Functional Resonance. *Int. J. Reliab. Qual. Saf. Eng.* **2018**, *25*, 1850001. [CrossRef]
69. Hollangel, E. *Functional Resonance Accident Model, Method and Examples*; Cognitive Systems Engineering Laboratory, University of Linöping: Linöping, Sweden, 2005.
70. Hill, R.; Hollnagel, E. Instructions for Use of the FRAM Model Visuliser (FMV). 2016. Available online: https://zerprize.co.nz/Content/FMV_instructions_2.1.pdf (accessed on 14 June 2021).
71. Hulme, A.; Stanton, N.; Walker, G.H.; Waterson, P.; Salmon, P.M. What do applications of systems thinking accident analysis methods tell us about accident causation? A systematic review of applications between 1990 and 2018. *Saf. Sci.* **2019**, *117*, 164–183. [CrossRef]
72. Woo, D.M.; Vicente, K.J. Sociotechnical systems, risk management, and public health: Comparing the North Battleford and Walkerton outbreaks. *Reliab. Eng. Syst. Saf.* **2003**, *80*, 253–269. [CrossRef]
73. Vicente, K.J.; Christoffersen, K. The Walkerton *E. coli* outbreak: A test of Rasmussen's framework for risk management in a dynamic society. *Theor. Issues Ergon. Sci.* **2006**, *7*, 93–112. [CrossRef]
74. Cassano-Piche, A.L.; Vicente, K.J.; Jamieson, G.A. A test of Rasmussen's risk management framework in the food safety domain: BSE in the UK. *Theor. Issues Ergon. Sci.* **2009**, *10*, 283–304. [CrossRef]
75. Newnam, S.; Goode, N. Do not blame the driver: A systems analysis of the causes of road freight crashes. *Accid. Anal. Prev.* **2015**, *76*, 141–151. [CrossRef]
76. Waterson, P. A systems ergonomics analysis of the Maidstone and Tunbridge Wells infection outbreaks. *Ergonomics* **2009**, *52*, 1196–1205. [CrossRef]
77. Stevens, N.J.; Salmon, P. Sand, surf and sideways: A systems analysis of beaches as complex roadway environments. *Saf. Sci.* **2016**, *85*, 152–162. [CrossRef]
78. Wang, W.; Liu, X.; Qin, Y.; Huang, J.; Liu, Y. Assessing contributory factors in potential systemic accidents using AcciMap and integrated fuzzy ISM—MICMAC approach. *Int. J. Ind. Ergon.* **2018**, *68*, 311–326. [CrossRef]
79. Jenkins, D.P.; Salmon, P.; Stanton, N.; Walker, G. A systemic approach to accident analysis: A case study of the Stockwell shooting. *Ergonomics* **2010**, *53*, 1–17. [CrossRef]
80. Akyuz, E. A hybrid accident analysis method to assess potential navigational contingencies: The case of ship grounding. *Saf. Sci.* **2015**, *79*, 268–276. [CrossRef]
81. Nayak, R.; Waterson, P. 'When Food Kills': A socio-technical systems analysis of the UK Pennington 1996 and 2005 *E. coli* O157 Outbreak reports. *Saf. Sci.* **2016**, *86*, 36–47. [CrossRef]
82. Kee, D.; Jun, G.T.; Waterson, P.; Haslam, R. A systemic analysis of South Korea Sewol ferry accident—Striking a balance between learning and accountability. *Appl. Ergon.* **2017**, *59*, 504–516. [CrossRef]
83. Lee, S.; Moh, Y.B.; Tabibzadeh, M.; Meshkati, N. Applying the AcciMap methodology to investigate the tragic Sewol Ferry accident in South Korea. *Appl. Ergon.* **2017**, *59*, 517–525. [CrossRef]

84. Ouyang, M.; Hong, L.; Yu, M.-H.; Fei, Q. STAMP-based analysis on the railway accident and accident spreading: Taking the China–Jiaoji railway accident for example. *Saf. Sci.* **2010**, *48*, 544–555. [CrossRef]
85. Kontogiannis, T.; Malakis, S. A systemic analysis of patterns of organizational breakdowns in accidents: A case from Helicopter Emergency Medical Service (HEMS) operations. *Reliab. Eng. Syst. Saf.* **2012**, *99*, 193–208. [CrossRef]
86. Rong, H.; Tian, J. STAMP-based HRA considering causality within a sociotechnical system: A case of Minuteman III missile accident. *Hum. Factors* **2015**, *57*, 375–396. [CrossRef]
87. Aps, R.; Fetissov, M.; Goerlandt, F.; Helferich, J.; Kopti, M.; Kujala, P. Towards STAMP Based Dynamic Safety Management of Eco-Socio-Technical Maritime Transport System. *Procedia Eng.* **2015**, *128*, 64–73. [CrossRef]
88. Altabbakh, H.; AlKazimi, M.A.; Murray, S.; Grantham, K. STAMP–Holistic system safety approach or just another risk model? *J. Loss Prev. Process Ind.* **2014**, *32*, 109–119. [CrossRef]
89. Kim, T.-E.; Nazir, S.; Ivar Øvergård, K. A STAMP-based causal analysis of the Korean Sewol ferry accident. *Saf. Sci.* **2016**, *83*, 93–101. [CrossRef]
90. Canham, A.; Jun, G.T.; Waterson, P.; Khalid, S. Integrating systemic accident analysis into patient safety incident investigation practices. *Appl. Ergon.* **2018**, *72*, 1–9. [CrossRef] [PubMed]
91. Düzgün, H.S.; Leveson, N. Analysis of soma mine disaster using causal analysis based on systems theory (CAST). *Saf. Sci.* **2018**, *110*, 37–57. [CrossRef]
92. Gong, Y.; Li, Y. STAMP-based causal analysis of China-Donghuang oil transportation pipeline leakage and explosion accident. *J. Loss Prev. Process Ind.* **2018**, *56*, 402–413. [CrossRef]
93. Yang, Z.; Lim, Y.; Tan, Y. An Accident Model with Considering Physical Processes for Indoor Environment Safety. *Appl. Sci.* **2019**, *9*, 4732. [CrossRef]
94. Kaya, G.K.; Ozturk, F.; Sariguzel, E.E. System-based risk analysis in a tram operating system: Integrating Monte Carlo simulation with the functional resonance analysis method. *Reliab. Eng. Syst. Saf.* **2021**, *215*, 107835. [CrossRef]
95. Lower, M.; Magott, J.; Skorupski, J. A System-Theoretic Accident Model and Process with Human Factors Analysis and Classification System taxonomy. *Saf. Sci.* **2018**, *110*, 393–410. [CrossRef]
96. Patriarca, R.; Di Gravio, G.; Costantino, F. A Monte Carlo evolution of the Functional Resonance Analysis Method (FRAM) to assess performance variability in complex systems. *Saf. Sci.* **2017**, *91*, 49–60. [CrossRef]
97. Bellini, E.; Coconea, L.; Nesi, P. A Functional Resonance Analysis Method Driven Resilience Quantification for Socio-Technical Systems. *IEEE Syst. J.* **2019**, *14*, 1234–1244. [CrossRef]
98. Slim, H.; Nadeau, S. A Proposal for a Predictive Performance Assessment Model in Complex Sociotechnical Systems Combining Fuzzy Logic and the Functional Resonance Analysis Method (FRAM). *Am. J. Ind. Bus. Manag.* **2019**, *09*, 1345–1375. [CrossRef]
99. Zinetullina, A.; Yang, M.; Khakzad, N.; Golman, B.; Li, X. Quantitative resilience assessment of chemical process systems using functional resonance analysis method and Dynamic Bayesian network. *Reliab. Eng. Syst. Saf.* **2020**, *205*, 107232. [CrossRef]
100. Kim, Y.C.; Yoon, W.C. Quantitative representation of the functional resonance analysis method for risk assessment. *Reliab. Eng. Syst. Saf.* **2021**, *214*, 107745. [CrossRef]
101. Studic, M.; Majumdar, A.; Schuster, W.; Ochieng, W.Y. A systemic modelling of ground handling services using the functional resonance analysis method. *Transp. Res. Part C Emerg. Technol.* **2017**, *74*, 245–260. [CrossRef]
102. De Carvalho, P.V.R. The use of Functional Resonance Analysis Method (FRAM) in a mid-air collision to understand some characteristics of the air traffic management system resilience. *Reliab. Eng. Syst. Saf.* **2011**, *96*, 1482–1498. [CrossRef]
103. Rosa, L.V.; Haddad, A.; De Carvalho, P.V.R. Assessing risk in sustainable construction using the Functional Resonance Analysis Method (FRAM). *Cogn. Technol. Work* **2015**, *17*, 559–573. [CrossRef]
104. Patriarca, R.; Bergström, J.; Di Gravio, G. Defining the functional resonance analysis space: Combining Abstraction Hierarchy and FRAM. *Reliab. Eng. Syst. Saf.* **2017**, *165*, 34–46. [CrossRef]
105. Huang, W.; Yin, D.; Xu, Y.; Zhang, R.; Xu, M. Using N-K Model to quantitatively calculate the variability in Functional Resonance Analysis Method. *Reliab. Eng. Syst. Saf.* **2021**, *217*, 108058. [CrossRef]
106. Merle, G.; Roussel, J.-M.; Lesage, J.-J.; Perchet, V.; Vayatis, N. Quantitative Analysis of Dynamic Fault Trees Based on the Coupling of Structure Functions and Monte Carlo Simulation. *Qual. Reliab. Eng. Int.* **2014**, *32*, 7–18. [CrossRef]
107. Gholamizadeh, K.; Zarei, E.; Omidvar, M.; Yazdi, M. Fuzzy Sets Theory and Human Reliability: Review, Applications, and Contributions. In *Linguistic Methods Under Fuzzy Information in System Safety and Reliability Analysis*; Springer: Berlin/Heidelberg, Germany, 2022; pp. 91–137.
108. Wang, Y.F.; Roohi, S.F.; Hu, X.M.; Xie, M. Investigations of Human and Organizational Factors in hazardous vapor accidents. *J. Hazard. Mater.* **2011**, *191*, 69–82. [CrossRef]
109. Phan, T.; Sahin, O.; Smart, J. System Dynamics and Bayesian Network Models for Vulnerability and Adaptation Assessment of a Coastal Water Supply and Demand System. In Proceedings of the 8th International Congress on Environmental Modelling and Software (iEMSs), Toulouse, France, 10–14 July 2016. [CrossRef]
110. Banda, O.A.V.; Goerlandt, F.; Salokannel, J.; van Gelder, P.H.A.J.M. An initial evaluation framework for the design and operational use of maritime STAMP-based safety management systems. *WMU J. Marit. Aff.* **2019**, *18*, 451–476. [CrossRef]
111. Balan, C.V.; Iordache, V.-M. Limitations of Systemic Accident Analysis Methods. *INCAS Bull.* **2016**, *8*, 167.
112. Iordache, V.-M.; Balan, C.V. Safety culture in modern aviation systems-civil and military. *Incas Bull.* **2016**, *8*, 135.

113. Manzur Tirado, A.M.; Brown, R.; Valdez Banda, O.A. *Risk and Safety Management of Autonomous Systems: A Literature Review and Initial Proposals for the Maritime Industry*; Aalto University: Espoo, Finland, 2019.
114. Sujan, M.A.; Huang, H.; Braithwaite, J. Learning from incidents in health care: Critique from a Safety-II perspective. *Saf. Sci.* **2017**, *99*, 115–121. [CrossRef]
115. Banda, O.A.V.; Hänninen, M.; Lappalainen, J.; Kujala, P.; Goerlandt, F. A method for extracting key performance indicators from maritime safety management norms. *WMU J. Marit. Aff.* **2016**, *15*, 237–265. [CrossRef]
116. Banda, O.A.V.; Goerlandt, F. A STAMP-based approach for designing maritime safety management systems. *Saf. Sci.* **2018**, *109*, 109–129. [CrossRef]
117. Hollnagel, E.; Wears, R.L.; Braithwaite, J. *From Safety-I to Safety-II: A White Paper. The Resilient Health Care Net*; University of Southern Denmark: Odense, Denmark; University of Florida: Gainesville, FL, USA; Macquarie University: Macquarie, Australia, 2015.
118. Lahtinen, J.; Banda OA, V.; Kujala, P.; Hirdaris, S. The Risks of Remote Pilotage in an Intelligent Fairway–preliminary considerations. In Proceedings of the International Seminar on Safety and Security of Autonomous Vessels, Helsinki, Finland, 17–18 September 2019.
119. Patterson, M.; Deutsch, E.S. Safety-I, Safety-II and resilience engineering. *Curr. Probl. Pediatric Adolesc. Health Care* **2015**, *45*, 382–389. [CrossRef] [PubMed]
120. Hollnagel, E. *The ETTO Principle: Efficiency-Thoroughness Trade-Off: Why Things That Go Right Sometimes Go Wrong*; CRC Press: Boca Raton, FL, USA, 2017.
121. Salihoglu, E.; Beşikçi, E.B. The use of Functional Resonance Analysis Method (FRAM) in a maritime accident: A case study of Prestige. *Ocean Eng.* **2020**, *219*, 108223. [CrossRef]
122. Hollnagel, E. *Safety-I and Safety-II: The Past and Future of Safety Management*; CRC Press: Boca Raton, FL, USA, 2018.
123. Martinetti, A.; Chatzimichailidou, M.M.; Maida, L.; van Dongen, L. Safety I–II, resilience and antifragility engineering: A debate explained through an accident occurring on a mobile elevating work platform. *Int. J. Occup. Saf. Ergon.* **2018**, *25*, 66–75. [CrossRef] [PubMed]
124. Leveson III, N. *Safety III: A Systems Approach to Safety and Resilience*; MIT: Cambridge, MA, USA, 2020.
125. Monat, J.P.; Gannon, T.F. Applying Systems Thinking to Engineering and Design. *Systems* **2018**, *6*, 34. [CrossRef]
126. Aps, R.; Fetissov, M.; Goerlandt, F.; Kujala, P.; Piel, A. Systems-Theoretic Process Analysis of Maritime Traffic Safety Management in the Gulf of Finland (Baltic Sea). *Procedia Eng.* **2017**, *179*, 2–12. [CrossRef]
127. Qiao, W.; Li, X.; Liu, Q. Systemic approaches to incident analysis in coal mines: Comparison of the STAMP, FRAM and "2–4" models. *Resour. Policy* **2019**, *63*, 101453. [CrossRef]
128. Johnson, C.; Almeida, I.M. An investigation into the loss of the Brazilian space programme's launch vehicle VLS-1 V03. *Saf. Sci.* **2008**, *46*, 38–53. [CrossRef]
129. Salmon, P.; Williamson, A.; Lenne, M.; Mitsopoulos-Rubens, E.; Rudin-Brown, C.M. Systems-based accident analysis in the led outdoor activity domain: Application and evaluation of a risk management framework. *Ergonomics* **2010**, *53*, 927–939. [CrossRef]
130. Salmon, P.; Cornelissen, M.; Trotter, M.J. Systems-based accident analysis methods: A comparison of Accimap, HFACS, and STAMP. *Saf. Sci.* **2012**, *50*, 1158–1170. [CrossRef]
131. Debrincat, J.; Bil, C.; Clark, G. Assessing organisational factors in aircraft accidents using a hybrid Reason and AcciMap model. *Eng. Fail. Anal.* **2013**, *27*, 52–60. [CrossRef]
132. Salmon, P.M.; Read, G.; Stanton, N.; Lenné, M.G. The crash at Kerang: Investigating systemic and psychological factors leading to unintentional non-compliance at rail level crossings. *Accid. Anal. Prev.* **2013**, *50*, 1278–1288. [CrossRef]
133. Salmon, P.M.; Goode, N.; Lenné, M.G.; Finch, C.F.; Cassell, E. Injury causation in the great outdoors: A systems analysis of led outdoor activity injury incidents. *Accid. Anal. Prev.* **2014**, *63*, 111–120. [CrossRef]
134. Salmon, P.; Goode, N.; Archer, F.; Spencer, C.; McArdle, D.; McClure, R. A systems approach to examining disaster response: Using Accimap to describe the factors influencing bushfire response. *Saf. Sci.* **2014**, *70*, 114–122. [CrossRef]
135. Underwood, P.; Waterson, P. Systems thinking, the Swiss Cheese Model and accident analysis: A comparative systemic analysis of the Grayrigg train derailment using the ATSB, AcciMap and STAMP models. *Accid. Anal. Prev.* **2014**, *68*, 75–94. [CrossRef]
136. Fan, Y.; Zhu, J.; Pei, J.; Li, Z.; Wu, Y. Analysis for Yangmingtan Bridge collapse. *Eng. Fail. Anal.* **2015**, *56*, 20–27. [CrossRef]
137. Hamim, O.F.; Hoque, S.; McIlroy, R.C.; Plant, K.L.; Stanton, N.A. Applying the AcciMap methodology to investigate the tragic Mirsharai road accident in Bangladesh. *MATEC Web Conf.* **2019**, *277*, 02019. [CrossRef]
138. Hulme, A.; Stanton, N.A.; Walker, G.H.; Waterson, P.; Salmon, P.M. Complexity theory in accident causation: Using AcciMap to identify the systems thinking tenets in 11 catastrophes. *Ergonomics* **2021**, *64*, 821–838. [CrossRef] [PubMed]
139. Igene, O.O.; Johnson, C.W.; Long, J. An evaluation of the formalised AcciMap approach for accident analysis in healthcare. *Cogn. Technol. Work* **2021**, *24*, 161–181. [CrossRef]
140. Zhu, D.; Yao, S. A Hazard Analysis Method for Software-Controlled Systems Based on System-Theoretic Accident Modeling and Process. In Proceedings of the 2018 IEEE 9th International Conference on Software Engineering and Service Science (ICSESS), Beijing, China, 23–25 November 2018; IEEE: Manhattan, NY, USA.
141. Sawaragi, T.; Horiguchi, Y.; Hina, A. Safety Analysis of Systemic Accidents Triggered by Performance Deviation. In Proceedings of the 2006 SICE-ICASE International Joint Conference, Busan, Korea, 18–21 October 2006; IEEE: Manhattan, NY, USA.

142. Herrera, I.; Woltjer, R. Comparing a multi-linear (STEP) and systemic (FRAM) method for accident analysis. *Reliab. Eng. Syst. Saf.* **2010**, *95*, 1269–1275. [CrossRef]
143. Frost, B.; Mo, J.P. System hazard analysis of a complex socio-technical system: The functional resonance analysis method in hazard identification. In Proceedings of the Australian System Safety Conference, Melbourne, Australia, 7–10 April 2014.
144. Tian, J.; Wu, J.; Yang, Q.; Zhao, T. FRAMA: A safety assessment approach based on Functional Resonance Analysis Method. *Saf. Sci.* **2016**, *85*, 41–52. [CrossRef]
145. Hosseinnia, B.; Khakzad, N.; Patriarca, R.; Paltrinieri, N. Modeling Risk Influencing Factors of Hydrocarbon Release Accidents in Maintenance Operations using FRAM. In Proceedings of the 2019 4th International Conference on System Reliability and Safety (ICSRS), Rome, Italy, 20–22 November 2019; IEEE: Manhattan, NY, USA.
146. O'Hara, J.K.; Baxter, R.; Hardicre, N. 'Handing over to the patient': A FRAM analysis of transitional care combining multiple stakeholder perspectives. *Appl. Ergon.* **2020**, *85*, 103060. [CrossRef]
147. Furniss, D.; Nelson, D.; Habli, I.; White, S.; Elliott, M.; Reynolds, N.; Sujan, M. Using FRAM to explore sources of performance variability in intravenous infusion administration in ICU: A non-normative approach to systems contradictions. *Appl. Ergon.* **2020**, *86*, 103113. [CrossRef]
148. Grabbe, N.; Kellnberger, A.; Aydin, B.; Bengler, K. Safety of automated driving: The need for a systems approach and application of the Functional Resonance Analysis Method. *Saf. Sci.* **2020**, *126*, 104665. [CrossRef]
149. França, J.E.M.; Hollnagel, E.; dos Santos, I.J.A.L.; Haddad, A.N. Analysing human factors and non-technical skills in offshore drilling operations using FRAM (functional resonance analysis method). *Cogn. Technol. Work* **2020**, *23*, 553–566. [CrossRef]
150. Salehi, V.; Hanson, N.; Smith, D.; McCloskey, R.; Jarrett, P.; Veitch, B. Modeling and analyzing hospital to home transition processes of frail older adults using the functional resonance analysis method (FRAM). *Appl. Ergon.* **2021**, *93*, 103392. [CrossRef]
151. de Carvalho, E.A.; Gomes, J.O.; Jatobá, A.; da Silva, M.F.; de Carvalho, P.V.R. Employing resilience engineering in eliciting software requirements for complex systems: Experiments with the functional resonance analysis method (FRAM). *Cogn. Technol. Work* **2021**, *23*, 65–83. [CrossRef]

sustainability

Article

Safety Risks Analysis: Moderating Effect of Risk Level on Mitigation Measures Using PLS-SEM Technique

Wong Chin Yew [1,*], Mal Kong Sia [2] and Own QianYi Janet [2]

1 Faculty of Social Sciences and Humanities, National University of Malaysia, Bangi 43600, Malaysia
2 Faculty of Built Environment, Tunku Abdul Rahman University of Management and Technology, Kuala Lumpur 53300, Malaysia
* Correspondence: vivienyew@ukm.edu.my

Abstract: The Malaysian construction sector registers higher fatal accidents than the manufacturing sector even though the latter has the highest cases of accidents. There is a need to implement effective safety risk management. The main objective of this study is to explore the moderating effect of risk level of accidents on mitigation measures implemented. For this purpose, the factors causing safety risks and the practical measures taken by contractors to mitigate these risks were identified, in addition to the operationalization of the likelihood and severity of accidents using suitable rating scales. Descriptive analysis shows that a fall-related accident is the most likely and the most severe safety risk at high risk level. Results from multivariate analysis using SmartPLS 4 show that safety risks have a significant positive relationship with mitigation measures, and risk level actually heightens this relationship. As a result, the practical measures implemented on construction sites to mitigate the impacts of accidents may be inadequate unless the moderating effect of risk level is considered during the planning, design, and management of construction safety. Therefore, mitigation measures taken by the contractors must take into account the types of factors causing safety risks, as well as the likelihood and severity of these factors.

Keywords: accidents; likelihood; mitigation measures; risk level; safety risks; severity; SmartPLS 4

Citation: Yew, W.C.; Sia, M.K.; Janet, O.Q. Safety Risks Analysis: Moderating Effect of Risk Level on Mitigation Measures Using PLS-SEM Technique. *Sustainability* **2023**, *15*, 1090. https://doi.org/10.3390/su15021090

Academic Editors: Esmaeil Zarei, Samuel Yousefi and Mohsen Omidvar

Received: 24 November 2022
Revised: 25 December 2022
Accepted: 4 January 2023
Published: 6 January 2023

1. Introduction

According to [1], risk is defined as the "*combination of the likelihood of occurrence of a work-related hazardous event or exposure(s) and the severity of injury and ill-health that can be caused by the event or exposures*". In addition, [2] defines "*a risk as the potential of a situation or event to impact on the achievement of specific objectives*". Risks are found in any business undertaking. As a result, incidents are bound to happen in any occupation of any sector which affects its smooth operation.

The construction industry is well-known for its complexity, dynamic nature, uniqueness, and diverse environments which could create uncertainty and challenges because the works involved nowadays could be high in the sky, deep underground, below water level, or across the sea which often involve adverse surroundings and situations. According to [3], risk may appear in any form and at any stage of the construction process. A construction site is thus full of hazards due to many people working in various activities and the use of heavy materials and moving machineries. Hence, the construction industry is highly prone to various factors which could cause safety risks. Therefore, implementing safety risk management in the industry is essential in anticipation of the unpredictable nature of safety risks with the objectives to mitigate or manage their impacts. Table 1 summarises the occupational accident statistics released by [4]. Among the industrial sectors listed, the construction sector registers the highest number of fatal accidents with an average of 89 cases per year over seven years compared to the manufacturing sector, even though the latter has the highest rate of accidents with an average of 3355 cases/year from 2015–2021, which is almost 14 times higher than the construction industry.

Table 1. Occupational accident statistics in Malaysia from 2015–2021.

Industrial Sector	Type	2015	2016	2017	2018	2019	2020	2021	Total from 2015–2021	Average over 7 Years	
										Type (per Year)	Sector (per Year)
Manufacturing	Death	46	68	68	62	73	73	48	438	62.6	3355
	NPD	1906	2173	1985	2969	4661	4202	4015	21911	3130.1	
	PD	89	74	125	197	214	231	206	1136	162.3	
Mining and Quarrying	Death	4	4	8	4	5	3	8	36	5.1	43.6
	NPD	32	20	37	34	52	35	44	254	36.3	
	PD	3	0	1	3	3	1	4	15	2.1	
Construction	Death	88	91	111	118	84	66	65	623	89.0	240.0
	NPD	138	126	123	106	227	137	147	1004	143.4	
	PD	11	5	6	8	15	3	5	53	7.6	
Agriculture, Forestry, Logging and Fishery	Death	31	23	23	26	43	43	16	205	29.3	763.7
	NPD	440	435	488	709	1111	916	939	5038	719.7	
	PD	9	9	11	14	22	20	18	103	14.7	
Utility (Electricity, Gas, Water and Sanitary Services)	Death	6	2	10	5	9	3	8	43	6.1	161.7
	NPD	86	68	90	168	245	214	198	1069	152.7	
	PD	4	4	4	NA	4	3	1	20	2.8	
Transport, Storage and Communication	Death	22	12	16	12	21	11	6	100	14.3	215.6
	NPD	107	113	105	124	359	294	281	1383	197.6	
	PD	2	2	1	1	9	6	5	26	3.7	
Wholesale and Retail Trade	Death	3	0	10	1	0	1	2	17	2.4	113.0
	NPD	102	107	86	69	85	126	182	757	108.1	
	PD	3	4	1	3	2	1	3	17	2.4	
Hotel and Restaurants	Death	0	3	3	1	5	2	0	14	2.0	127.1
	NPD	62	85	110	120	227	137	125	866	123.7	
	PD	0	2	1	2	3	1	1	10	1.4	
Financial, Insurance, Real Estate and Business Services	Death	14	14	16	22	16	8	17	107	15.3	232.3
	NPD	105	101	124	190	384	312	264	1480	211.4	
	PD	0	11	6	5	6	7	4	39	5.6	
Public Services and Statutory Bodies/Authorities	Death	0	6	2	9	3	3	4	27	3.9	73.7
	NPD	31	101	64	48	93	73	68	478	68.3	
	PD	1	3	0	1	3	1	2	11	1.6	
All industrial sectors combined	Death	214	223	267	260	259	213	174	1610	230.0	5322.9
	NPD	3009	3329	3212	4537	7444	6446	6263	34240	4891.4	
	PD	122	114	156	234	281	274	249	1430	204.3	

Note: NPD = Non-Permanent Disabilities; PD = Permanent Disabilities; NA = Not Available.

Common safety risks in the construction industry, such as fall from height, being struck by a moving object, or workers being buried in a landslide can be significantly reduced if not eliminated by introducing safety management. Therefore, it should be taken as a critical element for creating value and thus increasing a project's overall performance in terms of time, quality, and cost [5]. Realistically, achieving comprehensive and effective safety management is a challenge to all project managers because they must anticipate the risks that may occur and the resulting consequences. However, it has been found that the level of safety management practices in Malaysia construction companies is relatively low because of the lack of knowledge and understanding on the subject [5]. According to [6], most construction companies in Malaysia fail to implement a systematic process of risk management. It has been found that safety risks could be accepted, transferred, and mitigated by implementing a systematic process in safety management.

To address the lack of knowledge and understanding on safety management practices, this study set out to investigate the relationship between safety risks (SR) and the mitigation measures (MM) implemented to mitigate the impacts of these risks, as shown in Figure 1. It is hypothesised that there is a positive relationship between SR and MM. To achieve this purpose, this study aims to: (a) determine the likelihood and severity of commonly-occurred fatal construction accidents; (b) identify the factors causing safety risks; (c) identify the practical measures taken by contractors to mitigate or manage these safety risks; and

(d) explore the moderating effect of likelihood and severity, in terms of risk level, on the mitigation measures implemented using the PLS-SEM technique.

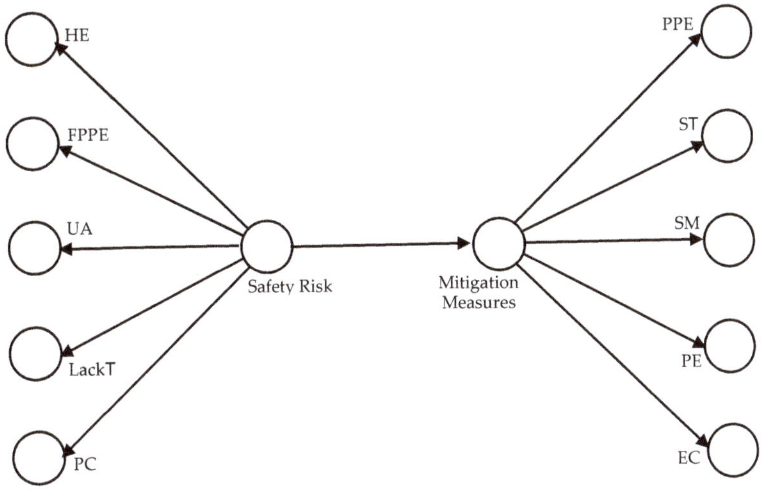

Figure 1. Path model on the relationship between safety risks and mitigation measures.

Likelihood and severity are expected to affect the relationship depicted in Figure 1. Inherent in any risk are the likelihood of an accident to happen and the severity of its impact when it happens. According to [7], risk increases when the probability of an incident occurring increases or the severity of injury increases. The more likely it is for an accident to occur, and the more severe the accident, the higher the risk level. Hence, these twin characteristics of risks, working hand in hand, will determine the risk level of an accident which is given by "risk level = likelihood × severity". The moderating effect of risk level on the relationship can be quantified by examining the R-squared value and path coefficient between these two constructs when risk level acts as a moderator between safety risks and mitigation measures.

2. Literature Review

Any industry which wants to succeed must operate safely, dependably, and on a long term basis [8]. Risks that have not been identified and managed will undoubtedly threaten a project's objectives, resulting in high cost and schedule overruns [9]. To accomplish this goal, the industry must first identify the dangers and assess the risks connected with them. If an industry could identify and categorise risks before the commencement of a project, they would be able to improve risk management and avoid any potential losses.

2.1. Commonly-Occurred Fatal Construction Accidents

Ref. [10] conducted a study in the United States based on the OSHA fatalities data from 1980, 1985, and 1990. They concluded that fall-related, struck-by, electrocution, and being caught in-between are the most common forms of accidents. Table 2 shows the statistics from 2016 to 2020 on the various types of fatal construction accidents provided by [11], where nearly 50% of the fatal accidents that happened were due to workers falling from height. Other studies have also revealed that fall-related accidents are the most common fatal construction accidents [12–15], including in China [16].

Table 2. Commonly-occurred fatal construction accidents.

Item	Type of Accident	Period Unavailable in [12]		From 2010–2015 in [13]		From 2013–2018 in [14]		From 2010–2018 in [15]		From 2016–2020 for this Study [11]	
		Cases	Percent	Cases	Percent	Cases	Percent	Cases	Percent	Cases	Percent
1	Falling from height	17	56.7%	56	43.4%	63	43.4%	304	38.2%	40	48.8%
2	Struck-by accident (e.g., moving object, moving vehicle, or by falling object)	4 + 2	20.0%	33	25.6%	49	33.8%	242	30.4%	25	30.5%
3	Fall into opening or drowning	2	6.7%	6	4.7%	8	5.5%	78	9.8%	5	6.1%
4	Buried	2	6.7%	8	6.2%	—	—	—	—	5	6.1%
5	Electrocution	3	10.0%	7	5.4%	7	4.8%	22	2.8%	3	3.7%
6	Road accident	—	—	—	—	—	—	—	—	1	1.2%
7	Caught in between	—	—	17	13.2%	11	7.6%	141	17.7%	1	1.2%
8	Fire or explosion	—	—	1	0.8%	2	1.4%	3	0.4%	1	1.2%
9	Insect pest	—	—	—	—	—	—	—	—	1	1.2%
10	Exposure to, or contact with, harmful substances	—	—	1	0.8%	1	0.7%	6	0.8%	—	—
11	Environmental factors	—	—	—	—	4	2.8%	—	—	—	—
Total		30	100%	129	100%	145	100%	796	100%	82	100%

2.1.1. Fall-Related Accident

Fall-related accidents are the most common type of safety risk not only in Malaysia but also in many other countries such as the United States, China, the United Kingdom, Spain, Korea, Singapore and Taiwan [15]. When compared to other forms of safety risks in the construction industry, fall-related accidents are believed to have the highest frequency of occurrence [17]. Any object that might cause a person to lose their balance and fall is considered to be a danger while working four feet or more above the ground. The majority of workplace accidents involve falling from a working platform, scaffolding, ladder, or structure. As a result, falls from height are still much more common in construction accidents than in other kinds of accidents [18].

2.1.2. Struck-by Accident

Being struck by any objects or equipment is known to be one of the factors that led to fatal injuries and deaths in the Malaysian construction industry from 2010 to 2018 [15]. A struck-by accident happens when a worker encounters any moving, dropping, or rolling material or object forcibly [19]. It shall also include incidents where the workers on-site or in public get hit by any falling material, moving vehicle, or machinery [20].

(a) Struck by a Swinging or Slipping Object

When materials are mechanically raised, there is a possibility that they may swing and harm the employees below. As the weight is lifted, the materials may swing, twist, or spin in their respective positions. This movement has the potential to catch employees off guard, and they may be struck by the swinging load. Windy circumstances are particularly dangerous since the weight will swing more widely. If the worker is hit from behind and falls to another level, the worker may receive even more severe injuries. This is dependent on where the worker is positioned and the power behind the weight [21].

(b) Struck by a Rolling Object

When an object is rolling, moving, or sliding on the same level as the worker, this is referred to as being struck by a rolling object. Incidents when the worker is hit or run over by a moving vehicle without being trapped beneath it, as well as incidents where the worker is struck by a sliding item or piece of equipment on the same level, are included under this category [22].

(c) Struck by Falling Object

Injuries sustained as a result of being struck by a falling object or equipment occurred when the source of the injury is falling from a higher to a lower level. This includes instances in which the injured person is crushed, pinned, or caught under an object falling from a higher to a lower level [22].

2.1.3. Drowning and Asphyxiation

Drowning is considered as the world's third highest factor causing fatal injury or death [15]. Drowning occurs when a person dies as a result of suffocation caused by a liquid that limits or blocks oxygen intake into the human body from the air, resulting in asphyxia [12]. Asphyxiation, on the other hand, is a situation comparable to drowning in which insufficient oxygen occurs in the human body as a result of poor breathing as a result of working in a confined space or drowning [23].

2.1.4. Buried

Accidents may happen when construction workers are found buried due to cave-in or collapse of earth during or after excavation work [24]. The author of [25] reported the occurrence of a gruesome work accident which led to the death of a construction worker after he was buried alive under a landslide.

2.1.5. Electrocution

Generally, an electrical hazard refers to the risk of getting burned, electrocution, shock, arc flash, or other injury due to exposure to a lethal amount of electrical energy. Burns could be defined as injuries due to contact or exposure to electricity, arc flash, or thermal contact, while shock often results when the human body reflex responds to the passage of electric current [26].

2.1.6. Road Accident

Road accident is one of the safety risks in the Malaysian construction industry. According to the Department of Occupational Safety and Health [11], a truck driver died in a road accident due to a malfunctioning blinker at a sharp bend. Road accidents could also happen due to the vehicle's brake failure and hydroplaning. "*Increasing number of highway construction zones*" in highway construction projects have disrupted regular traffic flows which could cause traffic safety problems and accidents [27].

2.1.7. Caught in-between Accidents

Caught in-between accidents occurred when two or more objects or components of an object are caught, squeezed, compressed, crushed, or pinched between one's body [13]. There are times when a construction worker is too focused on their own tasks and fails to see caught in-between hazards, such as standing between a heavy machine, such as a trailer and a forklift, or an immovable structure, such as a brick wall [28]. According to [29], incidents involving being squashed or crushed between rolling, sliding, or shifting things are also regarded as one of the most common forms of accidents in the construction industry.

2.1.8. Fire or Explosion

The potential danger of fire outbreak is particularly severe on many construction sites, especially during those high-risk activities such as hot work that generates heat, sparks or flame, or even overheating of the plants and equipment [30]. In fact, fire would easily break out with the presence of sufficient oxygen, fuel, and a source of ignition arising from hot work, overheating plant and equipment, smoking, faulty electrical installation, bonfires or arson [31].

The occurrence of explosions in construction sites, in fact, is not so frequent, but such risks will lead to significant consequences: not only defects on the structure but also the potential loss of a worker's life. There was a case of explosion in 2017 at a Malaysian MRT construction site caused by an old bomb from the Second World War, which resulted in the death of one construction worker while another two were critically injured [32].

2.1.9. Insect Pest

According to [11], there was one worker who died after been stung by hornets at the Sarawak construction site. The employer was required to conduct HIRARC to identify such a risk and provide risk control measures such as destroying the honeycomb to prevent the safety risk from happening.

2.2. Factors Influencing Safety Risks

Accidents may happen on a construction site due to many reasons [33]. There are many heavy plants, heavy materials, rough terrains, and people working at high places. As a result, a construction site is a high-risk place.

2.2.1. Human Errors

Human errors, no matter how minor, may occasionally have a domino effect, resulting in enormous economic or life loss [34]. Human errors are often related with improper attitude, inadequate tools used, body effort and lack of experience [35]. Human errors are considered to be the main cause of fall-related accidents. The contributing factors to fall-related accidents include human errors and inappropriate use of a control [36]. Workers' negligence in judgement accounts for approximately one-third of the fall accidents [37].

2.2.2. Failure to Use Personal Protective Equipment (PPE)

Every year, a large number of construction workers are killed or seriously harmed due to the improper usage and wearing of personal protective equipment (PPE) [38]. According to statistics from throughout the globe, 2 million individuals are predicted to be disabled each year as a result of work-related accidents, with 25% or more of those injuries occurring to the head, eyes, hands, and feet [39]. This is due to a lack of knowledge and use of safety equipment, such as hard safety helmets, which are only worn by 16% of those who have had occupational head injuries [40]. In addition, 23% of employees who had worn safety boots suffered from foot injuries. Moreover, 40 percent of those who had suffered from eye injuries had worn eye protection [39]. According to statistics, although there is no assurance that personal safety equipment can prevent incidents resulting in injuries from occurring, it may at least minimise the likelihood of such an incident occurring [41]. Ref. [42] believed that precise safety applications may help to minimise construction site accidents, as well as production costs, productivity development and profitability. Most significantly, he added, lives could be saved.

2.2.3. Unsafe Act and Site Condition

The major fundamental factors of accident cases are unsafe acts and site circumstances [43]. In total, 99% of construction safety risks are caused by either risky conduct or unsafe conditions, or both of these factors together [44]. These are regarded as the primary causes of all forms of construction safety risks. Unsafe activities are defined as the misuse of safety procedures, which increases the likelihood of an accident occurring on the construction site [13]. An unsafe site condition is a physical condition or environment that is surrounded by possible risks and might be the cause of a site accident [19]. The dangerous act mostly deals with hazardous equipment or unsafe methods, such as working without safety devices, equipment failure, inappropriate work process, worker knowledge level, and failure to follow work procedures [19]. Unsafe circumstances, on the other hand, include missing or inadequate guardrails on platforms, malfunctioning tools and equipment, fire dangers, a bad fire alarm system, a lack of housekeeping, poor climatic conditions, excessive noise, and insufficient light to operate.

site [65]. There is thus a need for foreign migrant workers to attend safety training package in order to address the higher accident rates than the local skilled workers [66].

The majority of foreign workers lack knowledge and awareness since they did not attend safety training. It is critical that the content of training uses more illustrations to explain it in order to increase worker safety awareness [67]. Workers will be able to understand and know how to manage the machine more effectively if the training techniques use animation to display and explain the processes of operating machines [68].

2.3.3. Safety Meeting

A safety meeting is one of the ways that will be used to offer an opportunity for all parties participating in the construction team to introduce and discuss the precautionary safety concerns linked to safety and health on the construction site. Before beginning work, a safety meeting must be held to ensure that all personnel are on the same page and may review the previously provided information [69]. A safety meeting is an important aspect of developing a workers' safety culture in order to reduce accidents on the construction site [45].

Before beginning a new project, kick-off meetings should be held to discuss the risks and hazards, how to select and utilise personal protective equipment (PPE), safety precautions, and safe work procedures that will be implemented at each stage of construction [70].

2.3.4. Proper Equipment

The construction company is responsible for supplying employees with suitable equipment and a safe working environment in order to properly implement the construction site safety culture among workers [40]. Poorly maintained equipment and machinery may result in significant injuries and fatalities. It is critical to offer suitable equipment and machinery that is in excellent working order. Machines must be serviced on a regular basis to guarantee proper operation. Even just a tiny piece of the tools also need to be handled well when carrying out the job in a construction site as it may extremely reduce the opportunity for injury or the fatality of a construction worker. Scaffolding, for example, must be built in the proper manner to provide construction workers safe access to the other level of the structure. As a result, the employer must provide enough equipment at all times while complying with OSHA's safety regulations [28].

2.3.5. Promote Effective Communication

Promoting effective communication on-site by all construction parties is needed to prevent accidents from happening. In order to avoid workplace accidents, workers, supervisors, managers, contractors, and everyone on-site should be encouraged to communicate with each other and with the employer [71]. Good and concise communication emphasising safety issues shall be practiced among everyone in the construction site so that any misfortune may be avoided [72]. Ref. [73] found that it is important to promote safety communication among construction workers because this will encourage workers to participate actively in providing and receiving safety information.

Poor and ineffective communication can be due to many factors. Ref. [74] identified 33 factors which are responsible for poor communication in the construction industry. Of the 30 factors identified, [51] has categorized them under four dimensions, namely, organizational and management factors, behavioural and cultural factors, project information factors, and technology and method factors. A high accident rate has been found to be one of the impacts of poor and ineffective communication by [50,74].

3. Methodology

This study used the quantitative research approach for data collection and analyses.

3.1. Research Design

In this study, partial least squares structural equation modelling (PLS-SEM) using SmartPLS 4 software [75] was employed as the multivariate analysis technique to explore the moderating effect of risk level on mitigation measures taken. Hence, a survey questionnaire is suitable for data collection as long as the measurement scales are equidistant. The basic conceptual model used for this study is shown in Figure 1: the safety risks construct is conceptualized as a second order hierarchical latent construct consisting of five categories of factors causing safety risks, and the mitigation measures construct is conceptualized as a second order hierarchical latent construct also consisting of five categories of mitigation measures.

Sampling

The respondents, purposively selected for this study, comprised personnel working in the construction industry from the Klang Valley, Malaysia. Three hundred (300) copies of questionnaires prepared in Google Forms were distributed through emails and WhatsApp messenger to the respondents from June 2022 to August 2022. A total of 83 completed questionnaires were received with no missing data, giving a response rate of nearly 28%.

3.2. Research Instrument

The questionnaire consists of four main sections with closed-ended questions as explained below.

3.2.1. Demographic Information

This section is designed to collect the demographic information of respondents such as education level, current practice and the total number of years of working experience, types of projects involved in, and familiarity with management of safety risks.

3.2.2. Likelihood and Severity of Commonly-Encountered Accidents

This section consists of a list of nine commonly-occurred fatal construction accidents listed in Table 2. The respondents were requested to rate the likelihood and severity of these safety risks based on their opinions and experiences according to 5-point rating scales as shown in Table 3.

Table 3. Rating scales for likelihood and severity of commonly-occurred accidents.

Likelihood of Commonly-Occurred Accidents			Severity of Commonly-Occurred Accidents		
Rating	**Likelihood**	**Definition**	**Rating**	**Severity**	**Definition**
1	Inconceivable	Has never occurred	1	Negligible	First aid, minor abrasions, cuts
2	Remote	Has not been known to occur after many years	2	Minor	Outpatient, medical leave not more than 4 days
3	Conceivable	Might occur sometimes in future	3	Serious	Hospitalized, medical leave 5 days or more
4	Possible	Chances to occur and not unusual	4	Major	Permanent disability, single fatality
5	Most likely	Happen extremely	5	Catastrophic	Numerous fatalities

3.2.3. Factors Influencing Safety Risks

This section contains 18 questions grouped under five categories of factors influencing safety risks. The respondents were requested to rate these factors measured on a 5-point Likert scale from '1 = strongly disagree', '2 = disagree', '3 = neutral', '4 = agree' to '5 = strongly agree' based on their opinion and experiences.

3.2.4. Practical Measures Taken

This section contains 20 questions grouped under five categories of practical measures taken to mitigate safety risks. The respondents were requested to rate the importance of these practical measures measured on a 5-point Likert scale from '1 = not important at all', '2 = slightly important', '3 = moderately important', '4 = important' to '5 = very important' based on their opinion and experiences.

4. Results

4.1. Descriptive Analysis

Table 4 presents the demographic information of the 83 respondents who participated in the questionnaire survey. Out of the 83 questionnaires received, 62 of the respondents are working in consultancy firms, while 21 respondents are from contractor companies. Of the 83 respondents, 62 of them have more than 2 years of working experience. In terms of educational background, 72 of them have at least a bachelor's degree and above. Of the 83 respondents who participated in the questionnaire survey, 77 of them indicated they are familiar with safety risks. In terms of projects involved, 36 respondents mentioned they are involved with main building works, whereas 47 of the respondents mentioned they are involved in infrastructure works, including highway and railway projects.

Table 4. Demographic information of respondents.

Item	Response Category	Frequency	Percentage (%)	Total Percentage (%)
Current practice	Consultant	62	74.7	100
	Contractor	21	25.3	
Types of projects involved	Highway project	17	20.5	100
	Infrastructure works	20	24.1	
	Main building works	36	43.4	
	Railway projects	10	12.0	
Educational level	SPM	1	1.2	100
	Diploma	10	12.0	
	Bachelor's Degree	60	72.3	
	Master's Degree	10	12.0	
	PhD	2	2.4	
Working experience in the construction industry	2 years or less	21	25.3	100
	3–6 years	24	28.9	
	7–10 years	24	28.9	
	11–14 years	9	10.8	
	15 years and above	5	6.0	
Familiarity with safety risks	Yes	77	92.8	100
	No	6	7.2	

4.1.1. Likelihood and Severity of Commonly-Occurred Fatal Construction Accidents

Table 5 displays the results for the likelihood of fatal construction accidents commonly happening in the Malaysian construction industry, with fall-related accidents having the highest mean value of 3.96, and insect pest as a safety risk having the lowest mean value of 3.18. The overall mean value is 3.57. In Table 5, the indicators for this construct have skewness values ranging from −0.108 to −1.072, and kurtosis values ranging from −1.322 to 0.292, showing that these indicators do not depart from the normality requirements according to Brown (cited in [76]).

Table 5. Likelihood of commonly-occurred fatal construction accidents (n = 83).

Construct	Indicator	Commonly-Occurred Accidents	Mean	Standard Deviation	Skewness	Kurtosis	Overall Mean
Likelihood	Likelihood1	Fall-related accident (human falling from height)	3.96	1.163	−1.072	0.292	3.57
	Likelihood2	Struck-by accident (struck by falling object, moving vehicle, rolling machinery)	3.54	0.979	−0.640	−0.179	
	Likelihood3	Drowning and Asphyxiation (insufficient oxygen)	3.31	1.352	−0.108	−1.322	
	Likelihood4	Buried (being buried under the landslide)	3.51	1.108	−0.511	−0.483	
	Likelihood5	Electrocution (getting burn, electrocution, shock, arc flash)	3.71	1.099	−0.866	0.099	
	Likelihood6	Road accident (hydroplaning, brake failure)	3.73	1.149	−0.746	−0.247	
	Likelihood7	Caught in-between accidents (caught, crushed, squeezed between two or more objects on site)	3.51	1.173	−0.526	−0.637	
	Likelihood8	Fire or explosion (fire outbreak, bomb explosion)	3.64	1.143	−0.652	−0.409	
	Likelihood9	Insect pest (for example: stung by hornets)	3.18	1.354	−0.368	−1.159	

Table 6 displays the results for the severity of the same fatal construction accidents commonly happening in the Malaysian construction industry, with fall-related accidents as the highest mean value of 4.18, and insect pest as a safety risk as the lowest mean value of 3.19. The overall mean value is 3.73. In Table 6, the indicators for this construct have skewness values ranging from −1.532 to −0.362, and kurtosis values ranging from −1.140 to 1.722, showing that these indicators, too, do not depart from the normality requirements according to Brown (cited in [76]).

Table 6. Severity of commonly-occurred fatal construction accidents (n = 83).

Construct	Indicator	Commonly-Occurred Accidents	Mean	Standard Deviation	Skewness	Kurtosis	Overall Mean
Severity	Severity1	Fall-related accident (human falling from height)	4.18	1.106	−1.532	1.722	3.73
	Severity2	Struck-by accident (struck by falling object, moving vehicle, rolling machinery)	3.72	0.860	−0.842	0.706	
	Severity3	Drowning and Asphyxiation (insufficient of oxygen)	3.57	1.139	−0.446	−0.561	
	Severity4	Buried (being buried under the landslide)	3.76	1.043	−0.886	0.583	
	Severity5	Electrocution (getting burn, electrocution, shock, arc flash)	3.87	0.985	−0.748	−0.013	
	Severity6	Road accident (hydroplaning, brake failure)	3.61	1.188	−0.817	−0.118	
	Severity7	Caught in-between accidents (caught, crushed, squeezed between two or more objects on site)	3.83	1.069	−0.944	0.326	
	Severity8	Fire or explosion (fire outbreak, bomb explosion)	3.82	1.038	−0.835	0.369	
	Severity9	Insect pest (for example: stung by hornets)	3.19	1.339	−0.362	−1.140	

4.1.2. Factors Influencing Safety Risks

Table 7 displays the results for the 18 indicators operationalizing the five categories of factors influencing safety risks, with unsafe act and site condition having the highest overall mean value of 4.31, and human error having the lowest overall mean value of 3.97. In Table 7, the indicators for this construct have skewness values ranging from −2.017 to

0.159, and kurtosis values ranging from −1.116 to 6.395, showing that these indicators do not depart from the normality requirements according to Brown (cited in [76]).

Table 7. Factors influencing safety risks (n = 83).

Construct	Indicator	Factors Influencing Safety Risks	Mean	Standard Deviation	Skewness	Kurtosis	Overall Mean
Human Error (HE)	HE1	Improper attitude	3.93	0.921	−0.238	−1.080	3.97
	HE2	Inadequate tools used	3.75	0.660	−0.199	0.097	
	HE3	Excessive physical exertion	4.18	0.587	−0.421	1.551	
	HE4	Lacks of experience	4.00	0.812	−0.140	−1.116	
Failure to use PPE (FPPE)	FPPE1	Failure to use safety helmets	4.52	0.722	−1.560	2.276	4.25
	FPPE2	Failure to use face protection	4.07	0.640	−0.921	2.723	
	FPPE3	Failure to use safety boots	4.23	0.831	−1.238	2.139	
	FPPE4	Failure to use eye protection	4.17	0.730	−0.852	1.166	
Unsafe act and site condition (UA)	UA1	Unsafe equipment	4.47	0.721	−1.190	0.746	4.31
	UA2	Unsafe methods	4.16	0.529	0.159	0.298	
	UA3	Hazardous environment	4.25	0.622	−0.229	−0.570	
	UA4	Improper work procedure	4.36	0.691	−0.847	0.469	
Lack of progressive training (LackT)	LackT1	Employer fail to offer sufficient training	4.34	0.928	−1.481	1.821	4.13
	LackT2	Lack of budget allocation on safety management	3.93	0.729	−2.017	6.395	
	LackT3	Lack of workforce due to subcontract work	4.12	0.929	−1.273	1.908	
Poor Communication (PC)	PC1	Language barrier	4.22	1.048	−1.557	2.244	4.14
	PC2	Miscommunication and misunderstanding	4.02	0.796	−1.084	2.249	
	PC3	Failure in conveying message	4.17	0.895	−1.284	2.260	

Note: 1 = strongly disagree; 2 = disagree; 3 = neutral; 4 = agree; 5 = strongly agree.

4.1.3. Mitigation Measures Taken

Table 8 displays the results for the 20 indicators operationalizing the five categories of mitigation measures, with proper equipment having the highest overall mean value of 4.47, and safety meeting having the lowest overall mean value of 4.30. In Table 8, the indicators for this construct have skewness values ranging from −3.752 to 1.025, and kurtosis values ranging from −0.993 to 19.0722, showing that only one indicator, that is PPE1, departs from the normality requirements according to Brown (cited in [76]).

4.2. *Structural Equation Modeling*

The Mann–Whitney U tests carried out earlier showed that there were no significant differences between the two subgroups, namely 62 respondents from consultant practices and 21 respondents from contractor companies for all the indicators of the twelve constructs. The raw data for these two subgroups were then combined to test the conceptual model shown in Figure 1. SmartPLS 4 software [75] was employed for partial least squares structural equation modelling (PLS-SEM) purposes. The 2-step procedure recommended by [77] was adopted for assessments of the measurement models and structural model.

4.2.1. Assessment of Measurement Models

The following are the quality criteria adopted for assessment of the measurement models in Figure 1:

1. Internal consistency reliability: A construct with high Cronbach's alpha value indicates the indicators have similar range and meaning [78];
2. Composite reliability (CR): Values greater than 0.60 are acceptable in exploratory study [79];
3. Indicator reliability: Loading values equal to and greater than 0.4 are acceptable if the sum of loadings results in higher loading scores, contributing to AVE scores of greater than 0.5 [80];

4. Convergent validity: In order to achieve adequate convergent validity, each construct should account for at least 50% of the average variance explained (AVE ≥ 0.50) [81–83];
5. Rho_A: The reliability of rho_A usually lies between Cronbach's alpha and composite reliability [84];
6. Discriminant validity: The square root of AVE of a construct should be larger than the correlations between the construct and other constructs in the model [82]. According to [85], HTMT$_{.90}$ value of 0.90 indicates that there is a problem of discriminant validity. Using cross loadings to assess discriminant validity, each indicator should load high on its own construct but low on other constructs. Cross loadings of <0.1 should be deleted [86].

Table 8. Practical measures taken to mitigate safety risks (n = 83).

Construct	Indicator	Practical Measures to Mitigate Safety Risks	Mean	Standard Deviation	Skewness	Kurtosis	Overall Mean
Personal Protective Equipment (PPE)	PPE1	Safety helmets	4.76	0.597	−3.752	19.072	4.45
	PPE2	Ear protection	4.16	0.689	−0.672	0.970	
	PPE3	High visibility clothing	4.39	0.641	−0.841	1.047	
	PPE4	Safety footwear	4.41	0.716	−1.612	4.911	
	PPE5	Safety harnesses	4.53	0.591	−1.207	2.406	
	PPE6	Training of PPE	4.45	0.590	−0.517	−0.632	
Safety and health training (ST)	ST1	Safety measures training	4.55	0.737	−1.317	0.162	4.35
	ST2	Machinery operator training	4.20	0.435	1.025	0.337	
	ST3	Working at height training	4.30	0.745	−0.555	−0.993	
Safety Meeting (SM)	SM1	Discuss the precautionary safety concerns	4.48	0.755	−1.245	0.565	4.30
	SM2	Communication between job groups	4.20	0.488	0.462	0.227	
	SM3	Report changes at the work site	4.36	0.531	0.081	−0.969	
	SM4	Update the existing safety plan and procedure	4.17	0.640	−0.163	−0.579	
Proper equipment (PE)	PE1	Supplying employees with suitable equipment	4.65	0.572	−1.419	1.081	4.47
	PE2	Safe working environment	4.30	0.535	0.124	−0.598	
	PE3	Machines serviced regularly	4.46	0.591	−0.925	1.881	
	PE4	Scaffolding with safe access	4.45	0.590	−0.517	−0.632	
Promote Effective Communication (EC)	EC1	Employee pay attention for safety briefing	4.36	0.790	−0.897	−0.293	4.33
	EC2	Construction parties communicate with each other	4.14	0.665	−0.169	−0.717	
	EC3	Rapid communication such as walkie-talkies	4.48	0.549	−0.380	−0.980	

Note: 1 = not important at all, 2 = slightly important, 3 = moderately important, 4 = important and 5 = very important.

Figure 1 is a higher component model (HCM) with two second-order hierarchical latent constructs. Hence, two-stage HCM analysis, a combination of repeated indicators approach and the use of latent variable scores is needed [87]. First stage analysis shows the following two indicators have to be removed for the lower order component models to achieve quality criteria, namely:

(a) Indicator PPE1 which has cross loadings of <0.10 needs to be removed for the PPE construct to achieve an AVE value > 0.50;
(b) Indicator HE3 which has cross loadings < 0.10 needs to be removed even though the AVE value for the HE construct > 0.50.

In the second stage analysis, the latent variable scores from the first stage serve as the manifest variables in the higher order component models. The AVE values for safety risks construct and mitigation measures construct together with the outer loadings, path coefficient, and p values are displayed in Figure 2.

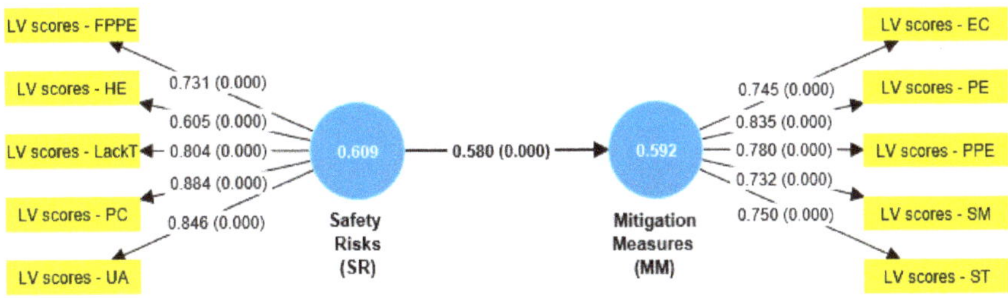

Figure 2. Higher order components with AVE values, path coefficients, and *p* values ($n = 83$).

The moderating effect of likelihood and severity is investigated with risk level as a second order hierarchical construct as shown in Figure 3, which gives the graphical output from first stage analysis. The AVE values of lower order constructs for safety risks, mitigation measures and risk level together with the outer loadings, path coefficient and *p* values are presented together.

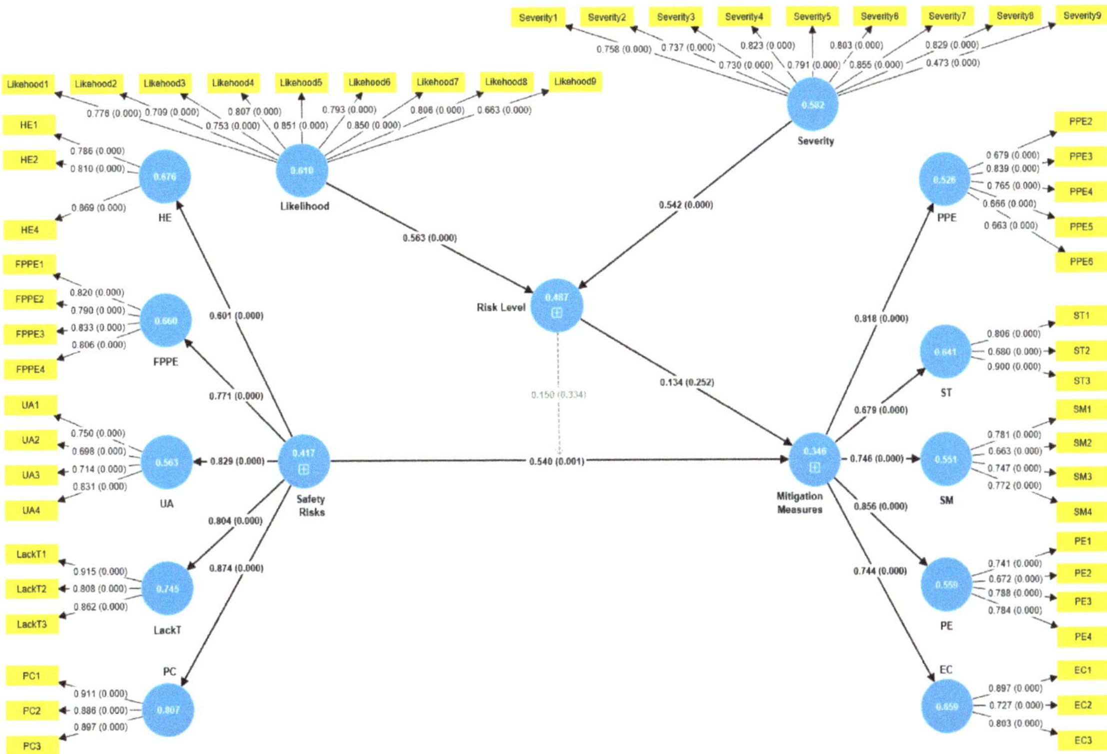

Figure 3. AVE values, path coefficients, and *p* values from first stage analysis ($n = 83$).

The assessment results for lower order components from first stage analysis are summarised in Table 9. Based on the quality criteria given earlier, the lower order components achieve convergent validity with AVE > 0.50 for all the twelve constructs, indicator reliability with outer loadings ranging from 0.473 to 0.915, as well as construct reliability and validity with Cronbach's alpha, Rho A and Rho C values well above 0.700.

Table 9. Assessment results for lower order components from first stage analysis (n = 83; CI = 95%).

Lower Order Component	Indicator	Outer Loadings	Construct Reliability and Validity			
			Cronbach's Alpha	Composite Reliability, Rho_A	Composite Reliability, Rho_C	AVE (≥0.50)
Likelihood of commonly-occurred accidents	Likehood1 Likehood2 Likehood3 Likehood4 Likehood5 Likehood6 Likehood7 Likehood8 Likehood9	0.776 0.709 0.753 0.807 0.851 0.793 0.850 0.806 0.663	0.919	0.923	0.933	0.610
Severity of commonly-occurred accidents	Severity1 Severity2 Severity3 Severity4 Severity5 Severity6 Severity7 Severity8 Severity9	0.758 0.737 0.730 0.823 0.791 0.803 0.855 0.829 0.473	0.907	0.915	0.925	0.582
Human Error (HE)	HE1 HE2 HE4	0.786 0.810 0.869	0.763	0.774	0.862	0.676
Failure to use Personal Protective Equipment (FPPE)	FPPE1 FPPE2 FPPE3 FPPE4	0.820 0.790 0.833 0.806	0.830	0.840	0.886	0.660
Unsafe act and site condition (UA)	UA1 UA2 UA3 UA4	0.750 0.698 0.714 0.831	0.741	0.753	0.837	0.563
Lack of progressive training (LackT)	LackT1 LackT2 LackT3	0.915 0.808 0.862	0.827	0.834	0.897	0.745
Poor Communication (PC)	PC1 PC2 PC3	0.911 0.886 0.897	0.880	0.880	0.926	0.807
Personal Protective Equipment (PPE)	PPE2 PPE3 PPE4 PPE5 PPE6	0.679 0.839 0.765 0.666 0.663	0.771	0.782	0.846	0.526
Safety and health training (ST)	ST1 ST2 ST3	0.806 0.680 0.900	0.711	0.734	0.841	0.641
Safety Meeting (SM)	SM1 SM2 SM3 SM4	0.781 0.663 0.747 0.772	0.726	0.726	0.830	0.551
Proper equipment (PE)	PE1 PE2 PE3 PE4	0.741 0.672 0.788 0.784	0.735	0.740	0.835	0.559
Promote effective communication (EC)	EC1 EC2 EC3	0.897 0.727 0.803	0.738	0.747	0.852	0.659

Table 10 shows the lower order components do not have any problem with discriminant validity because there is no $HTMT_{.90}$ value which is more than 0.90. In addition,

Table 11 shows the lower order components achieve satisfactory discriminant validity too because the square root of AVE (along the diagonal) is larger than the correlation (off diagonal) for all the lower order components.

Table 10. Discriminant validity for lower order components using HTMT ratio of correlation.

Indicator	EC	FPPE	HE	LackT	Likelihood	PC	PE	PPE	SM	ST	Severity	UA
EC												
FPPE	0.304											
HE	0.168	0.486										
LackT	0.485	0.517	0.440									
Likelihood	0.217	0.399	0.698	0.351								
PC	0.553	0.617	0.493	0.837	0.256							
PE	0.659	0.357	0.309	0.258	0.313	0.454						
PPE	0.783	0.286	0.335	0.406	0.350	0.494	0.865					
SM	0.650	0.241	0.329	0.285	0.438	0.456	0.758	0.531				
ST	0.451	0.406	0.436	0.506	0.330	0.704	0.751	0.507	0.684			
Severity	0.221	0.391	0.476	0.442	0.700	0.483	0.322	0.441	0.413	0.376		
UA	0.533	0.718	0.477	0.718	0.459	0.790	0.624	0.552	0.443	0.689	0.474	

Table 11. Discriminant validity for lower order components using Fornell and Larcker criterion.

Indicator	EC	FPPE	HE	LackT	Likelihood	PC	PE	PPE	SM	ST	Severity	UA
EC	0.812											
FPPE	0.247	0.812										
HE	0.139	0.405	0.822									
LackT	0.382	0.442	0.367	0.863								
Likelihood	0.179	0.348	0.598	0.305	0.781							
PC	0.450	0.541	0.416	0.716	0.229	0.898						
PE	0.491	0.292	0.239	0.172	0.248	0.366	0.748					
PPE	0.597	0.217	0.252	0.312	0.272	0.408	0.652	0.726				
SM	0.477	0.136	0.252	0.221	0.355	0.365	0.561	0.406	0.742			
ST	0.336	0.327	0.324	0.393	0.228	0.559	0.542	0.377	0.479	0.800		
Severity	0.183	0.344	0.411	0.382	0.638	0.431	0.253	0.365	0.335	0.278	0.763	
UA	0.399	0.577	0.376	0.574	0.355	0.648	0.452	0.406	0.317	0.504	0.381	0.750

In the second stage analysis for the moderating effect of risk level, the latent variable (LV) scores for safety risks, mitigation measures, and risk level from the first stage analysis serve as the manifest variables in the higher order components. The graphical output for second stage analysis is shown in Figure 4, where the AVE values for safety risks, mitigation measures, and risk level together with the outer loadings, path coefficient, and p values are presented.

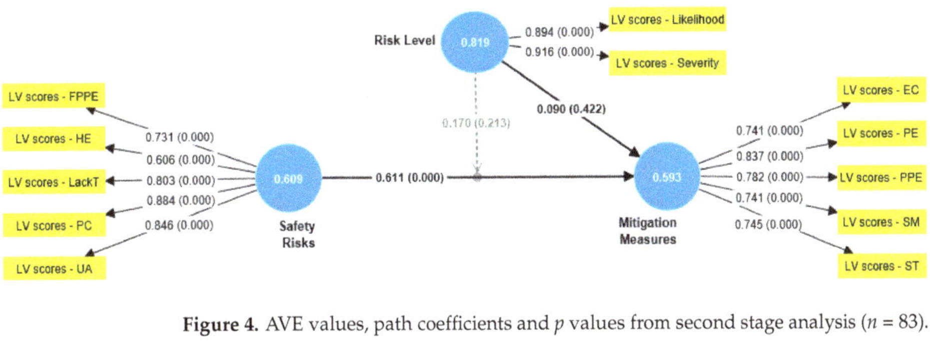

Figure 4. AVE values, path coefficients and p values from second stage analysis (n = 83).

The assessment results for higher order components from second stage analysis are summarised in Table 12. Based on the quality criteria given earlier, the higher order components achieve convergent validity with AVE > 0.50, indicator reliability with outer loadings ranging from 0.731 to 0.916, as well as construct reliability and validity with Cronbach's alpha, Rho A and Rho C values well above 0.700.

Table 12. Assessment results of higher order components (n = 83; CI = 95%).

Higher Order Component	Latent Variable Scores	Outer Loadings	R-Squared, R^2	Construct Reliability and Validity			
				Cronbach's Alpha	Composite Reliability, Rho_A	Composite Reliability, Rho_C	AVE (≥0.50)
Risk Level	LV Scores—Likelihood	0.894	—	0.779	0.786	0.900	0.819
	LV Scores—Severity	0.916					
Safety Risks	LV Scores—HE	0.606	—	0.837	0.882	0.885	0.609
	LV Scores—FPPE	0.731					
	LV Scores—UA	0.846					
	LV Scores—LackT	0.803					
	LV Scores—PC	0.884					
Mitigation Measures	LV Scores—PPE	0.782	0.368	0.829	0.834	0.879	0.593
	LV Scores—ST	0.745					
	LV Scores—SM	0.741					
	LV Scores—PE	0.837					
	LV Scores—EC	0.741					

Table 13 shows the higher order components have no problem with discriminant validity because there is no $HTMT_{.90}$ value which is more than 0.90. In addition, Table 14 shows the higher order components achieve satisfactory discriminant validity as well because the square root of AVE (along the diagonal) is larger than the correlation (off diagonal) for all the constructs.

Table 13. Discriminant validity of higher order components using HTMT ratio.

Higher Order Component	Mitigation Measures	Risk Level	Safety Risks	Risk Level × Safety Risks
Mitigation measures				
Risk level	0.481			
Safety risks	0.652	0.665		
Risk level × Safety risks	0.165	0.114	0.470	

Table 14. Discriminant validity of higher order components using Fornell and Larcker criterion.

Higher Order Component	Mitigation Measures	Risk Level	Safety Risks
Mitigation measures	0.770		
Risk level	0.386	0.905	
Safety risks	0.577	0.516	0.780

4.2.2. Assessment of Structural Model

The following are the criteria adopted to assess the higher component model shown in Figure 4:

Standardised root mean square residual (SRMR): A value less than 0.10 is considered a good fit [88].

Normed fit index (NFI): A value above 0.9 usually represents acceptable fit [89].

The assessment results for model fit of higher component model are summarised in Table 15, showing the higher component model has a good fit with SRMR = 0.094. However, the NFI value < 0.90.

Table 15. Assessment results of structural model (n = 83; CI = 95%).

Item	Original Sample	Saturated Model Sample Mean	95%	99%	Estimated Sample	Estimated Model Sample Mean	95%	99%
SRMR (≤0.10)	0.094	0.071	0.088	0.099	0.095	0.073	0.093	0.104
d_ULS	0.696	0.403	0.604	0.760	0.700	0.429	0.681	0.840
d_G	0.328	0.258	0.380	0.452	0.328	0.328	0.385	0.458
Chi-square	153.132				153.258			
NFI (≥0.90)	0.698				0.697			

4.3. Moderating Effect of Risk Level

Table 16 summarises the results obtained in second stage analyses for Figures 2 and 4. Without risk level as a moderator in Figure 2, the main effect between safety risks and mitigation measures is $\beta = 0.580$ ($p < 0.001$) as shown in Figure 2, with $R^2 = 0.337$ ($p < 0.001$) and effect size = 0.508 ($p < 0.05$). With risk level as a moderator in Figure 4, the simple effect is $\beta = 0.611$ ($p < 0.001$), with $R^2 = 0.368$ ($p < 0.001$) and effect size = 0.339 ($p > 0.05$). The path coefficient β increased to 0.611 from 0.580, which is an increase of 5.3% and the R^2 value increased to 0.368 from 0.337, which is an increase of 9.2%. The strength of the relationship between safety risks and mitigation measures increases when the risk level increases in the presence of risk level as a moderator. This is illustrated by the interaction plot shown in Figure 5. The effect size is 0.038 ($p > 0.05$), which is small according to [90].

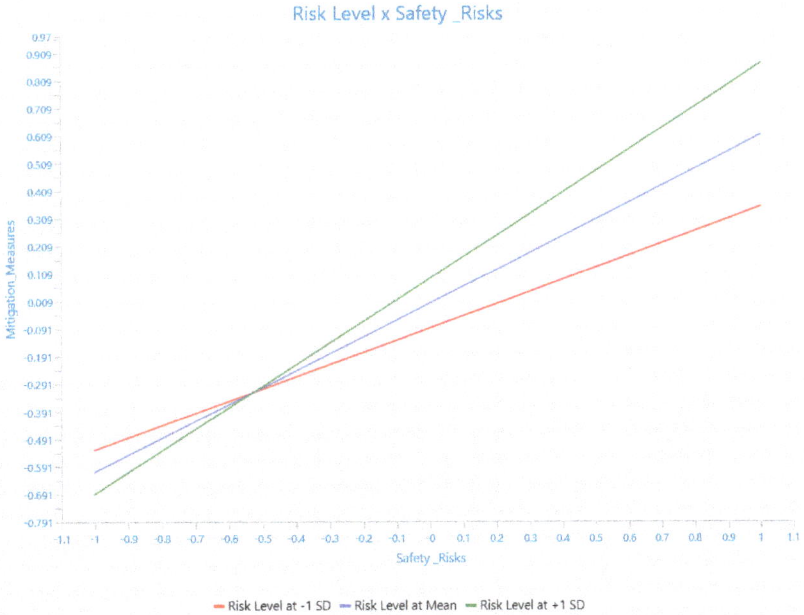

Figure 5. Interaction plot showing the moderating effect of risk level.

Table 16. Moderating effect of risk level on mitigation measures (n =83, CI = 95%).

Case	Safety Risks	Mitigation Measures		Path	Path Coefficient	Total Effect	Effect Size, f-Square
	AVE	R^2	AVE				
Figure 2 Base model	0.609, p = 0.000	0.337, p = 0.000	0.592, p = 0.000	Safety risks → MM	0.580, p = 0.000.	0.580, p = 0.000.	0.508, p = 0.031.
Figure 4 Risk level as a moderating construct	0.609. p = 0.000	0.368, p = 0.000	0.593, p = 0.000	Safety risks → MM	0.611, p = 0.000.	0.611, p = 0.000.	0.339, p = 0.099
				Risk level → MM	0.090, p = 0.422	0.090, p = 0.422	0.009, p = 0.788
				Risk level × Safety risks → MM	0.170, p = 0.213	0.170, p = 0.213	0.038, p = 0.475

5. Discussion

Table 17 displays the results for composite reliability and validity of the initial conceptual model which consists of all the indicators identified for this research which are presented in Tables 5–8. The results show that the measurement instrument used for data collection has a high internal consistency reliability with all the values well above 0.707. With the deletion of two indicators, namely HE3 and PPE1, the results in Table 9 show that the measurement instrument is further improved with good indicator reliability, adequate convergent validity, and adequate discriminant validity.

Table 17. Composite reliability and validity of lower order components.

Construct	Indicators in Construct		Cronbach's Alpha	Composite Reliability, Rho_A	Composite Reliability, Rho_C(>0.50 but <0.90)	Average Variance Extracted Values, AVE
	Number	Reference				
Likelihood of commonly-occurred accidents	9	Table 5	0.919	0.938	0.931	0.602
Severity of commonly-occurred accidents	9	Table 6	0.907	0.944	0.921	0.572
Human Error (HE)	4	Table 7	0.747	0.751	0.840	0.569
Failure to use Personal Protective Equipment (FPPE)	4		0.830	0.839	0.886	0.660
Unsafe act and site condition (UA)	4		0.741	0.754	0.837	0.563
Lack of progressive training (LackT)	3		0.827	0.833	0.897	0.745
Poor Communication (PC)	3		0.880	0.880	0.926	0.807
Personal Protective Equipment (PPE)	6	Table 8	0.776	0.781	0.843	0.475
Safety and health training (ST)	3		0.711	0.736	0.841	0.642
Safety Meeting (SM)	4		0.726	0.726	0.830	0.551
Proper equipment (PE)	4		0.735	0.741	0.835	0.559
Promote effective communication (EC)	3		0.738	0.750	0.852	0.659

The risk levels for the nine commonly-occurred fatal construction accidents presented in Tables 5 and 6 were calculated and the values are presented in Table 18. The results show that fall-related accident is at a high risk level, confirming the finding from earlier studies which mentioned falling from height is the number one killer in the Malaysian construction industry [11–16]. All the other types of safety risks are in the medium risk levels, with insect pest as a safety risk having the lowest risk level score. Table 18 also summarises the ranking for all the nine commonly-occurred safety risks measured in terms of mean likelihood score, mean severity score and risk level. The results show that fall-related accident remains at the top, signifying that fall-related accidents are highly risky and the most likely and severe risk to happen at the construction site; fatal accidents due to insect pest is at the bottom of the ranking, confirming the data from [11].

Table 18. Risk levels of commonly-occurred fatal construction accidents.

Item	Commonly-Occurred Accidents	Likelihood		Severity		Risk Level (Mean Likelihood × Mean Severity)		
		Mean Value	Rank	Mean Value	Rank	Score	Rank	Description
1	Fall-related accident (human falling from height)	3.96	1	4.18	1	16.6	1	High risk (≥15)
5	Electrocution (getting burn, electrocution, shock, arc flash)	3.71	3	3.87	2	14.4	2	4 < Medium risk < 15
7	Caught in-between accidents (caught, crushed, squeezed between two or more objects on site)	3.51	6	3.83	3	13.4	5	4 < Medium risk < 15
8	Fire or explosion (fire outbreak, bomb explosion)	3.64	4	3.82	4	13.9	3	4 < Medium risk < 15
4	Buried (being buried under the landslide)	3.51	6	3.76	5	13.2	6	4 < Medium risk < 15
2	Struck by accident (struck by falling object, moving vehicle, rolling machinery)	3.54	5	3.72	6	13.2	6	4 < Medium risk < 15
6	Road accident (hydroplaning, brake failure)	3.73	2	3.61	7	13.5	4	4 < Medium risk < 15
3	Drowning and Asphyxiation (insufficient of oxygen)	3.31	7	3.57	8	11.8	7	4 < Medium risk < 15
9	Insect pest (for example: stung by hornets)	3.18	8	3.19	9	10.1	8	4 < Medium risk < 15

Table 19 summarises that rankings for all the 18 indicators or factors of the five construction safety risks mitigation measures. The results show that 15 factors have mean scores well above 4, with the top 5 factors being failure to use safety helmets, unsafe equipment, improper work procedure, employers failing to provide sufficient training, and hazardous environment. Only three factors have mean scores slightly lower than 4, and they are improper attitude, lack of budget allocation on safety management, and inadequate tools used. In terms of overall mean value, the ranking for the five construction safety risks in descending order are: unsafe act and site condition, failure to use PPE, poor communication, lack of progressive training, and human error.

Table 19. Ranking of factors influencing safety risks.

Construct	Indicator	Factors Influencing Safety Risks	Mean		Overall Mean	
			Value	Rank	Value	Rank
Human Error (HE)	HE1	Improper attitude	3.93	3	3.97	5
	HE2	Inadequate tools used	3.75	4		
	HE3	Excessive physical exertion	4.18	1		
	HE4	Lacks of experience	4.00	2		
Failure to use PPE (FPPE)	FPPE1	Failure to use safety helmets	4.52	1	4.25	2
	FPPE2	Failure to use face protection	4.07	4		
	FPPE3	Failure to use safety boots	4.23	2		
	FPPE4	Failure to use eye protection	4.17	3		
Unsafe act and site condition (UA)	UA1	Unsafe equipment	4.47	1	4.31	1
	UA2	Unsafe methods	4.16	4		
	UA3	Hazardous environment	4.25	3		
	UA4	Improper work procedure	4.36	2		
Lack of progressive training (LackT)	LackT1	Employers fail to offer sufficient training	4.34	1	4.13	4
	LackT2	Lack of budget allocation on safety management	3.93	3		
	LackT3	Lack of workforce due to subcontract work	4.12	2		
Poor Communication (PC)	PC1	Language barrier	4.22	1	4.14	3
	PC2	Miscommunication and misunderstanding	4.02	3		
	PC3	Failure in conveying message	4.17	2		

Note: 1 = strongly disagree; 2 = disagree; 3 = neutral; 4 = agree; 5 = strongly agree.

Table 20 summarises the rankings for all the 20 indicators or factors of the five construction mitigation measures. The results show that all the 20 factors have mean scores well above 4, with the top 6 factors being the use of safety helmet, supplying employees with suitable equipment, safety measures training, safety harnesses, discussion on the precautionary safety concerns, and rapid communication such as walkie-talkies. All the five construction mitigation measures have overall mean values between 4.30–4.47. The construct on proper equipment has the highest overall mean score, and the measurement indicators correspond to control the risks at the source and to design of safe work systems to minimise risks.

Table 20. Ranking of practical measures taken to mitigate safety risks.

Construct	Indicator	Practical Measures to Mitigate Safety Risks	Mean		Overall Mean	
			Value	Rank	Value	Rank
Personal Protective Equipment (PPE)	PPE1	Safety helmets	4.76	1	4.45	2
	PPE2	Ear protection	4.16	6		
	PPE3	High visibility clothing	4.39	5		
	PPE4	Safety footwear	4.41	4		
	PPE5	Safety harnesses	4.53	2		
	PPE6	Training of PPE	4.45	3		
Safety and health training(ST)	ST1	Safety measures training	4.55	1	4.35	3
	ST2	Machinery operator training	4.20	3		
	ST3	Working at height training	4.30	2		
Safety meeting (SM)	SM1	Discuss the precautionary safety concerns	4.48	1	4.30	5
	SM2	Communication between job groups	4.20	3		
	SM3	Report changes at the work site	4.36	2		
	SM4	Update the existing safety plan and procedure	4.17	4		
Proper equipment (PE)	PE1	Supplying employees with suitable equipment	4.65	1	4.47	1
	PE2	Safe working environment	4.30	4		
	PE3	Machines serviced regularly	4.46	2		
	PE4	Scaffolding with safe access	4.45	3		
Promote effective communication (EC)	EC1	Employee pay attention for safety briefing	4.36	3	4.33	4
	EC2	Construction parties communicate with each other	4.14	3		
	EC3	Rapid communication such as walkie-talkies	4.48	1		

Note: 1 = not important at all, 2 = slightly important, 3 = moderately important, 4 = important and 5 = very important.

The goodness of fit (GoF) for the path model can be determined manually by using the formula GoF = [(mean R^2) × (mean AVE)]$^{1/2}$ [91]. Based on the R^2 value of 0.368 and mean AVE value of 0.674 for risk level, safety risks and mitigation measures constructs in Table 12, the GoF for the path model is found to be $(0.368 \times 0.674)^{1/2} = 0.497$, which is greater than 0.36 for large fit [92]. It can be concluded that the GoF for the model shown in Figure 4 is large for global PLS model validity.

6. Conclusions

This study investigated the moderating effect of risk level on mitigation measures implemented due to the numerous factors causing safety risks. In Table 16, the results for Figure 2 show that safety risks have a significant positive relationship with mitigation measures with $\beta = 0.580$, and $p < 0.001$. The effect size is large with $f^2 = 0.508$, and $p < 0.05$.

The following conclusions can be made from the results for Figure 4 in Table 16:

Safety risks has significant positive relationship with mitigation measures with $\beta = 0.611$, $p < 0.001$. The effect size is medium with $f^2 = 0.339$, $p > 0.05$.

Risk level has a positive but insignificant relationship with mitigation measures with $\beta = 0.090$, $p > 0.05$. The effect size is negligible with $f^2 = 0.009$, $p > 0.05$.

The interaction term, risk level × safety risks has a positive but insignificant relationship with mitigation measures with $\beta = 0.170$, $p > 0.05$. The effect size is small with $f^2 = 0.038$, $p > 0.05$.

The results for Figure 4 show that the relationship between safety risks and mitigation measures increases in the presence of risk level as a moderator with path coefficient β = 0.611, and $p < 0.001$. The interaction plot in Figure 5 actually illustrates that the relationship between safety risks and mitigation measures is further heightened in the presence of risk level as a moderator. Because of the positive moderating effect (β = 0.170), the relationship between safety risks and mitigation measures becomes stronger with higher levels of risk level. Even though the effect size of the interaction term ($f^2 = 0.038$) is small, under severe situations such as incidents that are categorized as 'Acts of God', the sudden surge in risk level would result in an immediate change in β value. Therefore, it is imperative to consider these extreme situations in the planning, design, and management of construction safety because the consequential impacts of these sudden and unexpected incidents could be disastrous, thereby disrupting the continuity of construction works.

It is important to note that uncertainty and severity are intrinsic/inherent properties of safety risks. Mitigation measures are put into place to eliminate the likelihood of safety risks from happening, and to reduce the severity and impacts of these safety risks when they actually happen, which could lead to the loss of lives and hence emotional sufferings, damage to property, disruptions to on-going works, stop-work orders, liquidated ascertained damages and litigation cases. The mitigation measures implemented should always consider the moderating effect of risk level of safety risks which may cause the practical measures implemented on construction sites to be inadequate. The effect of risk level is higher when either the likelihood or severity, or both, are higher. Therefore, mitigation measures taken by the contractors must always take into account the types of factors causing safety risks, as well as the uncertainty or likelihood and severity of these factors for the sustainability of development projects. According to [7], the likelihood of incidents and their severity could be reduced by conducting effective pre-job safety analyses.

The findings from this study have practical values in view of Section 15 in the Occupational Safety and Health Act 1994, which states "*it shall be the duty of every employer and every self-employed person to ensure, so far as is practicable, the safety, health and welfare at work of all his employees*". The term 'practicable' should consider the following aspects, namely: "*(a) the severity of the hazard or risk in question, (b) the state of knowledge about the hazard or risk and any way of removing or mitigating the hazard or risk, (c) the availability and suitability of ways to remove or mitigate the hazard or risk, and (d) the cost of removing or mitigating the hazard or risk*" [93]. In this study, numerous factors which influence or cause safety risks were identified and presented in Table 7; some of the practical measures which can be implemented to mitigate or manage these safety risks in order to reduce their impacts were presented in Table 8. The author of [94] asserted that all the factors that influence safety on construction projects should be identified and categorized in order to prepare a construction accident causation framework which maps out these factors in terms of originating influences, shaping factors and immediate factors so that a comprehensive plan for training, awareness and monitoring can be prepared. The mitigation measures implemented should be able to manage or mitigate the impacts from accidents which are categorized as high-risks. This study also has academic value in applying the PLS-SEM method to analyse the data collected from the Malaysian construction industry. For generalization purposes, further research with larger sample size using the same technique should be replicated to provide additional evidence on the effects of likelihood and severity on mitigation measures taken for safety risks.

Author Contributions: Conceptualization, M.K.S. and W.C.Y.; methodology, M.K.S. and W.C.Y.; validation, W.C.Y. and M.K.S.; formal analysis, M.K.S. and W.C.Y.; investigation, M.K.S. and O.Q.J.; resources, M.K.S. and O.Q.J.; data curation, M.K.S. and O.Q.J.; writing—original draft preparation, M.K.S., W.C.Y. and O.Q.J.; writing—review and editing, W.C.Y. and M.K.S.; funding acquisition, W.C.Y. All authors have read and agreed to the published version of the manuscript.

Funding: MPOB-UKM Endowment Chair, Research Grant number: EP-2019-054, financed the APC.

Institutional Review Board Statement: Not applicable.

Informed Consent Statement: All subjects who took part in the study provided informed consent.

Data Availability Statement: On request, the corresponding author will provide the data that back up the study's conclusions.

Acknowledgments: Special appreciation to the research assistant as well as the participants who contributed considerable time and effort to the success of this study.

Conflicts of Interest: The authors declare no conflict of interest.

References

1. *ISO 45001:2018*; Occupational Health and Safety Management Systems—Requirements with Guidance for Use. International Organization for Standardization: Geneva, Switzerland, 2018.
2. What Is Risk Management? Available online: https://www.apm.org.uk/resources/ (accessed on 14 November 2022).
3. Chinn, S. 8 Construction Risk Facing the Construction Industry and How to Reduce. Available online: https://esub.com/blog/8-construction-risks-facing-the-construction-industry-and-how-to-reduce/ (accessed on 1 March 2022).
4. Department of Occupational Safety and Health. Occupational Accident Statistics. Available online: https://www.dosh.gov.my/index.php/statistic-v/occupational-accident-statistics (accessed on 7 November 2022).
5. Kang, B.G.; Fazlie, M.A.; Goh, B.H.; Song, M.K.; Zhang, C. Current practice of risk management in the Malaysia construction industry–the process and tools/techniques. *Int. J. Struct. Civ. Eng. Res.* **2015**, *4*, 371–377. [CrossRef]
6. Qazi, A.; Quigley, J.; Dickson, A.; Kirytopoulos, K. Project complexity and risk management (PROCRIM): Towards modelling project complexity driven risk paths in construction projects. *Int. J. Proj. Manag.* **2016**, *34*, 1183–1198. [CrossRef]
7. Colin, M. Incident prevention tools: Incident investigations and pre-job safety analyses. *Int. J. Min. Sci. Technol.* **2017**, *27*, 635–640.
8. Purohit, D.P.; Siddiqui, N.A.; Nandan, A.; Yadav, B.P. Hazard identification and risk assessment in construction industry. *Int. J. Appl. Eng. Res.* **2018**, *13*, 7639–7667.
9. Goh, C.S.; Rahman, H.A. The identification and management of major risks in the Malaysian construction industry. *J. Constr. Dev. Ctries.* **2013**, *18*, 19–32.
10. Hinze, J.; Russell, D.B. Analysis of fatalities recorded by OSHA. *J. Constr. Eng. Manag.* **1995**, *121*, 209–214. [CrossRef]
11. Fatal Accident Case. Available online: https://www.dosh.gov.my/index.php/fatal-accident-case-1 (accessed on 28 February 2022).
12. Williams, O.S.; Hamid, R.A.; Misnan, M.S. Analysis of fatal building construction accidents: Cases and causes. *J. Multidiscip. Eng. Sci. Technol.* **2017**, *4*, 8030–8040.
13. Hamid, A.R.A.; Azmi, M.R.A.N.; Aminudin, E.; Jaya, R.P.; Zakaria, R.; Zawawi, A.M.M.; Yahya, K.; Haron, Z.; Yunus, R.; Saar, C.C. Causes of fatal construction accidents in Malaysia. *IOP Conf. Ser. Earth Environ. Sci.* **2019**, *220*, 012044. [CrossRef]
14. Halim, N.N.A.A.; Jaafar, M.H.; Kamaruddin, M.A.; Kamaruzaman, N.A.; Singh, P.J. The causes of Malaysian construction fatalities. *J. Sustain. Sci. Manag.* **2020**, *15*, 236–256. [CrossRef]
15. Darda'u Rafindadi, A.; Napiah, M.; Othman, I.; Mikic, M.; Al-Ashmori, Y.Y. Rate of occurrence of fatal accidents in Malaysian construction industry after BIM implementation. *Int. J. Eng. Manag. Res.* **2020**, *10*, 49–76. [CrossRef]
16. Shao, B.; Hu, Z.; Liu, Q.; Chen, S.; He, W. Fatal accident patterns of building construction activities in China. *Saf. Sci.* **2019**, *111*, 253–263. [CrossRef]
17. Zahrizan, Z.; Mohamen Ali, N.; Haron, A.T.; Marshall-Ponting, A.; Hamid, Z.A. Exploring the adoption of building information modelling (BIM) in the Malaysian construction industry: A qualitative approach. *Int. J. Res. Eng. Technol.* **2013**, *2*, 384–395.
18. Zaini, N.Z.M.; Salleh, M.A.M.; Hasmori, M.F.; Abas, N.H. Effect of accident due to fall from height at construction sites in Malaysia. *IOP Conf. Ser. Earth Environ. Sci.* **2020**, *498*, 012106. [CrossRef]
19. Hamid, A.R.A.; Majid, M.Z.A.; Singh, B. Causes of accidents at construction sites. *Malays. J. Civ. Eng.* **2008**, *18*, 242–259.
20. Hinze, J.; Pedersen, C.; Fredley, J. Identifying root causes of construction injuries. *J. Constr. Eng. Manag.* **1998**, *124*, 67–71. [CrossRef]
21. Ahmed, S. Causes and effects of accident at construction site: A study for the construction industry in Bangladesh. *Int. J. Sustain. Constr. Eng. Technol.* **2019**, *10*, 18–40.
22. Fatal Construction Work Accident: Injuries from Being Struck by Object. Available online: https://www.kenallenlaw.com/2020/12/fatal-construction-work-accident-injuries-from-being-struck-by-object/ (accessed on 8 March 2022).
23. Ballowe, P.L. *Fatal Construction Accidents Categorized as "Other"*; University of Florida: Gainesville, FL, USA, 2008.
24. Tözer, K.D.; Çelik, T.; Gürcanli, G.E. Classification of construction accidents in Northern Cyprus. *Tek. Dergi* **2018**, *29*, 8295–8316. [CrossRef]
25. Zack, J. Construction Worker Buried Alive under Landslide. Available online: https://www.thestar.com.my/news/nation/2020/07/21/construction-worker-buried-alive-under-landslide (accessed on 2 March 2020).
26. Occupational Safety and Health Administration. *Construction Focus Four: Electrocution Hazards*; Department of Labor: Washington, DC, USA, 2011.
27. Li, Y.; Bai, Y. Comparison of characteristics between fatal and injury accidents in the highway construction zones. *Saf. Sci.* **2008**, *46*, 646–660. [CrossRef]

28. Lee, B.L.C.; Chen, J.C.; Fo, K.W. Accidents in construction sites: A study on the causes and preventive approaches to mitigate accident rate. *INTI J.* **2018**, *1*, e2600-7920.
29. Occupational Safety and Health Administration. *Construction Focus Four: Caught-In or—Between Hazards*; Department of Labor: Washington, DC, USA, 2011.
30. Ayob, A.; Shaari, A.A.; Zaki, M.F.M.; Munaaim, M.A.C. Fatal occupational injuries in the Malaysian construction sector–causes and accidental agents. *IOP Conf. Ser. Earth Environ. Sci.* **2018**, *140*, 012095. [CrossRef]
31. Health and Safety Executive. *Fire Safety in Construction—Guidance for Clients, Designers and Those Managing and Carrying Out Construction Work Involving Significant Fire Risks*; HSE: Sudbury, ON, Canada, 2010.
32. Zack, J. Construction Worker Succumbs to Injuries after Bandar Malaysia Explosion. Available online: https://www.thestar.com.my/news/nation/2017/10/11/construction-worker-succumbs-to-injuries-after-bandar-malaysia-explosion (accessed on 2 March 2020).
33. Rahman, H.A.; Wang, C.; Loo, S.C. *Risk Management in Construction Industry*; University of Malaya Press: Kuala Lumpur, Malaysia, 2015.
34. Jiskani, I.M.; da Silva, J.M.N.; Chalgri, S.R.; Behrani, P.; Lu, X.; Manda, E. Mine health and safety: Influence of psychosocial factors on musculoskeletal disorders among miners in Pakistan. *Int. J. Min. Miner. Eng.* **2020**, *11*, 152–167. [CrossRef]
35. Dekker, S.W.A. Reconstructing human contributions to accidents: The new view on error and performance. *J. Saf. Res.* **2002**, *33*, 371–385. [CrossRef] [PubMed]
36. Huang, X.; Hinze, J. Analysis of construction worker fall accidents'. *J. Constr. Eng. Manag.* **2003**, *129*, 262–271. [CrossRef]
37. Haron, A.T. Organisational Readiness to Implement Building Information Modelling: A Framework for Design Consultants in Malaysia. Ph.D. Thesis, University of Salford, Manchester, UK, 2013.
38. Goh, K.C.; Goh, H.H.; Omar, M.F.; Toh, T.C.; Zin, A.A.M. Accidents preventive practice for high-rise construction. In *MATEC Web of Conferences*; EDP Sciences: Les Ulis, France, 2016; Volume 47, p. 04004.
39. Herzberg, C.; Rödel, H.; Engelmann, E. New shielding protective equipment for live working. *Int. J. Cloth. Sci. Technol.* **2013**, *13*, 301–308. [CrossRef]
40. Aziz, N.A.A.; Nordin, R.M.; Ismail, Z.; Yunus, J. Identifying Factors and Impacts of Occupational Safety and Health (OSH) towards Work Accident in Achieving Sustainable Construction. Available online: http://www.mgbc.org.my/Downloads/08052018_MUIGBC2018_E-PROCEEDING_PAPER/20180508_1_Theory_Technology_Practice_of_Green_Building/3_Identifying_Factors_and_Impacts_of_Occupational_Safety_and_Helath_Towards_Work_Accident.pdf (accessed on 3 March 2022).
41. Hrynyk, R.; Frydrych, I. Study on textile assemblies with aluminized basalt fabrics destined for protective gloves. *Int. J. Cloth. Sci. Technol.* **2015**, *27*, 705–719. [CrossRef]
42. Ahmad, R. Best Practices in Safety Management for Conventional Civil Construction Industry in Malaysia. Ph.D. Thesis, University Teknologi Malaysia, Johor Bahru, Malaysia, 2008.
43. Holt, A.S.J. *Principles of Construction Safety*; Blackwell Publishing: Hoboken, NJ, USA, 2005.
44. Ridley, J.; Channing, J. *Safety at Work*; Elsevier Ltd.: Oxford, UK, 2008.
45. Tan, C.K.; Razak, N.A. Case studies on the safety management at construction site. *J. Sustain. Sci. Manag.* **2014**, *9*, 90–108.
46. Ahn, S.; Kim, T.; Park, Y.J.; Kim, J.M. Improving effectiveness of safety training at construction worksite using 3D BIM simulation. *Adv. Civ. Eng.* **2020**, *2020*, 2473138. [CrossRef]
47. Hassan, W.; Razi, A.; Qamar, R.; Jaffir, R.; Suhail, S. The effect of training on employee retention. *Glob. J. Manag. Bus. Res.* **2013**, *12*, 17–20.
48. Jones, K. 6 Keys to Better Construction Communication. Available online: https://www.constructconnect.com/blog/6-keys-better-construction-communication (accessed on 29 March 2022).
49. Tam, C.M. Use of the internet to enhance construction communication: Total information transfer system. *Int. J. Proj. Manag.* **1999**, *17*, 107–111. [CrossRef]
50. Gamil, Y.; Abdul Rahman, I. Identification of causes and effects of poor communication in construction industry: A theoretical review. *Emerg. Sci. J.* **2017**, *1*, 239–247. [CrossRef]
51. Gamil, Y.; Abd Rahman, I. Studying the relationship between causes and effects of poor communication in construction projects using PLS-SEM approach. *J. Facil. Manag.* 2021; *ahead of print*. [CrossRef]
52. Hoezen, M.E.L.; Reymen, I.M.M.J.; Dewulf, G.P.M.R. The problem of communication in construction. In Proceedings of the Adaptables 2006, International Conference on Adaptable Building Structures, Eindhoven, The Netherlands, 3–5 July 2006.
53. Valitherm, A. Communication barrier in Malaysia construction sites. *Int. J. Educ. Res.* **2014**, *2*, 1–10.
54. International Labour Office. *Guidelines on Occupational Safety and Health Management Systems ILO-OSH 2001*, 2nd ed.; International Labour Office: Geneva, Switzerland, 2009.
55. The Importance of Personal Protective Equipment (PPE). Available online: https://www.danielsinsuranceinc.com/the-importance-of-personal-protective-equipment-ppe/ (accessed on 9 March 2022).
56. Healthy Working Lives. Personal Protective Equipment (PPE). Available online: https://www.healthyworkinglives.scot/workplace-guidance/safety/personal-protective-equipment/Pages/types-of-ppe.aspx (accessed on 9 March 2022).
57. Mitchell, S.J.; Braithwaite, G.R. Perceptions of safety and offshore helicopter travel. *Int. J. Energy Sect. Manag.* **2008**, *2*, 479–498. [CrossRef]

58. Smith, T.D.; DeJoy, D.M. Safety climate, safety behaviors and line-of-duty injuries in the fire service. *Int. J. Emerg. Serv.* **2014**, *3*, 49–64. [CrossRef]
59. Cooper, M. Current issues in health and safety training in the UK. *J. Eur. Ind. Train.* **1998**, *22*, 354–361. [CrossRef]
60. Cheng, E.W.; Li, H.; Fang, D.P.; Xie, F. Construction safety management: An exploratory study from China. *Constr. Innov.* **2004**, *4*, 229–241. [CrossRef]
61. Riaz, Z.; Parn, E.; Edwards, D.; Arslan, M.; Shen, C.; Mora, F. BIM and sensor-based data management system for construction safety monitoring. *J. Eng. Des. Technol.* **2017**, *15*, 738–753. [CrossRef]
62. Low, S.P.; Sua, C.S. The maintenance of construction safety: Riding on ISO 9000 quality management systems. *J. Qual. Maint. Eng.* **2000**, *6*, 28–44.
63. Lakhiar, M.T.; Lakhiar, M.T.; Abdullah, A.H. Occupational health and safety performance in high-rise building projects in Pakistan: A systematic literature review. *Oper. Res. Eng. Sci. Theory Appl.* **2021**, *4*, 99–114. [CrossRef]
64. Buniya, M.K.; Othman, I.; Durdyev, S.; Sunindijo, R.Y.; Ismail, S.; Kineber, A.F. Safety program elements in the construction industry: The case of Iraq. *Int. J. Environ. Res. Public Health* **2021**, *18*, 411. [CrossRef]
65. Marhani, M.A.; Adnan, H.; Baharuddin, H.E.; Esa, M.R.; Hassan, A.A. Dependency of foreign workers in Malaysian construction industry. *Built Environ. J.* **2012**, *9*, 39–50.
66. Vignoli, M.; Nielsen, K.; Guglielmi, D.; Mariani, M.G.; Patras, L.; Peiró, J.M. Design of a safety training package for migrant workers in the construction industry. *Saf. Sci.* **2021**, *136*, 105124. [CrossRef]
67. Ismail, R.B.; Palliyaguru, R.; Karunasena, G.; Othman, N.A. Health and safety (H&S) challenges confronted by foreign workers in the Malaysian construction industry: A background study. In Proceedings of the WCS 2018: Built Asset Sustainability: Rethinking Design, Construction and Operations, The 7th World Construction Symposium 2018, Galadari Hotel, Colombo, Sri Lanka, 29 June–1 July 2018; pp. 277–2873.
68. Mahdi, M.Z.; Jamaludin, K.R. Occupational stress management among SU-30MKM maintenance personnel in Gong Kedak Air Base. In Proceedings of the 1st Master Project Symposium on Occupational Safety and Health, Universiti Teknologi Malaysia, Kuala Lumpur, Malaysia, 13–15 December 2016.
69. The Importance of Safety Meetings. Available online: https://smallbusiness.chron.com/importance-safety-meetings-78270.html (accessed on 29 March 2022).
70. Olson, R.; Varga, A.; Cannon, A.; Jones, J.; Gilbert-Jones, I.; Zoller, E. Toolbox talks to prevent construction fatalities: Empirical development and evaluation. *Saf. Sci.* **2016**, *86*, 122–131. [CrossRef]
71. Asnicar, S. Need for Effective Communication in Construction. Available online: https://www.whitecardonline.com.au/blog/white-card/need-for-effective-communication-in-construction/ (accessed on 29 March 2022).
72. 10 Essential Construction Communication Skills for Your Team. Available online: https://www.bigrentz.com/blog/communication-in-construction (accessed on 29 March 2022).
73. Pandit, B.; Albert, A.; Patil, Y.; Al-Bayati, A.J. Fostering safety communication among construction workers: Role of safety climate and crew-level cohesion. *Int. J. Environ. Res. Public Health* **2019**, *16*, 71. [CrossRef] [PubMed]
74. Suleiman, A. Causes and effects of poor communication in the construction industry in the MENA region. *J. Civ. Eng. Manag.* **2022**, *28*, 365–376. [CrossRef]
75. SmartPLS 4. Available online: https://www.smartpls.com (accessed on 31 October 2022).
76. Griffin, M.M.; Steinbrecher, T.D. Large-scale datasets in special education research. *Int. Rev. Res. Dev. Disabil.* **2013**, *45*, 155–183.
77. Henseler, J.; Ringle, C.M.; Sinkovics, R.R. The use of partial least squares path modeling in international marketing. In *New Challenges to International Marketing*; Emerald Publishing Limited: Bingley, UK, 2009; Volume 20, pp. 277–319.
78. Cronbach, L.J. Test validation. In *Educational Measurement*, 2nd ed.; Thorndike, R., Ed.; American Council on Education: Washington, DC, USA, 1971.
79. Gefen, D.; Straub, D.W.; Boudreau, M.C. Structural equation modeling and regression: Guidelines for research practice. *Commun. Assoc. Inf. Syst.* **2000**, *4*, 1–79. [CrossRef]
80. Hulland, J. Use of partial least squares (PLS) in strategic management research: A review of four recent studies. *Strateg. Manag. J.* **1999**, *20*, 195–204. [CrossRef]
81. Bagozzi, R.P.; Yi, Y. On the evaluation of structural equation models. *J. Acad. Mark. Sci.* **1988**, *16*, 74–94. [CrossRef]
82. Fornell, C.; Larcker, D.F. Evaluating structural equation models with unobservable variables and measurement error. *J. Mark. Res.* **1981**, *18*, 39–50. [CrossRef]
83. Hair, J.F.; Gabriel, M.; Patel, V. AMOS covariance-based structural equation modeling (CB-SEM): Guidelines on its application as a marketing research tool. *Braz. J. Mark.* **2014**, *13*, 44–55.
84. Henseler, J.; Ringle, C.M.; Sarstedt, M. Testing measurement invariance of composites using partial least squares. *Int. Mark. Rev.* **2016**, *33*, 405–431. [CrossRef]
85. Gold, A.H.; Malhotra, A.; Segars, A.H. Knowledge management: An organizational capabilities perspective. *J. Manag. Inf. Syst.* **2001**, *18*, 185–214. [CrossRef]
86. Chin, W.W. The partial least squares approach for structural equation modeling. In *Modern Methods for Business Research*; Marcoulides, G.A., Ed.; Lawrence Erlbaum Associates Publishers: Mahwah, NJ, USA, 1998; pp. 295–336.
87. Hair, J.F., Jr.; Hult, G.T.M.; Ringle, C.M.; Sarstedt, M. *A Primer on Partial Least Squares Structural Equation Modeling (PLS-SEM)*, 2nd ed.; Sage Publications: Thousand Oaks, CA, USA, 2017.

88. Hu, L.T.; Bentler, P.M. Fit indices in covariance structure modeling: Sensitivity to underparameterized model misspecification. *Psychol. Methods* **1998**, *3*, 424–453. [CrossRef]
89. Bentler, P.M.; Bonett, D.G. Significance tests and goodness of fit in the analysis of covariance structures. *Psychol. Bull.* **1980**, *88*, 588–606. [CrossRef]
90. Cohen, J. *Statistical Power Analysis for the Behavioral Sciences*, 2nd ed.; Lawrence Erlbaum: Hillsdale, MI, USA, 1988.
91. Coughlan, J.M.M. Structural equation modelling: Guidelines for determining model fit. *Electron. J. Bus. Res. Methods* **2008**, *6*, 53–60.
92. Wetzels, M.; Odekerken-Schröder, G.; Van Oppen, C. Using PLS path modeling for assessing hierarchical construct models: Guidelines and empirical illustration. *MIS Q.* **2009**, *33*, 177–195. [CrossRef]
93. Department of Occupational Safety and Health. *Guidelines on Occupational Safety and Health Act 1994 (Act 514)*; Ministry of Human Resources: Johor Bahru, Malaysia, 2006.
94. Muñoz-La Rivera, F.; Mora-Serrano, J.; Oñate, E. Factors Influencing Safety on Construction Projects (fSCPs): Types and Categories. *Int. J. Environ. Res. Public Health* **2021**, *18*, 0884. [CrossRef]

sustainability

MDPI

Article

Occupational Risk Assessment for Flight Schools: A 3,4-Quasirung Fuzzy Multi-Criteria Decision Making-Based Approach

Muhammet Gul [1,*] and Muhammet Fatih Ak [2]

[1] Department of Transportation and Logistics, Istanbul University, 34320 Istanbul, Turkey
[2] Department of Industrial Engineering, Antalya Bilim University, 07190 Antalya, Turkey; fatih.ak@antalya.edu.tr
* Correspondence: muhammetgul@istanbul.edu.tr; Tel.: +90-212-473-7000 (ext. 19254)

Abstract: The concept of occupational risk assessment is related to the analysis and prioritization of the hazards arising in a production or service facility and the risks associated with these hazards; risk assessment considers occupational health and safety (OHS). Elimination or reduction to an acceptable level of analyzed risks, which is a systematic and proactive process, is then put into action. Although fuzzy logic-related decision models related to the assessment of these risks have been developed and applied a lot in the literature, there is an opportunity to develop novel occupational risk assessment models depending on the development of new fuzzy logic extensions. The 3,4-quasirung fuzzy set (3,4-QFS) is a new type of fuzzy set theory emerged as an extension of the Pythagorean fuzzy sets and Fermatean fuzzy sets. In this approach, the sum of the cube of the degree of membership and the fourth power of the degree of non-membership must be less than or equal to 1. Since this new approach has a wider space, it can express uncertain information in a more flexible and exhaustive way. This makes this type of fuzzy set applicable in addressing many problems in multi-criteria decision making (MCDM). In this study, an occupational risk assessment approach based on 3,4-quasirung fuzzy MCDM is presented. Within the scope of the study, the hazards pertaining to the flight and ground training, training management, administrative and facilities in a flight school were assessed and prioritized. The results of existing studies were tested, and we considered both Pythagorean and Fermatean fuzzy aggregation operators. In addition, by an innovative sensitivity analysis, the effect of major changes in the weight of each risk parameter on the final priority score and ranking of the hazards was evaluated. The outcomes of this study are beneficial for OHS decision-makers by highlighting the most prioritized hazards causing serious occupational accidents in flights schools as part of aviation industry. The approach can also be suggested and adapted for production and service science environments where their occupational health & safety are highly required.

Keywords: 3,4-quasirung fuzzy set; multi-criteria decision making; risk assessment; Pythagorean fuzzy set; Fermatean fuzzy set; flight school; occupational health and safety; transportation; system safety; soft computing; uncertainty analysis

Citation: Gul, M.; Ak, M.F. Occupational Risk Assessment for Flight Schools: A 3,4-Quasirung Fuzzy Multi-Criteria Decision Making-Based Approach. *Sustainability* **2022**, *14*, 9373. https://doi.org/10.3390/su14159373

Academic Editors: Esmaeil Zarei, Samuel Yousefi and Mohsen Omidvar

Received: 19 April 2022
Accepted: 28 July 2022
Published: 31 July 2022

1. Introduction

Occupational risk assessment is a process that covers the evaluation, ranking, and classification of hazards arising within a production or service system and the risks associated with these hazards from occupational health & safety (OHS) perspective [1]. This process determines whether the emerging hazards and the risks associated with these hazards are at an acceptable level and takes the necessary measures with a proactive approach [2]. While the primary purpose of the occupational risk assessment is to protect the employee from the dangers that arise in and around the workplace, the safety of business operations is also a secondary objective within the scope of the occupational risk assessment [3]. As in the manufacturing industry, in all other industries, harmony and good management of the workplace environment, resources and employees is necessary for the activities to be carried out according to OHS principles [4]. Occupational risk assessment is an essential component of a coherent safety management system and organizations seeking applicable, fast, and practical risk assessment.

Risk assessment with an OHS perspective is performed with some particular quantitative, qualitative, or hybrid methods that are a combination of these two. Many risk assessment methods are mentioned in the content of IEC 31010:2019, an important standard of ISO [5]. These methods are used for some purposes such as defining the risks, determining the source, cause, and trigger elements of the risk, allowing to choose between options and understanding the consequences of risk and probability. Multi-criteria decision-making (MCDM) is one of the methods mentioned by ISO within the scope of this standard. It is a sub-branch of Operations Research rather than a specific approach and consists of many different methods. MCDM provides an innovative perspective that allows selection, ranking, or classification among alternatives by considering multiple criteria in decision making. It is frequently used in risk analysis studies conducted with an OHS perspective. In this context, MCDM is used in integration with many well-known concepts such as fuzzy logic, data analytics, and artificial intelligence/expert systems [6]. MCDM, integrated with fuzzy logic, constitutes an important slice of the OHS risk assessment literature and contributes to the OHS risk assessment literature, especially by eliminating some of the drawbacks of traditional qualitative and quantitative risk assessment methods mentioned in IEC 31010:2019. These disadvantages have been emphasized many times in the literature [6] For example, in methods such as the risk matrix method, Fine–Kinney method, Failure Mode and Effect Analysis (FMEA), Event Tree Analysis (ETA), Fault Tree Analysis (FTA), Bow-tie analysis and Hazard and Operability Analysis (HAZOP), risk parameters do not have importance weights, and the evaluation is not done precisely due to the numerical scale defined for the parameters, logical problems and the insufficient number of parameters are some of the drawbacks [7–13].

Since the fuzzy logic theory was first proposed by Zadeh [14], many versions have been developed and integrated with many MCDM methods [15]. The 3,4-quasirung fuzzy set (3,4-QFS) is a new extension of fuzzy set theory [16]. It is proposed as an extension of the Pythagorean fuzzy sets [17] and Fermatean fuzzy sets [18]. Another study used fuzzy sets among major accidents in human reliability analysis [19]. Pouyakian et al. used fuzzy MCDM to assist in obtaining an optimum allocation of control measures [20]. In this version, the sum of the cube of the degree of membership and the fourth power of the degree of non-membership must be less than or equal to one. Since this new approach has a wider space, it can express uncertain information more flexibly and exhaustively in decision-making problems such as occupational risk assessment. Therefore, in this study, an occupational risk assessment approach based on 3,4-quasirung fuzzy MCDM is provided.

The aviation industry is one of the industries that have grown in recent years, while it plays a crucial role and is essential for developing countries. When the aviation sector and flight school processes are examined, minor and major differences are observed after the COVID-19 pandemic. According to the figures of the Turkey Directorate General of Civil Aviation's 2020 annual bulletin of safety incidents, although it is seen that the number of traffic movements and the number of safety incidents decreased in 2020 compared to the data of the previous two years, it is obvious that there will be an increase again in these days when the effects of COVID-19 have decreased and the return to normal has been experienced. While the aircraft traffic movement across Turkey was 1,544,169 in 2018 and 1,556,417 in 2019, it decreased to 855,833 in 2020. In 2020, there is a 45% decrease in traffic movement compared to the previous year. While there was an increase in traffic movements in January and February of 2020 compared to 2019, there was a serious decrease due to the subsequent COVID-19 restrictions. Although the traffic movement has increased again with the reduction of restrictions since June, it is seen that it is far behind 2019. The total number of incident reports made during the year decreased from 2319 in 2018 and 2736 in

2019 to 2073 in 2020. With the effect of the decrease in the number of traffic movements, the number of incident reports decreased by 24% in 2020 [21].

Flight schools, which produce professional teams for the aviation industry, are one of the most important pillars of the sector. It is important to develop an appropriate risk assessment process by considering the activities carried out in these schools from an OHS perspective. Delikhoon et al. [22] mentioned that systems thinking accident analysis models can be utilized in different studies to increase the system's sustainability of aviation safety. In 1998, both a flight instructor and a student died in an accident on Lake Manitoba. In August 2008, a C172 crashed in Toronto during an aviation training flight, killing one person and seriously injuring two. In the accident that took place in Istanbul in 2020, a piloting undergraduate student was rescued with injuries. In 2022, 2 pilots lost their lives as a result of the crash of a single-engine training plane near Bursa Yunuseli Airport. The examples given are only examples of the accidents that occurred before and after the pandemic in flight schools and processes, and it is observed that there are a large number of fatal and serious accidents. At first glance, it may seem like there is nothing in common between these accidents. Observations and accident analyses reveal the lack of a feasible and comprehensive risk assessment. Since both the flight and ground training and training management activities, which are among the activities carried out by the flight school, contain various risks and the existence of administrative and facility-related hazards reinforces this need. Flight instructors are responsible for understanding and taking precautions against a wide variety of risks, both for themselves and their students. A consistent and comprehensive quantitative risk assessment before flight training can systematically help you determine if the risk level is too high, and provide an opportunity to reduce or reject risk before it is too late. For these reasons, the risk assessment model proposed in this study was applied in a flight school risk prioritization process and it was emphasized that it should include common features based on expert opinions to be applicable in flight school risk assessment processes as well.

The 3,4-QFSs are superior to Pythagorean and Fermatean fuzzy sets in the MCDM domain, but it has not yet been applied to the occupational risk assessment. Therefore, this study remedies the gap and also improves the traditional risk assessment techniques' limitations, thereby more accurately transforming expert opinions into computable quantitative data. The characteristics and objectives of this paper are (1) to offer a risk assessment method for the OHS field, (2) to use a new 5-point 3,4-QF linguistic scale in the approach, and (3) to apply the proposed approach in a flight school risk assessment process. Along with this real case application in a flight school, a comparative study is also performed to confirm its adaptability to any other sector's OHS process and its applicability.

2. Research Background

Since this research is an occupational risk assessment study based on 3,4-QF-MCDM, initially the recent occupational risk assessment studies based on fuzzy MCDM are reviewed, then a summary of the newly proposed 3,4-QFS theory is given. Finally, the research gaps and main contributions of the study in terms of research methodology and application viewpoints are presented.

In recent years, there has been an increase and development in the application of the combination of MCDM & fuzzy set to the field of occupational risk assessment, due to the proposal of new methods in the field of MCDM and the gradual development of fuzzy logic extensions. Since fuzzy MCDM has produced remedies for the deficiencies of traditional risk assessment approaches such as "weighting of risk parameters" and "prioritizing hazards more sensitively" and has succeeded in improving it continuously. Starting with Zadeh's initial fuzzy theory [14,23], triangular and trapezoidal fuzzy numbers [24], then intuitionistic fuzzy number [25], type-2 fuzzy number [26], hesitant fuzzy number [27], Pythagorean fuzzy number [17], Picture fuzzy sets [28], Spherical fuzzy numbers [29], Fermatean fuzzy number [18], q-rung fuzzy numbers [30], and finally 3,4-quasirung fuzzy numbers [16] are proposed and are ready to be implemented to many real-world problems.

Many traditional occupational risk assessment methods have been made more effective by jointly using with fuzzy MCDM. To cite recent studies, Marhavilas et al. [31] conducted a study integrating Decision Risk-Matrix (also known as risk matrix) and HAZOP methods with the Fuzzy Analytical Hierarchy Process (FAHP). They used the study to identify and prioritize potential hazards at a sour crude oil processing facility. Celik and Gul [32] performed a two-dimensional occupational risk assessment via BWM-MARCOS integration under interval type-2 fuzzy sets for dam safety. While two risk parameters (severity and occurrence) are weighted interval type-2 fuzzy BWM, hazards are prioritized via interval type-2 fuzzy MARCOS method. Another classical method, the Fine–Kinney method, is often integrated with fuzzy MCDM. A fundamental book on the subject, Gul et al. [33], includes many approaches applied to different cases and provided their Python codes in modeling. Similarly, there are many studies combining this method with fuzzy MCDM [34–39]. Another important traditional method is FMEA. Many disadvantages of FMEA such as the lack of weight of risk parameters, loss of information in evaluating failure modes, not taking into account the relationship between failure modes in the calculation of risk priority number, different scores of the parameters giving the same risk priority number, and not considering additional parameters other than three parameters have been eliminated by its usage with fuzzy MCDM [40–45].

On the other hand, almost all of the fuzzy set extensions mentioned above have been applied in occupational risk assessment in an integrated manner with MCDM methods [20,46–52]. Mohandes et al. [48] developed a five-dimensional-safety risk assessment model to improve construction safety. They used FAHP as a weighting tool for the five dimensions, and FTOPSIS to obtain a precise prioritized ranking system for the identified safety risks. Liu et al. [46] developed a new occupational risk assessment model by integrating picture fuzzy sets and the Alternative Queuing Method (AQM) to assess and rank the risk of occupational hazards for corrective actions. Gul et al. [53] proposed a Fermatean fuzzy TOPSIS-based approach for occupational risk assessment in manufacturing. Ak et al. [54] studied occupational health, safety, and environmental risk assessment in the textile production industry through a Bayesian BWM-VIKOR approach.

The 3,4-QF-MCDM can express a wider field, imprecise information in decision-making more flexible, applicable, and detailed [16]. The adequacy and suitability of the proposed model are verified by solving a numerical problem concerned with the occupational risk assessment pertaining to the flight and ground training, training management, administration, and facilities in a flight school. When the fuzzy logic-based MCDM methods in the literature are examined, it is seen that more consistent decisions can be obtained and more consistent models can be modeled in OHS with the 3,4-QF-MCDM study. It provides a higher degree of consistency to risk prioritization. The main advantage of 3,4-QFS is that it allows decision-makers to take advantage of additional areas such as flexibility, and reduction of uncertainty when applying to MCDM problems [16]. Occupational health and safety risk analysis studies require a detailed examination of the effectiveness in decision-making processes due to the uncertainties in the scope and detail. The literature has revealed that more detailed and flexible decision-making processes can be performed with 3,4-QF-MCDM [16].

3. Research Method

3.1. Preliminaries on 3,4-QFSs

Before moving on to the detailed notation adapted from [16], it is useful to define the 3,4-QFS. For the universal set U, a 3,4-QFS (3,4Q) is defined as $3{,}4Q = \{\langle d, f_{3,4Q}(d), h_{3,4Q}(d) | d \in U \rangle\}$. Here, $f_{3,4Q} : U \to [0,1]$ and $h_{3,4Q} : U \to [0,1]$ represent membership and non-membership degree by satisfying the condition of $0 \leq (f_{3,4Q}(d))^3 + (h_{3,4Q}(d))^4 \leq 1$.

The term of $\psi_{3,4Q}(d) = \sqrt[12]{1 - (f_{3,4Q}(d))^3 - (h_{3,4Q}(d))^4}$ is the hesitancy degree. 3,4-QFSs can describe inexact data more precisely than Pythagorean and Fermatean fuzzy sets. In addition, 3,4-QFS allows the decision maker to take advantage of more space in the use of membership and non-membership values when handling the MCDM problem. Therefore, there are some decision-making situations that can be handled with 3,4-QFSs, but cannot be expressed with Pythagorean and Fermatean fuzzy numbers and their corresponding linguistic terms. As an example, suppose a decision maker sets a satisfaction degree of 0.8 and a dissatisfaction degree of 0.8. We cannot handle this situation with Pythagorean and Fermatean fuzzy sets since $0.8^2 + 0.8^2 > 1$ and $0.8^3 + 0.8^3 > 1$. On the other hand, this can be expressed with 3,4-QFSs ($0.8^3 + 0.8^4 < 1$). In such decision-making problems, 3,4-QFSs are more useful to process uncertain information and better reflect this uncertainty [16]. A comparison of the spaces of all three fuzzy set versions is given in Figure 1.

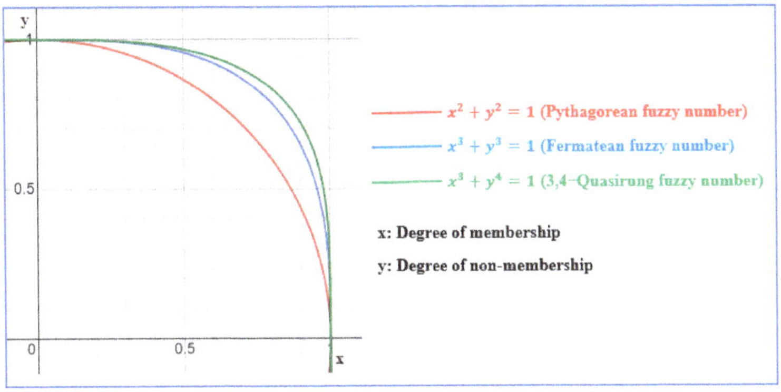

Figure 1. Comparison of the spaces of Pythagorean, Fermatean and 3,4-Quasirung fuzzy numbers.

In order to be used in the MCDM approach used for this study, the score and accuracy functions should be formulated for this type of fuzzy extension. According to the Seikh and Mandal [16], the following equations are suggested:

The score function Φ for the 3,4-QFS $\delta = (f_\delta, h_\delta)$ is formulized as in Equation (1).

$$\Phi(\delta) = \frac{1 + {f_\delta}^3 - {h_\delta}^4}{2},\ \Phi(\delta)\epsilon[0,1] \tag{1}$$

The score function Θ for the 3,4-QFS $\delta = (f_\delta, h_\delta)$ is formulized as in Equation (2).

$$\Theta(\delta) = \frac{{f_\delta}^3 + {h_\delta}^4}{2}, \Theta(\delta)\epsilon[0,1] \tag{2}$$

For more detailed theorems which the score and accuracy functions have satisfied, one can be referred the study [16]. Some basic arithmetic operations of 3,4-QFSs are given in Equations (3)–(6).

Let $A = (f_A, h_A)$ and $Z = (f_Z, h_Z)$ be two 3,4-QF numbers.

$$A \oplus Z = \left(\sqrt[3]{f_A^3 + f_Z^3 - f_A^3 f_Z^3}, h_A h_Z\right) \tag{3}$$

$$A \otimes Z = \left(f_A f_Z, \sqrt[4]{h_A^4 + h_Z^4 - h_A^4 h_Z^4}\right) \tag{4}$$

$$\lambda A = \left(\sqrt[3]{1 - \left(1 - f_A^3\right)^\lambda}, {h_A}^\lambda\right) \tag{5}$$

$$A^\lambda = \left({f_A}^\lambda, \sqrt[4]{1 - \left(1 - h_A^4\right)^\lambda}\right) \tag{6}$$

Some aggregation operators are needed to combine the evaluations of the decision makers and to inject the crisp criterion weights (the weights of the risk parameters obtained with BWM for this problem) into the calculations in the form of 3,4-QF numbers. These are the 3,4-Quasirung fuzzy weighted averaging aggregation operator (3,4-QFWA) and the 3,4-Quasirung fuzzy weighted geometric aggregation operator (3,4-QFGA). Formulas and calculation details are given in Equations (7) and (8).

The aggregated value of a number of 3,4-QF numbers $3,4Q_r = (f_{3,4Q_r}, h_{3,4Q_r})$, $r = 1,2,\ldots,k$ is calculated with the arithmetic operator as in Equation (7).

$$3,4-\text{QFWA}(3,4Q_1, 3,4Q_2, \ldots, 3,4Q_k) = \oplus_{r=1}^{k} \varrho_r * 3,4Q_r = \left(\sqrt[3]{1-\prod_{r=1}^{k}(1-f_{3,4Q_r}{}^{3})^{\varrho_r}}, \prod_{r=1}^{k} h_{3,4Q_r}{}^{\varrho_r}\right) \quad (7)$$

Here $\varrho = (\varrho_1, \varrho_2, \ldots, \varrho_k)^T$ is the weight vector of $3,4Q_r = (f_{3,4Q_r}, h_{3,4Q_r})$, $r = 1,2,\ldots,k$. $\varrho_r > 0$ and $\sum_{r=1}^{k} \varrho_r = 1$.

The aggregated value of a number of 3,4-QF numbers $3,4Q_r = (f_{3,4Q_r}, h_{3,4Q_r})$, $r = 1,2,\ldots,k$ is calculated with the geometric operator as in Equation (8).

$$3,4-\text{QFGA}(3,4Q_1, 3,4Q_2, \ldots, 3,4Q_k) = \oplus_{r=1}^{k} 3,4Q_r{}^{\varrho_r} = \left(\prod_{r=1}^{k} f_{3,4Q_r}{}^{\varrho_r}, \sqrt[4]{1-\prod_{r=1}^{k}(1-h_{3,4Q_r}{}^{4})^{\varrho_r}}\right) \quad (8)$$

To provide an easy understanding of the readers, one small example is given in the following to show how the QFWA is computed. Let $A1 = (0.5, 0.2)$, $A2 = (0.8, 0.3)$, $A3 = (0.8, 0.3)$, $A4 = (0.7, 0.3)$, $A5 = (0.4, 0.2)$, and $A6 = (0.4, 0.8)$ be six values provided under 3,4-QF numbers which are the ratings of an alternative regarding six different decision criteria. Let the weights of these six criteria be as follows, respectively: 0.2, 0.1, 0.3, 0.15, 0.15 and 0.1. With QFWA, the membership value and non-membership value of this alternative are calculated as follows:

$$\begin{aligned} &membership\ value \\ &= (1 \\ &-(((1-0.5^3)^{0.2}) * ((1-0.8^3)^{0.1}) * ((1-0.8)^{0.3}) * ((1-0.7^3)^{0.15}) * ((1-0.4^3)^{0.15}) \\ &*((1-0.4^3)^{0.1})))^{\frac{1}{3}} = 0.6876 \end{aligned}$$

$$non-membership\ value = 0.2^{0.2} * 0.3^{0.1} * 0.3^{0.3} * 0.3^{0.15} * 0.2^{0.15} * 0.8^{0.1} = 0.2871$$

Then, finally the score function Φ for this alternative is computed as follows:

$$\Phi = \frac{1 + 0.6876^3 - 0.2871^4}{2} = 0.6591$$

The similar procedure is followed for the QFGA computations.

3.2. Development of 3,4-QF MCDM-Based Occupational Risk Assessment Model

In this study, we propose an occupational risk assessment study based on 3,4-QF MCDM. The structure of the OHS risk assessment problem dealt with in this study is suitable for 3,4-QF-MCDM. For a 3,4-QF-MCDM problem, (1) evaluation criteria, (2) alternatives, (3) criterion weights, and (4) performance values obtained by evaluating alternatives against criteria are required. For the OHS risk assessment problem discussed in the study, these four components are planned as follows: Evaluation criteria in an occupational risk assessment study are the parameters that are effective in defining the risk. In this study, we consider six risk parameters unlike the traditional risk assessment methods such as the risk matrix method, Fine−Kinney method, and FMEA as follows: (1) Probability: The frequency of occurrence of the hazard [11,48,54], (2) Severity: The degree of hazard that the risk will pose on personnel, machinery-equipment, environment and continuity of production/service [11,42,48,54], (3) Detectability: The detectability of the risk with the eye or any digital device [11,48], (4) Cost: Percentage of the total annual budget determined by the company for OHS measures [42,55], (5) Sensitivity to not using personal protective equipment: To what extent the use of personal protective equipment

affects the severity of the risk [56], (6) Applicability of preventive measures: Opportunities for preventive measures and their degree of applicability [55,57]. The second component considered as an alternative is the hazards and associated risks identified in the context of OHS in the observed flight school. Criterion weights represent the relative importance weights of six risk parameters and were calculated with Best-Worst Method (BWM) [58]. The performance values obtained by evaluating the alternatives according to the criteria refer to the value obtained by scoring each hazard according to six different risk parameters. These ratings were made for different decision makers with relatively the same level of experience, using a 5-point 3,4 quasirung fuzzy linguistic scale. This scale was first proposed and used by the authors in this study. Here, the values named as criteria in a usual MCDM problem and specified as "risk parameter" in our problem consist of real numbers. These values were obtained by applying the BWM method. The details of the BWM method have not been given here. Already, the steps of the traditional BWM method can be followed by Rezaei [58]. The values expressing the performance values of the alternatives against the criteria and showing the score given as a result of the evaluation of each hazard by the expert according to each risk parameter for our study are expressed with 3,4-QF numbers. For our problem, let $H = \{H_1, H_2, \ldots, H_m\}$ be a set of hazards emerged at the observed case study facility and $RP = \{RP_1, RP_2, \ldots, RP_n\}$ be the set of risk parameters considered. The number of risk parameters for this study is six. Let $\varrho = \{\varrho_1, \varrho_2, \ldots, \varrho_n\}$ be the weight vector of risk parameters obtained via BWM where $\varrho_j(j = 1,2,\ldots,n)$, $\varrho_j > 0$, $\sum_{j=1}^{n} \varrho_j = 1$. Let $A = (\alpha_{ij})_{mxn} = \left(\left(f_{3,4Q_{ij}}, h_{3,4Q_{ij}}\right)\right)_{mxn}$ be the 3,4-QF decision matrix. Here, $\alpha_{ij} = \left(f_{3,4Q_{ij}}, h_{3,4Q_{ij}}\right)$ shows assessment of an expert on the hazard H_i with respect to risk parameter RP_j. It should be noted that $\left(f_{3,4Q_{ij}}\right)^3 + \left(h_{3,4Q_{ij}}\right)^4 \leq 1$ and $f_{3,4Q_{ij}} \in [0,1]$ and $h_{3,4Q_{ij}} \in [0,1]$. In our proposed occupational risk assessment model, both 3,4-QFWA and 3,4-QFGA operators are tested to find the priority values and orders of each hazard. The steps of the suggested model are adapted from Seikh and Mandal [16]'s study as in the following:

Step 1: Determine components of the occupational risk assessment model: the risk parameters; hazard list; OHS experts who participate in the assessment (with their expertise coefficient).

Step 2: In this second step, OHS experts make their individual assessments on the hazards with respect to risk parameters, using the scale as suggested by the authors. It is a new 5-point 3,4-QF linguistic scale and given in Table 1. Individual assessments of experts are aggregated with the operators of 3,4-QFWA and/or 3,4-QFGA. Experts' coefficients are assumed to be equal in terms of experience in the sector. Here, we introduce the aggregated decision matrix as $B = (\beta_{ij})_{mxn} = \left(\left(f_{3,4Q_{ij}}, h_{3,4Q_{ij}}\right)\right)_{mxn}$.

Table 1. 5-point 3,4-QF linguistic scale used for assessing hazards with respect to risk parameters.

Linguistic Term of Risk Parameter						Corresponding 3,4-QF Number	
RP1	**RP2**	**RP3**	**RP4**	**RP5**	**RP6**	**Membership Degree**	**Non-Membership Degree**
Very Low	Needs first aid	Easy	Very low cost	Negligible	Totally possible	0.11	0.99
Low	Minor injury	Highly possible	Lower costs	Low	Highly possible	0.44	0.95
Medium	Serious injury	Sometimes possible	Moderate cost	Medium	Medium	0.69	0.82
High	Fatality	Highly difficult	High cost	High	Low possibility	0.92	0.51
Very High	Many fatalities	Extremely difficult	Very high cost	Maximum	Not possible at all	1.00	0.00

Note: RP1: Probability; RP2: Severity; RP3: Detectability; RP4: Cost; RP5: Sensitivity to not using personal protective equipment; RP6: Applicability of preventive measures.

Step 3: Normalize the aggregated decision matrix $B = (\beta_{ij})_{mxn} = \left(\left(f_{3,4Q_{ij}}, h_{3,4Q_{ij}}\right)\right)_{mxn}$ into a new matrix named by $C = (\gamma_{ij})_{mxn} = \left(\left(f_{3,4Q_{ij}}, h_{3,4Q_{ij}}\right)\right)_{mxn}$ the following two rules in Equation (9):

$$\gamma_{ij} = \begin{cases} \left(f_{3,4Q_{ij}}, h_{3,4Q_{ij}}\right), \ if \ H_j \ is \ a \ benefit \ risk \ parameter \\ \left(h_{3,4Q_{ij}}, f_{3,4Q_{ij}}\right), \ if \ H_j \ is \ a \ cost \ risk \ parameter \end{cases} \tag{9}$$

Step 4: Determine weights of risk parameters $\varrho_j (j = 1, 2, \ldots, n)$ via Rezaei's BWM [58]. For the computations, two pairwise comparison matrix is required as called Best-to-Others and Others-to-Worst. Then, optimal weights for each risk parameter is computed by solving the mathematical optimization model. Also, the consistency of matrices should be checked by the conditions in Rezaei [58].

Step 5: Compute the information $\zeta_k (k = 1, 2, \ldots, m)$ of the hazard $H_k (k = 1, 2, \ldots, m)$ via one of the Equations (10) and (11).

$$\zeta_k = 3,4 - \text{QFWA}(\gamma_{k1}, \gamma_{k2}, \ldots, \gamma_{kn}) = \oplus_{j=1}^{n} \varrho_j \gamma_{kj} = \left(\sqrt[3]{1 - \prod_{j=1}^{n} \left(1 - \left(f_{3,4Q_{kj}}{}^{\gamma}\right)^3\right)^{\varrho_j}}, \prod_{j=1}^{n} \left(h_{3,4Q_{kj}}{}^{\gamma}\right)^{\varrho_j} \right) \tag{10}$$

$$\zeta_k = 3,4 - \text{QFGA}(\gamma_{k1}, \gamma_{k2}, \ldots, \gamma_{kn}) = \otimes_{j=1}^{n} \left(\gamma_{kj}\right)^{\varrho_j} = \left(\prod_{j=1}^{n} \left(f_{3,4Q_{kj}}{}^{\gamma}\right)^{\varrho_j}, \sqrt[4]{1 - \prod_{j=1}^{n} \left(1 - \left(h_{3,4Q_{kj}}{}^{\gamma}\right)^4\right)^{\varrho_j}} \right) \tag{11}$$

Step 6: Compute the score function $\Phi(\zeta_k)$ for each hazard with Equation (1).

Step 7: If the scores of $\Phi(\zeta_k)$ for $(k = 1, 2, \ldots, m)$ be all distinct, then the most serious hazard (the most priority one) is H_k if $\Phi(\zeta_k) = \max_{1 \le l \le m} \{\Phi(\zeta_l)\}$.

Step 8: If there exists more than one hazard, $\Phi(\zeta_k)$ $(k = 1, 2, \ldots, m)$ are equal, we consider accuracy values of each hazard $\Theta(\zeta_k)$ via Equation (2).

- If $\Phi(\zeta_k)$ provides maximum value for one particular hazard, then this hazard has the highest priority and is the most serious/riskiest.
- If $\Phi(\zeta_k)$ provides maximum value for more than one particular hazard, then the most serious/riskiest hazard is one which has the highest $\Theta(\zeta_k)$ value.
- If the $\Theta(\zeta_k)$ values are equal for two or more than two hazards, the decision maker is free to select one of them. Both are possible and have the same priority orders.

4. Method Implementation and Results

4.1. Case Study Description

In this section, we applied the 3,4 QF MCDM-based OHS risk assessment in a real case study concerned with the occupational risk assessment pertaining to the flight and ground training, training management, administrative, and facilities in a flight school to verify the validity and effectiveness of the proposed method. In direct proportion to the development of aviation, the demand for airplanes and pilots is increasing. The demand for flight schools, a total number of flight schools as well, has been increasing in recent years due to meet it. The flight school, where the study was carried out, started its training activities as a flight school in order to meet the pilot needs of the rapidly developing civil aviation industry. The flight school has the authorization to give Modular ATPL(A), ATPL(A) Integrated, and Multi Pilot License (MPL) Integrated into flight training.

Flight school activities contain occupational hazards and related risks in different categories in terms of occupational health and safety. Especially during the training phase, the possibility and effects of risk require a more detailed examination and a proactive approach. The processes in which occupational hazards and related risks occur in these activities are as follows: flight training process, ground services training process, managerial training processes, facility and related training processes, and training management process. Use of equipment, Perception, Task management, Communication, and Personnel actions are the five highest serious incidents according to European Aviation Safety Agency Report [59].

In this study, we consider six risk parameters unlike the traditional risk assessment methods such as risk matrix method, Fine−Kinney method, FMEA as follows: (1) Probability, (2) Severity, (3) Detectability, (4) Cost, (5) Sensitivity to not using personal protective equipment, (6) Applicability of preventive measures. Scale for six parameters can be seen in Tables 2–7. Probability refers to the frequency of occurrence of the hazard. Quantitative value and qualitative value of the probability parameter, related explanations are given in Table 2.

Table 2. Ratings of probability.

Quantitative Value	Qualitative Value	Description of Parameter
1	Very low	Hardly ever
2	Low	Once a year
3	Medium	Once in a month
4	High	Once a week
5	Very high	Every day (very often)

Table 3. Ratings of severity.

Quantitative Value	Qualitativevalue	Description of Parameter
1	Very Light	No loss of working hours, first aid required
2	Light	No lost workdays, outpatient treatment
3	Serious	Minor injury, treatment in bed
4	Very serious	Serious injury, loss of limb, occupational disease
5	Disaster	One or more deaths

Table 4. Ratings of detectability.

Quantitative Value	Qualitative Value	Description of Parameter
1	Very high	Risk can be detected very quickly and easily.
2	High	Risk can be detected quickly and easily.
3	Medium	Risk can be detected with reasonable time and experience.
4	Low	Determining the risk is very time-consuming and difficult.
5	Very low	Identifying the risk is almost impossible.

Table 5. Ratings of cost.

Quantitative Value	Qualitative Value	Description of Parameter
1	Very low cost	Between 0% and 20% of the total annual budget is allocated to OHS measures.
2	Lower costs	Between 21% and 40% of the total annual budget is allocated to OHS measures.
3	Moderate cost	Between 41% and 60% of the total annual budget is allocated to OHS measures.
4	High cost	Between 61% and 80% of the total annual budget is allocated to OHS measures.
5	Very high cost	Between 81% and 100% of the total annual budget is allocated to OHS measures.

Table 6. Ratings of sensitivity to not using personal protective equipment.

Quantitative Value	Qualitative Value	Description of Parameter
1	Negligible	Risk can be avoided without using PPE.
2	Low	The use of PPE can reasonably reduce the risk.
3	Moderate	The use of PPE reduces the risk.
4	High	It is necessary to use PPE to reduce the risk.
5	Maximum	PPE must be used.

Table 7. Ratings of applicability of preventive measures.

Quantitative Value	Qualitative Value	Description of Parameter
1	Quite possible	Opportunities for preventive measures and their applicability are entirely possible.
2	High	Opportunities for preventive measures and their feasibility are high.
3	Moderate	Opportunities for preventive measures and their applicability are moderate.
4	High	Opportunities for preventive measures and their viability are low.
5	Practically impossible	Opportunities for preventive measures and their applicability are not possible.

Severity refers to the degree of hazard that the risk will pose on personnel, machinery-equipment, environment and continuity of production/service. Quantitative value and qualitative value of the severity parameter, related explanations are given in Table 3.

Detectability refers to the detectability of the risk with the eye or any digital device. Quantitative value and qualitative value of the detectability parameter, related explanations are given in Table 4.

Cost refers to the percentage of the total annual budget determined by the company for OHS measures. Quantitative value and qualitative value of the cost parameter, related explanations are given in Table 5.

Sensitivity to not using personal protective equipment refers to what extent the use of personal protective equipment affects the severity of the risk. Quantitative value and qualitative value of the PPE parameter, related explanations are given in Table 6.

Applicability of preventive measures refers to opportunities for preventive measures and their degree of applicability. Quantitative value and qualitative value of the PPE parameter, related explanations are given in Table 7.

Risks and related processes within the scope of flight school activities are 1-Flight Training, 2-Ground Training, 3-Administrative Process, 4-Training Management, 5-Facilities. Five basic processes and risks in the processes are listed in Table 8.

In this study, an aviation management specialist, 2 assistant professor trainers, and 2 pilot trainers evaluated the risks in the process on 6 determined parameters. Instructors have more than 10 years of teaching and piloting experience. In this study, which includes risk analysis and evaluation, the fact that the experts have industry experience makes the findings valuable. The inclusion of experts with field experience and piloting training practice in the determination process of the parameters provided a more detailed and consistent evaluation of the problems, hazards, and related risk situations experienced in the flight school processes. In addition, a format has been created that will allow the use of both the content of the study and the method applied by other flight schools. In the application of the 3,4-QF MCDM-based occupational risk assessment model, the provision of literature-supported content and integration of expert opinions have allowed a comprehensive and consistent assessment of the dangers and risks inherent in flight schools. A consistent and comprehensive quantitative risk assessment before flight training can systematically help you determine if the risk level is too high, and provide an opportunity to reduce or reject risk before it is too late. Flight instructors are responsible for understanding and taking action against a wide range of risks, both for themselves and their students. The study provides the opportunity to apply and use risk assessments specific to flight schools, with information and evaluations obtained

from instructors who have flight instructor experience, work at different flight schools and continue their actual training. This study creates a baseline for risk assessing processes of flight education and brings attention to the decisions makers on the highest priority risks.

Table 8. Description of hazards, associated risks and related process.

Hazard ID	Hazard	Related Risk	Process
H1	Lack of flight safety	Mid-air collusion	Flight Training
H2	Mechanical: Engine	Engine fails in flight	Flight Training
H3	Mechanical: Control Mechanism	Flight Control Mechanism Malfunction	Flight Training
H4	Mechanical: Landing Gear	Landing gear not deployed	Flight Training
H5	Inadequate preflight planning	Smoke, fire, and fumes	Flight Training
H6	Mismanagement of fuel	Critical level of fuel	Flight Training
H7	Mechanical: Control Mechanism	System malfunction	Flight Training
H8	Misjudgment of distance and speed	Excursion	Flight Training
H9	Misjudgment of distance and speed	Incursion	Flight Training
H10	Improper in-flight decision	Abandoned take-off	Flight Training
H11	Improper in-flight decision	Emergency declaration	Flight Training
H12	Lack of flight safety	Forced landing off track	Flight Training
H13	Improper in-flight decisions	Hard landing	Flight Training
H14	Failure to maintain directional control	Landing on the wrong runway	Flight Training
H15	Inadequate preflight planning	Tire damage and blowouts	Flight Training
H16	Lack of flight safety	Runway Crossing Incursion	Flight Training
H17	Failure to see and avoid objects or obstructions.	Bird Strike	Flight Training
H18	Improper in-flight decision	Getting lost in flight (individual flight)	Flight Training
H19	Physiological factors	Pilot Incapacitation	Flight Training
H20	Violation of aviation safety rules	NOTAM	Flight Training
H21	Violation of aviation safety rules	Worksite Violation	Flight Training
H22	Lack of flight safety	Disobey ATC instructions	Flight Training
H23	Work environment factors	FOD on runway	Flight Training
H24	Inadequate preflight controls	Planning with a lack of Instructor Authorization: Ground training	Ground Training
H25	Inadequate preflight controls	Lack of training of trainers certificate: Ground training	Ground Training
H26	Inadequate preflight controls	Availability of staff/teachers who were recruited without registration	Administrative Process
H27	Insufficient practical training	Ensuring the integration of theoretical training and flight training	Training Management
H28	Improper in-flight decisions	Uncertainty of communication in emergency situations, course of action in incidents or accidents	Flight Training
H29	Mechanical	Injury in the candidate selection process	Facilities
H30	Human error	Injury in the candidate selection process	Facilities
H31	Violation of aviation safety rules	Candidate restricted area entry and simulator use	Flight Training
H32	Weakness of communication in education	Distrust between candidate and instructor	Administrative Process
H33	COVID-19 virus	COVID-19 transmission risk	Administrative Process
H34	COVID-19 virus	Online course due to pandemic risk	Training Management

Table 8. *Cont.*

Hazard ID	Hazard	Related Risk	Process
H35	COVID-19 virus	Continuation of flight activity in the pandemic	Flight Training
H36	COVID-19 virus	Delay of the normalization process due to the pandemic	Administrative Process
H37	COVID-19 virus	The risk of mass transmission in theoretical trainings made face-to-face due to the pandemic	Training Management
H38	COVID-19 virus	Risk of virus transmission from headphones	Training Management
H39	COVID-19 virus	The need to give online lessons to students during the full closure of the pandemic	Training Management
H40	COVID-19 virus	Continuation of flight activity during the full closure of the pandemic	Flight Training

4.2. Results of 3,4-QF MCDM-Based Occupational Risk Assessment Model

In order to demonstrate the applicability of the adapted approach to the field of occupational risk assessment, the step-by-step implementation results of the approach detailed in Section 3.2 is presented below. Since detailed information is given in the previous sub section about the preparation stage before the occupational risk assessment and the components needed, it is useful to start with the steps in which direct numerical calculations are made. This corresponds to the second step of the steps given in Section 3.2. In this step, the evaluations of 40 hazards according to 6 risk parameters and the scale in Table 1. The risk parameters were taken from 4 decision-making expert participants. These evaluations taken are aggregated using both the operators given in Equations (7) and (8). It should be noted here that the expert weights are taken equally as 0.25. Considering that the geometric mean, which is one of the applied operators, reduces the information loss relatively less, 3,4-QFGA was preferred in the calculation. The aggregated decision matrix as B is computed as in Table 9. In the third step, the normalized aggregated decision matrix is the same as the aggregated decision matrix, since all risk parameters are evaluated as "benefit" and the scale is prepared accordingly. Fourth step is on the determination of six risk parameter weights via BWM method. It is a recently proposed MCDM method based on pairwise comparison [54]. It requires less pairwise comparisons when compared to the most known and applied pairwise comparison-based MCDM method "Analytic Hierarchy Process".

It also provides a more consistent assessment of the subjective judgments of experts. Therefore, we used BWM to determine the importance weights of *RP1—probability, RP2—severity, RP3—detectability, RP4—cost, RP5—sensitivity to not using personal protective equipment* and *RP6—applicability of preventive measures* parameters. With the Saaty's 1-9 scale on the Best-to-Others and Others-to-Worst evaluations (the OHS experts from the facility make this assessment in a group consensus), we computed the weights of six risk parameters by using the BWM solver (developed by Rezaei) as shown in Figure 2. Moreover, the consistency has been checked and found valid and consistent. With the proposed risk assessment method, it will be possible to minimize the uncertainty of hazards and risks, analyze, evaluate and examine them consistently. For the implementation of the model, it is necessary to report in detail the experienced and possible cases and to ensure data reliability. The flight school will be able to benefit from the proposed risk assessment method during the curriculum formation, theoretical and practical training process.

Table 9. The aggregated decision matrix.

Hazard	Aggregated Value in 3,4-QF Number											
	RP1		RP2		RP3		RP4		RP5		RP6	
H1	0.11	0.99	1.00	0.00	0.77	0.78	0.83	0.71	0.55	0.91	0.69	0.82
H2	0.22	0.98	0.92	0.51	0.86	0.65	0.89	0.63	0.55	0.91	0.62	0.87
H3	0.44	0.95	1.00	0.00	0.86	0.65	0.83	0.71	0.55	0.91	0.62	0.87
H4	0.22	0.98	0.92	0.51	0.86	0.65	0.83	0.71	0.55	0.91	0.49	0.93
H5	0.11	0.99	1.00	0.00	0.86	0.65	0.83	0.71	0.62	0.87	0.62	0.87
H6	0.11	0.99	0.92	0.51	0.77	0.78	0.66	0.87	0.39	0.94	0.16	0.99
H7	0.11	0.99	0.55	0.91	0.86	0.65	0.83	0.71	0.28	0.96	0.49	0.93
H8	0.11	0.99	0.92	0.51	0.86	0.65	0.66	0.87	0.28	0.96	0.44	0.95
H9	0.11	0.99	0.94	0.48	0.86	0.65	0.74	0.81	0.28	0.96	0.49	0.93
H10	0.22	0.98	0.92	0.51	0.86	0.65	0.66	0.87	0.28	0.96	0.11	0.99
H11	0.11	0.99	0.92	0.51	0.92	0.51	0.98	0.36	0.28	0.96	0.44	0.95
H12	0.31	0.97	0.94	0.48	0.77	0.78	0.83	0.71	0.28	0.96	0.44	0.95
H13	0.25	0.97	0.92	0.51	0.86	0.65	0.66	0.87	0.28	0.96	0.49	0.93
H14	0.11	0.99	0.86	0.65	0.86	0.65	0.66	0.87	0.28	0.96	0.11	0.99
H15	0.11	0.99	0.92	0.51	0.86	0.65	0.66	0.87	0.28	0.96	0.49	0.93
H16	0.55	0.91	0.69	0.82	0.77	0.78	0.33	0.95	0.28	0.96	0.44	0.95
H17	0.55	0.91	1.00	0.00	0.77	0.78	0.66	0.87	0.28	0.96	0.49	0.93
H18	0.11	0.99	0.49	0.93	0.77	0.78	0.83	0.71	0.28	0.96	0.44	0.95
H19	0.11	0.99	0.92	0.51	0.86	0.65	0.83	0.71	0.28	0.96	0.31	0.97
H20	0.22	0.98	0.80	0.73	0.77	0.78	0.66	0.87	0.28	0.96	0.44	0.95
H21	0.31	0.97	0.69	0.82	0.77	0.78	0.66	0.87	0.28	0.96	0.44	0.95
H22	0.11	0.99	0.74	0.78	0.86	0.65	0.33	0.95	0.28	0.96	0.44	0.95
H23	0.16	0.99	0.69	0.82	0.86	0.65	0.33	0.95	0.28	0.96	0.44	0.95
H24	0.11	0.99	0.11	0.99	0.77	0.78	0.33	0.95	0.28	0.96	0.11	0.99
H25	0.11	0.99	0.11	0.99	0.77	0.78	0.33	0.95	0.28	0.96	0.11	0.99
H26	0.11	0.99	0.11	0.99	0.77	0.78	0.33	0.95	0.28	0.96	0.11	0.99
H27	0.11	0.99	0.11	0.99	0.77	0.78	0.33	0.95	0.28	0.96	0.44	0.95
H28	0.11	0.99	0.11	0.99	0.77	0.78	0.33	0.95	0.28	0.96	0.44	0.95
H29	0.11	0.99	0.69	0.82	0.77	0.78	0.33	0.95	0.28	0.96	0.69	0.82
H30	0.11	0.99	0.69	0.82	0.77	0.78	0.33	0.95	0.28	0.96	0.69	0.82
H31	0.11	0.99	0.11	0.99	0.77	0.78	0.33	0.95	0.28	0.96	0.44	0.95
H32	0.86	0.65	0.11	0.99	0.77	0.78	0.33	0.95	0.28	0.96	0.44	0.95
H33	0.86	0.65	0.44	0.95	0.77	0.78	0.33	0.95	0.28	0.96	0.62	0.87
H34	0.69	0.82	0.44	0.95	0.77	0.78	0.33	0.95	0.28	0.96	0.44	0.95
H35	0.44	0.95	0.44	0.95	0.77	0.78	0.33	0.95	0.28	0.96	0.44	0.95
H36	0.44	0.95	0.44	0.95	0.77	0.78	0.33	0.95	0.28	0.96	0.69	0.82
H37	0.44	0.95	0.44	0.95	0.77	0.78	0.33	0.95	0.28	0.96	0.69	0.82
H38	0.44	0.95	0.44	0.95	0.77	0.78	0.33	0.95	0.28	0.96	0.86	0.65
H39	0.44	0.95	0.44	0.95	0.77	0.78	0.33	0.95	0.28	0.96	0.69	0.82
H40	0.44	0.95	0.44	0.95	0.77	0.78	0.33	0.95	0.28	0.96	0.62	0.87

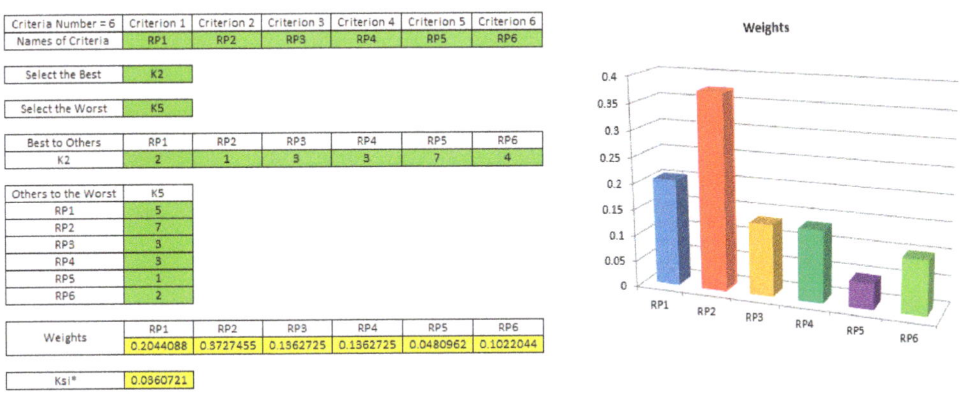

Figure 2. Weight determination of risk parameters via BWM. * note: the optimal value of Ksi.

In the fifth step, we have computed the information $\zeta_k (k = 1 \text{ to } 40)$ of each hazard via the Equation (11). Then, we have computed the score function $\Phi(\zeta_k)$ for each hazard. The results are demonstrated in Figure 3.

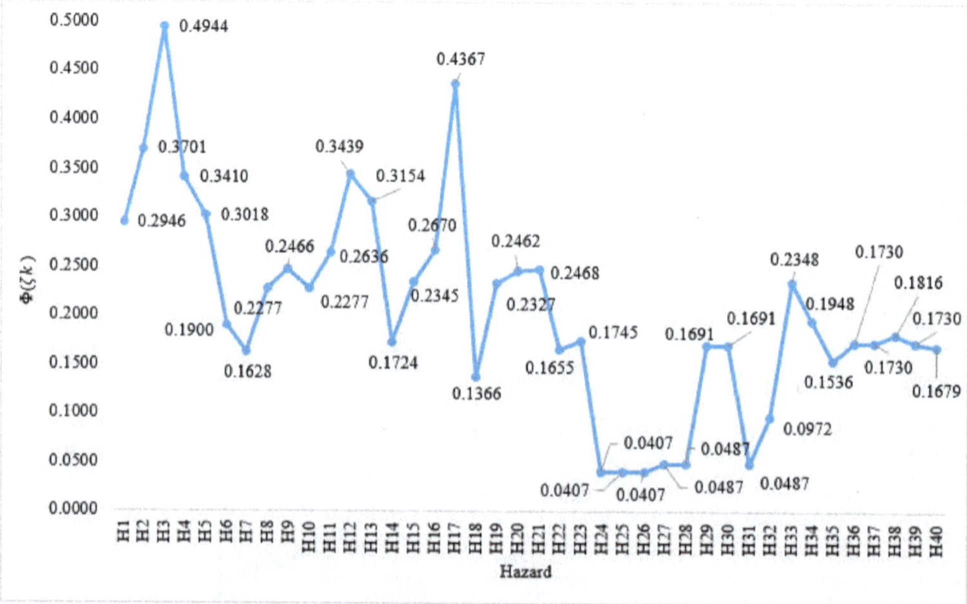

Figure 3. $\Phi(\zeta_k)$ values for each hazard.

According to the calculation results, the hazards H3 (0.4944) and H17 (0.4367) have the highest Φ values, and they are in the first and second place. These are control failure and bird strike hazards, respectively. These are followed by H2 (0.3701), H12 (0.3439), and H4 (0.3410), respectively. These are related to engine failure, forced landing and failure of landing gear. It's important to see that all of these five top priorities relate to flight training. To identify hazards with the same Φ value, H10 & H8 (0.227), H36 & H37 & H39 (0.1730), H29 & H30 (0.1691), H27 & H28 & H31 (0.0487) and H24 & H25 & H26 (0.0407), the Θ values of the relevant groups were examined. Since it was seen that the Θ values of all these five groups, which were looked for in order to rank within themselves, were the same (Figure 4), it was observed that there was no difference between their rankings. The final rankings are presented in Table 10. According to the numerical results of the priority scores of each emerged hazard in the flight school, the most important hazards and associated risks are related to flight training such as control failure, engine failure and bird strike. However, some of the flight training hazards that we will prioritize secondary are: forced landing, landing gear not deployed, hard landing, fire and smoke, mid-air collision, fuel criticality, and emergency declaration.

For H3, H17 and H2 hazards, training should be planned in a structure that will include interpersonal activities such as optimizing the human-machine interface, building and maintaining effective teams, problem solving, decision making and maintaining situational awareness. In terms of flight school training, human factor-related errors will be integrated into the curriculum, and practical and theoretical knowledge will be developed. Crew Resource Management, Line Oriented Flight Training and Threat and Error Management have been developed and mandated by the International Civil Aviation Organization (ICAO). Safety management regulations are supplemented by ICAO [60] with manuals such as; ICAO Bird Strike Information System Manual, Air Traffic Services Planning Manual.

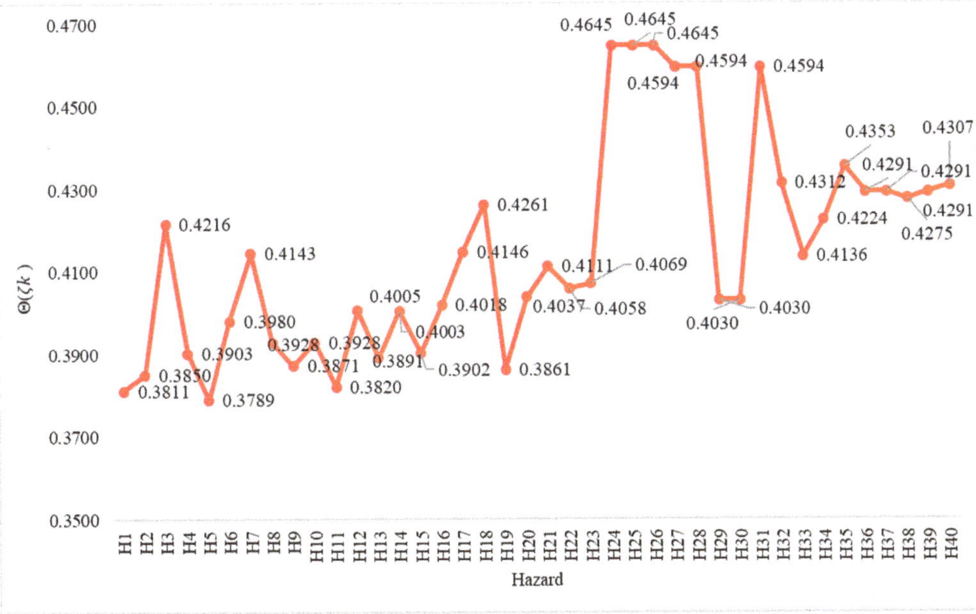

Figure 4. $\Theta(\zeta_k)$ values for each hazard.

Table 10. Final priority ranking of each hazard.

Ranking Order	Hazard	Ranking Order	Hazard
1	H3	17	H8; H10
2	H17	18	H34
3	H2	19	H6
4	H12	20	H38
5	H4	21	H23
6	H13	22	H36; H37; H39
7	H5	23	H14
8	H1	24	H29; H30
9	H16	25	H40
10	H11	26	H22
11	H21	27	H7
12	H9	28	H35
13	H20	29	H18
14	H33	30	H32
15	H15	31	H27; H28; H31
16	H19	32	H24; H25; H26

4.3. Comparative Study and Discussion

In this section, the numerical results obtained by applying the 3,4-QF MCDM-based occupational risk assessment model proposed in the article and numerical results obtained by solving the same problem with the Pythagorean and Fermatean fuzzy aggregation operators were compared. While the applied Fermatean fuzzy weighted geometric (FFWG) operator is adapted from the works of Senapati and Yager [61] and Zhou et al. [62], the Pythagorean fuzzy weighted geometric (PFWG) operator is used as in [63,64] These results are given in Figure 5. According to the results obtained with all three collection operators, the top five hazards have not changed. These are H3, H17, H2, H12 and H4 respectively.

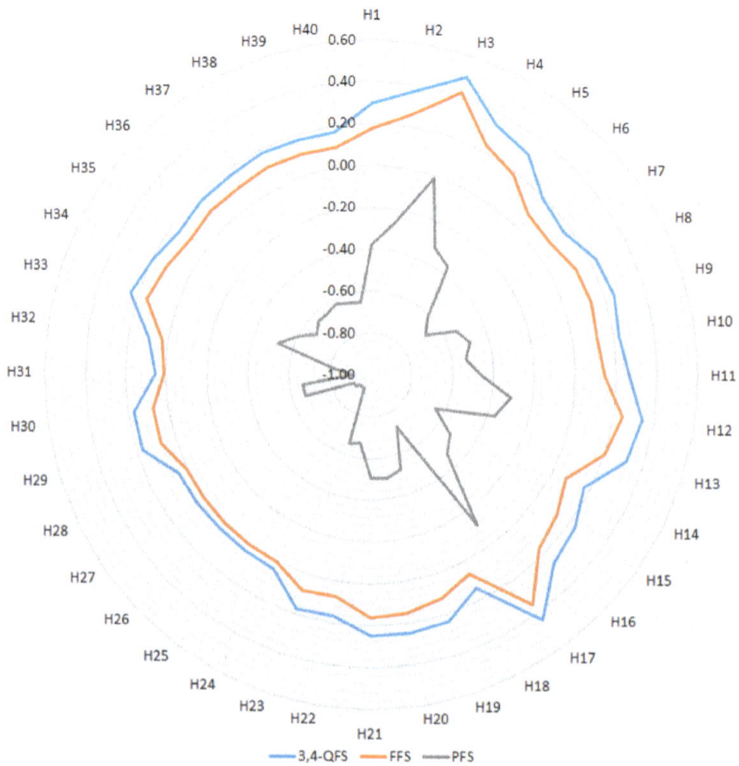

Figure 5. Comparison of final scores.

In fact, the Pearson correlation coefficient between the final scores of each hazard solved with these three operators was also found to be quite high (Figure 6). Note that the values obtained from the problem solved with PFWG are in the range of $[-1, 1]$. Because, score function values can be negative in Pythagorean fuzzy set [65]. Similarly, the final scores of the last eight hazards (H18, H32, H27, H28, H31, H24, H25 and H26) are the same with respect to all three aggregation operators based MCDM models, and the same measures can be arranged for the control measures to be taken for these eight least serious hazards.

A sensitivity analysis is also needed for implementation. Sensitivity analysis is the process of determining how changes in risk parameter weights will affect the final scores of hazards. In many occupational risk assessment studies, this is an extra study. As a matter of fact, it is an analysis that strengthens the robustness of the applied approach. In this sensitivity analysis, one of the risk parameters was defined as the major parameter and the others as the minor parameters. By highlighting the weight of the major parameter and keeping the other minor parameters at the same weight, it can be observed how much the results are affected by the major parameter. Three different scenarios listed in Table 11 are discussed in this section. According to the first scenario, the major parameter is selected as one of the six risk parameters one by one respectively (with a weight value of 0.20), while the other parameters are determined as minor parameters (all the same and with a weight value of 0.16).

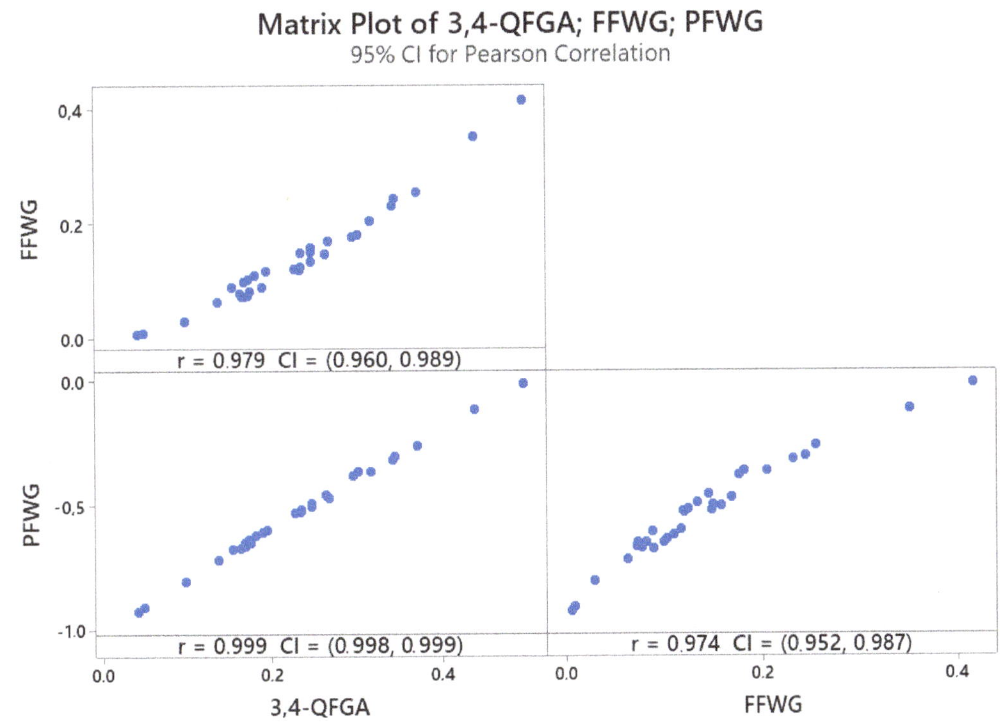

Figure 6. Pearson correlation analysis results.

Table 11. Scenario design of sensitivity analysis.

Scenario #	Weight of Major Risk Parameter	Weight of Minor Risk Parameter
Scenario 1	0.20	0.16
Scenario 2	0.40	0.12
Scenario 3	0.60	0.08

According to the results of scenario 1 as provided in Figure 7, it is seen that the H17 hazard is affected by the "RP1—Probability" risk parameter. The frequency of occurrence of the H17 risk appears to be a priority hazard when very significant. Additionally, when the weight of the "RP2—Severity" parameter is the highest, the hazard H17 falls one step back and takes less priority. Instead, H4 becomes a priority hazard. Overall speaking, H3 is the highest priority hazard, followed by H2, H4, H5 and H17. In addition, another striking result is that the hazard H1 is not affected by none of the risk parameters' weight increase.

Risk parameter	1st	2nd	3rd	4th	5th	6th	7th	8th	9th	10th	...	31th	32th	33th	34th	35th	36th	37th	38th	39th	40th
RP1	H3	H2	H17	H4	H5	H1	H33	H12	H13	H16	...	H18	H6	H22	H14	H27	H28	H31	H24	H25	H26
RP2	H3	H2	H4	H17	H5	H1	H12	H13	H33	H11	...	H19	H22	H14	H32	H27	H28	H31	H24	H25	H26
RP3	H3	H2	H4	H5	H17	H1	H12	H13	H33	H11	...	H6	H22	H32	H14	H27	H28	H31	H24	H25	H26
RP4	H3	H2	H4	H5	H17	H1	H12	H13	H11	H33	...	H23	H32	H22	H14	H27	H28	H31	H24	H25	H26
RP5	H3	H2	H4	H5	H17	H1	H12	H13	H33	H11	...	H6	H32	H22	H14	H27	H28	H31	H24	H25	H26
RP6	H3	H2	H4	H5	H17	H1	H12	H13	H33	H38	...	H22	H10	H6	H14	H27	H28	H31	H24	H25	H26

H1	H2	H3	H4	H5	H6	H7	H8	H9	H10	H11	H12	H13	H14	H15	H16	H17	H18	H19	H20
H21	H22	H23	H24	H25	H26	H27	H28	H29	H30	H31	H32	H33	H34	H35	H36	H37	H38	H39	H40

Figure 7. Sensitivity analysis on the results by Scenario 1.

The results of scenario 2 are given in Figure 8. Accordingly, when compared to Scenario 1 (Figure 7), it is seen that changes in risk parameter weights have more impact on the priority rankings of hazards.

Risk parameter	1st	2nd	3rd	4th	5th	6th	7th	8th	9th	10th	...	31th	32th	33th	34th	35th	36th	37th	38th	39th	40th
RP1	H33	H3	H17	H34	H32	H16	H2	H12	H38	H4	...	H18	H6	H22	H14	H27	H28	H31	H24	H25	H26
RP2	H3	H2	H17	H5	H4	H1	H12	H13	H11	H9	...	H40	H18	H35	H32	H27	H28	H31	H24	H25	H26
RP3	H3	H2	H4	H5	H17	H1	H13	H11	H12	H33	...	H18	H6	H32	H14	H27	H28	H31	H24	H25	H26
RP4	H3	H2	H4	H5	H1	H12	H11	H17	H13	H19	...	H35	H23	H32	H22	H27	H28	H31	H24	H25	H26
RP5	H3	H2	H5	H4	H17	H12	H13	H33	H11	H16	...	H18	H32	H22	H14	H27	H28	H31	H24	H25	H26
RP6	H3	H2	H1	H5	H38	H4	H17	H33	H36	H37	...	H22	H6	H27	H28	H31	H10	H14	H24	H25	H26

H1	H2	H3	H4	H5	H6	H7	H8	H9	H10	H11	H12	H13	H14	H15	H16	H17	H18	H19	H20
H21	H22	H23	H24	H25	H26	H27	H28	H29	H30	H31	H32	H33	H34	H35	H36	H37	H38	H39	H40

Figure 8. Sensitivity analysis on the results by Scenario 2.

According to the sensitivity analysis result of Scenario 3 presented in Figure 9, it is seen that the ranking result is similar to the previous one.

Risk parameter	1st	2nd	3rd	4th	5th	6th	7th	8th	9th	10th	...	31th	32th	33th	34th	35th	36th	37th	38th	39th	40th
RP1	H33	H32	H34	H17	H3	H16	H38	H36	H37	H39	...	H18	H6	H22	H14	H27	H28	H31	H24	H25	H26
RP2	H3	H17	H2	H5	H1	H4	H12	H13	H11	H9	...	H39	H40	H35	H32	H27	H28	H31	H24	H25	H26
RP3	H3	H2	H4	H5	H11	H13	H17	H1	H9	H15	...	H30	H18	H6	H32	H27	H28	H31	H24	H25	H26
RP4	H3	H2	H11	H4	H5	H1	H12	H19	H7	H17	...	H35	H23	H32	H22	H27	H28	H31	H24	H25	H26
RP5	H3	H5	H2	H4	H1	H17	H6	H12	H13	H33	...	H18	H32	H22	H14	H27	H28	H31	H24	H25	H26
RP6	H38	H3	H1	H2	H5	H36	H37	H39	H33	H29	...	H19	H27	H28	H31	H6	H10	H14	H24	H25	H26

H1	H2	H3	H4	H5	H6	H7	H8	H9	H10	H11	H12	H13	H14	H15	H16	H17	H18	H19	H20
H21	H22	H23	H24	H25	H26	H27	H28	H29	H30	H31	H32	H33	H34	H35	H36	H37	H38	H39	H40

Figure 9. Sensitivity analysis on the results by Scenario 3.

5. Conclusions

In this study, an occupational risk assessment approach based on 3,4-QF MCDM was proposed as the first attempt in the literature. Risk parameters, which are one of the basic components of occupational risk assessment studies, are modeled via six different parameters, different from classical risk analysis methods with two or three parameters. The weights of these parameters were obtained by [58]. The evaluations of the hazards arising in the workplace environment depending on each risk parameter were made by OHS experts and aggregated by the 3,4-QFGA operator. Comparative and sensitivity analyzes were also performed to consolidate the results of the approach.

5.1. Summary of Findings

According to the results of the risk parameter weight values determined by the BWM model, the most important parameter for this occupational risk assessment is the "severity" parameter with a weight value of 0.37. This is followed by "probability" with a significance weight of 0.20. These two parameters are followed by "detectability" and "cost" with weight values of about 0.14. The two least important parameters are "applicability of preventive measures (0.10)" and "sensitivity to not using PPE (0.05)", respectively. According to 3,4-QF MCDM risk assessment model, the most important hazards and associated risks are stemmed from the processes of flight training such as control failure, engine failure and bird strike. Moreover, secondary flight training hazards are forced landing, landing gear not deployed, hard landing, fire and smoke, mid-air collision, fuel criticality, and emergency declaration. According to the comparison analysis, there is not a significant difference between the results of the model solved with the other two types of fuzzy version-based aggregation operators (FFWG and PFWG) and the results of the current model. According to the results of the sensitivity analysis, it is seen that the H33 hazard, which is the hazard related to facilities, has the highest priority in increasing the weight of the *RP1—probability* risk parameter. This result appeared in both Scenario 2 and Scenario 3. A similar case shows that in the case of Scenario 3, H38 is the top priority hazard where the *RP6—applicability of preventive measures* parameter has a weight of 0.60 and each of the other parameters has a weight of 0.08.

5.2. Research Contributions

This study has made the following contributions from both a methodological and practical perspective.

- A new extension 3,4-QFS with a broader space than the Fermatean and Pythagorean fuzzy numbers has been adapted for the first time to an occupational risk assessment study. The proposed 3,4-QF-MCDM based approach uses more risk parameters than conventional risk assessments and calculates their weight values with Rezaei's BWM method.
- In addition, with the developed 3,4-QF scale, each hazard can be evaluated according to the relevant risk parameter, and the subjective judgments given by all the experts participating in the evaluation are aggregated with the 3,4-QFGA operator.
- Experts with field experience and pilot training practice were included in the process of determining risk assessment parameters which allows for a more detailed and consistent evaluation of problems, hazards, and related risk situations in the flight school processes. It made model more sustainable and applicable model. An innovative sensitivity analysis was conducted to analyze how the change in the weights of the parameters used in the flight school occupational risk assessment affected the priority score and, of course, the order of each hazard. In this respect, it is considered to make an important methodological contribution.
- Risk assessment for flight schools, which constitute the education pillar of the aviation industry, is undoubtedly extremely important in terms of serious hazards it contains. In this context, the occupational risk assessment study carried out in a flight school in order to test the applicability of the model contributes to the application as it is an adaptable model.

5.3. Limitations and Future Remarks

Since the proposed fuzzy set extension is still new, it is seen that this set has not yet been integrated into the MCDM methods that are widely applied in the field of occupational risk assessment. For future studies, it is planned to develop new risk assessment approaches such as 3,4-QFS-based TOPSIS and VIKOR. In addition, an approach can be suggested in which each risk parameter can be modeled how the production or service facility will be affected by some future states. With this approach, it can be considered how the risk parameter weights change in response to possible states and this change can be modeled with a fuzzy stratified MCDM structure.

Author Contributions: Data curation, M.F.A.; Methodology, M.G.; Software, M.G.; Validation, M.G.; Writing—original draft, M.G. and M.F.A.; Writing—review & editing, M.G. and M.F.A. All authors have read and agreed to the published version of the manuscript.

Funding: This research received no external funding.

Institutional Review Board Statement: Not applicable.

Informed Consent Statement: Not applicable.

Data Availability Statement: Not applicable.

Conflicts of Interest: The authors declare no conflict of interest.

References

1. European Commission; Directorate-General for Employment, Social Affairs and Inclusion. *Guidance on Risk Assessment at Work*; Publications Office: Luxembourg, 1996.
2. Cox, L.A., Jr. *Risk Analysis Foundations, Models, and Methods*; Springer Science & Business Media: Berlin, Germany, 2012; Volume 45.
3. European Agency for Safety and Health at Work. *Directive 89/391/EEC—OSH "Framework Directive"*; 1989; Available online: https://osha.europa.eu/en/legislation/directives/the-osh-framework-directive/1 (accessed on 30 July 2022).
4. Pinto, A. QRAM a Qualitative Occupational Safety Risk Assessment Model for the construction industry that incorporate uncertainties by the use of fuzzy sets. *Saf. Sci.* **2014**, *63*, 57–76. [CrossRef]
5. *IEC 31010:2019*; Risk Management—Risk Assessment Techniques, 2nd ed. International Organization for Standardization (ISO): Geneva, Switzerland, 2019. Available online: https://www.iso.org/standard/72140.html (accessed on 25 May 2022).
6. Gul, M. A review of occupational health and safety risk assessment approaches based on multi-criteria decision-making methods and their fuzzy versions. *Hum. Ecol. Risk Assess. Int. J.* **2018**, *24*, 1723–1760. [CrossRef]
7. Markowski, A.S.; Mannan, M.S. Fuzzy risk matrix. *J. Hazard. Mater.* **2008**, *159*, 152–157. [CrossRef] [PubMed]
8. Markowski, A.S.; Siuta, D. Fuzzy logic approach for identifying representative accident scenarios. *J. Loss Prev. Process Ind.* **2018**, *56*, 414–423. [CrossRef]
9. Markowski, A.S.; Mannan, M.S.; Bigoszewska, A. Fuzzy logic for process safety analysis. *J. Loss Prev. Process Ind.* **2009**, *22*, 695–702. [CrossRef]
10. Kabir, S. An overview of fault tree analysis and its application in model based dependability analysis. *Expert Syst. Appl.* **2017**, *77*, 114–135. [CrossRef]
11. Liu, H.C. FMEA using uncertainty theories and MCDM methods. In *FMEA Using Uncertainty Theories and MCDM Methods*; Springer: Singapore, 2016; pp. 13–27.
12. Dunjó, J.; Fthenakis, V.; Vílchez, J.A.; Arnaldos, J. Hazard and operability (HAZOP) analysis. A literature review. *J. Hazard. Mater.* **2010**, *173*, 19–32. [CrossRef]
13. Marhavilas, P.K.; Filippidis, M.; Koulinas, G.K.; Koulouriotis, D.E. The integration of HAZOP study with risk-matrix and the analytical-hierarchy process for identifying critical control-points and prioritizing risks in industry—A case study. *J. Loss Prev. Process Ind.* **2019**, *62*, 103981. [CrossRef]
14. Zadeh, L.A. Fuzzy sets. *Inf. Control* **1965**, *8*, 338–353. [CrossRef]
15. Ozdemir, Y.; Basligil, H.; Ak, M.F. Airport safety risk evaluation based on fuzzy Anp and fuzzy ahp. In *Airport Safety Risk Evaluation Based on Fuzzy Anp and Fuzzy Ahp*; World Scientific: Singapore, 2016.
16. Seikh, M.R.; Mandal, U. Multiple attribute decision-making based on 3, 4-quasirung fuzzy sets. *Granular Computing* **2022**, 1–14. [CrossRef]
17. Yager, R.R. Pythagorean membership grades in multicriteria decision making. *IEEE Trans. Fuzzy Syst.* **2014**, 22, 958–965. [CrossRef]
18. Senapati, T.; Yager, R.R. Fermatean fuzzy sets. *J. Ambient Intell. Humaniz. Comput.* **2020**, *11*, 663–674. [CrossRef]
19. Gholamizadeh, K.; Zarei, E.; Omidvar, M.; Yazdi, M. Fuzzy Sets Theory and Human Reliability: Review, Applications, and Contributions. In *Linguistic Methods under Fuzzy Information in System Safety and Reliability Analysis*; Springer: Berlin/Heidelberg, Germany, 2022; pp. 91–137.
20. Pouyakian, M.; Khatabakhsh, A.; Yazdi, M.; Zarei, E. Optimizing the Allocation of Risk Control Measures Using Fuzzy MCDM Approach: Review and Application. In *Linguistic Methods under Fuzzy Information in System Safety and Reliability Analysis*; Yazdi, M., Ed.; Studies in Fuzziness and Soft Computing; Springer: Cham, Switzerland, 2022; Volume 414.
21. Turkey Directorate General of Civil Aviation (DGCA). 2020 Annual Bulletin of Safety İncidents. 2020. Available online: https://web.shgm.gov.tr/documents/sivilhavacilik/files/mevzuat/sektorel/bulten/2020/EOYB-2020.pdf (accessed on 30 July 2022).

22. Delikhoon, M.; Zarei, E.; Banda, O.V.; Faridan, M.; Habibi, E. Systems Thinking Accident Analysis Models: A Systematic Review for Sustainable Safety Management. *Sustainability* **2022**, *14*, 5869. [CrossRef]
23. Zadeh, L.A. Fuzzy sets. In *Fuzzy Sets, Fuzzy Logic, and Fuzzy Systems*; World Scientific: Singapore, 1996; pp. 394–432.
24. Pedrycz, W. Why triangular membership functions? *Fuzzy Sets Syst.* **1994**, *64*, 21–30. [CrossRef]
25. Atanassov, K.T. Intuitionistic fuzzy sets. In *Intuitionistic Fuzzy Sets*; Physica: Heidelberg, Germany, 1999; pp. 1–137.
26. Karnik, N.N.; Mendel, J.M. Operations on type-2 fuzzy sets. *Fuzzy Sets Syst.* **2001**, *122*, 327–348. [CrossRef]
27. Torra, V. Hesitant fuzzy sets. *Int. J. Intell. Syst.* **2010**, *25*, 529–539. [CrossRef]
28. Cuong, B.C.; Kreinovich, V. Picture fuzzy sets. *J. Comput. Sci. Cybern.* **2014**, *30*, 409–420.
29. Kutlu Gündoğdu, F.; Kahraman, C. Spherical fuzzy sets and spherical fuzzy TOPSIS method. *J. Intell. Fuzzy Syst.* **2019**, *36*, 337–352. [CrossRef]
30. Yager, R.R. Generalized orthopair fuzzy sets. *IEEE Trans. Fuzzy Syst.* **2016**, *25*, 1222–1230. [CrossRef]
31. Marhavilas, P.K.; Filippidis, M.; Koulinas, G.K.; Koulouriotis, D.E. An expanded HAZOP-study with fuzzy-AHP (XPA-HAZOP technique): Application in a sour crude-oil processing plant. *Saf. Sci.* **2020**, *124*, 104590. [CrossRef]
32. Celik, E.; Gul, M. Hazard identification, risk assessment and control for dam construction safety using an integrated BWM and MARCOS approach under interval type-2 fuzzy sets environment. *Autom. Constr.* **2021**, *127*, 103699. [CrossRef]
33. Gul, M.; Mete, S.; Serin, F.; Celik, E. *Fine–Kinney-Based Fuzzy Multi-Criteria Occupational Risk Assessment*; Springer: Cham, Switzerland, 2021.
34. Wang, W.; Ding, L.; Liu, X.; Liu, S. An interval 2-Tuple linguistic Fine-Kinney model for risk analysis based on extended ORESTE method with cumulative prospect theory. *Inf. Fusion* **2022**, *78*, 40–56. [CrossRef]
35. Wang, W.; Jiang, W.; Han, X.; Liu, S. An extended gained and lost dominance score method based risk prioritization for Fine-Kinney model with interval type-2 fuzzy information. *Hum. Ecol. Risk Assess. Int. J.* **2022**, *28*, 154–183. [CrossRef]
36. Çalış Boyacı, A.; Selim, A. Assessment of occupational health and safety risks in a Turkish public hospital using a two-stage hesitant fuzzy linguistic approach. *Environ. Sci. Pollut. Res.* **2022**, *29*, 36313–36325. [CrossRef] [PubMed]
37. Can, G.F.; Toktas, P. An Advanced Stochastic Risk Assessment Approach Proposal Based on KEMIRA-M, QFD and Fine-Kinney Hybridization. *Int. J. Inf. Technol. Decis. Mak.* **2021**, *20*, 431–468. [CrossRef]
38. Dagsuyu, C.; Oturakci, M.; Essiz, E.S. A New Fine-Kinney Method Based on Clustering Approach. *Int. J. Uncertain. Fuzziness Knowl. Based Syst.* **2020**, *28*, 497–512. [CrossRef]
39. Tang, J.; Liu, X.; Wang, W. A hybrid risk prioritization method based on generalized TODIM and BWM for Fine-Kinney under interval type-2 fuzzy environment. *Hum. Ecol. Risk Assess. Int. J.* **2021**, *27*, 954–979. [CrossRef]
40. Khalilzadeh, M.; Ghasemi, P.; Afrasiabi, A.; Shakeri, H. Hybrid fuzzy MCDM and FMEA integrating with linear programming approach for the health and safety executive risks: A case study. *J. Model. Manag.* **2021**, *16*, 1025–1053. [CrossRef]
41. Shahri, M.M.; Jahromi, A.E.; Houshmand, M. Failure Mode and Effect Analysis using an integrated approach of clustering and MCDM under pythagorean fuzzy environment. *J. Loss Prev. Process Ind.* **2021**, *72*, 104591. [CrossRef]
42. Jafarzadeh Ghoushchi, S.; Soleimani Nik, M.; Pourasad, Y. Health Safety and Environment Risk Assessment Using an Extended BWM-COPRAS Approach Based on G-Number Theory. *Int. J. Fuzzy Syst.* **2022**, *24*, 1888–1908. [CrossRef]
43. Valipour, M.; Yousefi, S.; Jahangoshai Rezaee, M.; Saberi, M. A clustering-based approach for prioritizing health, safety and environment risks integrating fuzzy C-means and hybrid decision-making methods. *Stoch. Environ. Res. Risk Assess.* **2021**, *36*, 919–938. [CrossRef]
44. La Fata, C.M.; Giallanza, A.; Micale, R.; La Scalia, G. Improved FMECA for effective risk management decision making by failure modes classification under uncertainty. *Eng. Fail. Anal.* **2022**, *135*, 106163. [CrossRef]
45. Karamustafa, M.; Cebi, S. Extension of safety and critical effect analysis to neutrosophic sets for the evaluation of occupational risks. *Appl. Soft Comput.* **2021**, *110*, 107719. [CrossRef]
46. Liu, R.; Liu, Z.; Liu, H.C.; Shi, H. An improved alternative queuing method for occupational health and safety risk assessment and its application to construction excavation. *Autom. Constr.* **2021**, *126*, 103672. [CrossRef]
47. Liu, R.; Zhu, Y.J.; Chen, Y.; Liu, H.C. Occupational health and safety risk assessment using an integrated TODIM-PROMETHEE model under linguistic spherical fuzzy environment. *Int. J. Intell. Syst.* **2021**, *36*, 6814–6836. [CrossRef]
48. Mohandes, S.R.; Durdyev, S.; Sadeghi, H.; Mahdiyar, A.; Hosseini, M.R.; Banihashemi, S.; Martek, I. Towards enhancement in reliability and safety of construction projects: Developing a hybrid multi-dimensional fuzzy-based approach. *Eng. Constr. Archit. Manag.* **2022**. [CrossRef]
49. Mohandes, S.R.; Sadeghi, H.; Fazeli, A.; Mahdiyar, A.; Hosseini, M.R.; Arashpour, M.; Zayed, T. Causal analysis of accidents on construction sites: A hybrid fuzzy Delphi and DEMATEL approach. *Saf. Sci.* **2022**, *151*, 105730. [CrossRef]
50. La Fata, C.M.; Giallanza, A.; Micale, R.; La Scalia, G. Ranking of occupational health and safety risks by a multi-criteria perspective: Inclusion of human factors and application of VIKOR. *Saf. Sci.* **2021**, *138*, 105234. [CrossRef]
51. Lolli, F.; Coruzzolo, A.M.; Alessandro, G.; Balugani, E.; Butturi, M.A.; Marinello, S.; Marinelli, S. The Dynamic, Individual and Integrated Risk Assessment: A Multi-criteria Approach Using Big Data. In Proceedings of the International Conference on Applied Human Factors and Ergonomics, New York, NY, USA, 25–29 July 2021; Springer: Cham, Switzerland, 2021; pp. 207–215.
52. Zarei, E.; Ramavandi, B.; Darabi, A.H.; Omidvar, M. A framework for resilience assessment in process systems using a fuzzy hybrid MCDM model. *J. Loss Prev. Process Ind.* **2021**, *69*, 104375. [CrossRef]

53. Gul, M.; Lo, H.W.; Yucesan, M. Fermatean fuzzy TOPSIS-based approach for occupational risk assessment in manufacturing. *Complex Intell. Syst.* **2021**, *7*, 2635–2653. [CrossRef]
54. Ak, M.F.; Yucesan, M.; Gul, M. Occupational health, safety and environmental risk assessment in textile production industry through a Bayesian BWM-VIKOR approach. *Stoch. Environ. Res. Risk Assess.* **2022**, *36*, 629–642. [CrossRef]
55. Di Bona, G.; Silvestri, A.; Forcina, A.; Petrillo, A. Total efficient risk priority number (TERPN): A new method for risk assessment. *J. Risk Res.* **2018**, *21*, 1384–1408. [CrossRef]
56. Grassi, A.; Gamberini, R.; Mora, C.; Rimini, B. A fuzzy multi-attribute model for risk evaluation in workplaces. *Saf. Sci.* **2009**, *47*, 707–716. [CrossRef]
57. Chen, J.K. Prioritization of corrective actions from utility viewpoint in FMEA application. *Qual. Reliab. Eng. Int.* **2017**, *33*, 883–894. [CrossRef]
58. Rezaei, J. Best-worst multi-criteria decision-making method. *Omega* **2015**, *53*, 49–57. [CrossRef]
59. European Aviation Safety Agency. *Annual Safety Review 2020*; Publications Office: Luxembourg, 2020.
60. ICAO. International Civil Aviation Organization. *Global Aviation Safety Plan 2020–2022*. Available online: https://www.icao.int/Meetings/anconf13/Documents/Doc_10004_GASP_2020_2022_Edition.pdf (accessed on 13 October 2021).
61. Senapati, T.; Yager, R.R. Fermatean fuzzy weighted averaging/geometric operators and its application in multi-criteria decision-making methods. *Eng. Appl. Artif. Intell.* **2019**, *85*, 112–121. [CrossRef]
62. Zhou, L.P.; Wan, S.P.; Dong, J.Y. A Fermatean Fuzzy ELECTRE Method for Multi-Criteria Group Decision-Making. *Informatica* **2022**, *33*, 181–224. [CrossRef]
63. Yager, R.R. Pythagorean fuzzy subsets. In Proceedings of the 2013 Joint IFSA World Congress and NAFIPS Annual Meeting (IFSA/NAFIPS), Edmonton, AB, Canada, 24–28 June 2013; pp. 57–61.
64. Peng, X.; Selvachandran, G. Pythagorean fuzzy set: State of the art and future directions. *Artif. Intell. Rev.* **2019**, *52*, 1873–1927. [CrossRef]
65. Peng, X.D.; Yang, Y.; Song, J.; Jiang, Y. Pythagorean fuzzy soft set and its application. *Comput. Eng.* **2015**, *41*, 224–229.

sustainability

MDPI

Article

Flight Training Risk Identification and Assessment Based on the HHM-RFRM Model

Hong Sun [1], Fangquan Yang [2], Peiwen Zhang [3,*] and Yunxiang Zhao [4]

1 Flight Technology and Flight Safety Research Base, Civil Aviation Flight University of China, Guanghan 618307, China
2 School of Airport, Civil Aviation Flight University of China, Guanghan 618307, China
3 School of Economics and Management, Civil Aviation Flight University of China, Guanghan 618307, China
4 Flight Training Standards Branch, Civil Aviation Flight University of China, Guanghan 618307, China
* Correspondence: zhangpeiwen@cafuc.edu.cn

Abstract: Due to the unavoidable operational risks and insufficient risk management capabilities of beginner pilots in flight training, the challenge of risk control in aviation schools has become increasingly prominent. To ensure the safety of flight training in aviation schools and to reduce costs and increase revenue, the essential prerequisite for improving efficiency is risk management. Therefore, it is necessary to explore risk identification and assessment methods. This paper adopts the holographic modeling (HHM) method and risk filtering, rating and management (RFRM) theory. First, the HHM idea is used to construct a risk identification framework (HHM-PAVE) for flight training. Second, based on the dual criteria, multiple criteria and cloud model (CM) in the RFRM approach, an improved risk assessment matrix-cloud model (IPC-CM) is proposed and combined with the N-K model and Bayes' theorem to propose a coupled risk scenario hazard measurement model (CR-HM) based on the HHM-RFRM approach in risk assessment. In the assessment process, the impact of risk factors on system stability as well as the uncertainty problem and coupling–risk quantification problem in expert assessment are considered to obtain scientific and objective quantitative assessment results. Finally, the risk identification and assessment experiments were conducted using HHM-RFRM on the flight training. The results show that the method can more accurately identify critical risk factors in a flight training system and provide a new perspective for risk prevention and control.

Keywords: safety engineering; flight training; HHM-RFRM; risk identification; risk assessment

Citation: Sun, H.; Yang, F.; Zhang, P.; Zhao, Y. Flight Training Risk Identification and Assessment Based on the HHM-RFRM Model. *Sustainability* **2023**, *15*, 1693. https://doi.org/10.3390/su15021693

Academic Editors: Esmaeil Zarei, Samuel Yousefi and Mohsen Omidvar

Received: 14 December 2022
Revised: 10 January 2023
Accepted: 12 January 2023
Published: 16 January 2023

1. Introduction

Safety is a top priority for the aviation industry. Aviation safety has significantly improved from the development of the global aviation industry in the past seventy years. From 2017 through 2021, the total number of accidents, the real accident rate and the number of fatalities continued to decrease. However, the overall risk of death increased to 0.23 in 2021 due to the rise in fatal accidents in turboprops, and various types of accidents still occur. Aviation Safety Network (ASN) data [1] indicated that 453 accidents have occurred worldwide since 2020, causing widespread public concern as well as loss of life and damage to property. Since 2010, 57% of the total accidents have been caused by pilots. From the early training of pilots and throughout pilots' lifecycles, pilot risk control capability is lacking, and the risk control and management of flight training in flight schools are becoming increasingly prominent. Therefore, it is crucial to perform comprehensive and effective risk identification and assessment of risks in flight training, which is the key to risk management for flight schools and pilots.

Risk management has always been an active area of research. It has penetrated all walks of life. Evaluation methods have been developed and evolved in cross-discipline integration. For example, Wenjun Zhang et al. [2] used the HHM-RFRM model in ship navigation safety to analyze navigation risk management from the perspective of risk coupling. In addition, many studies [3,4] in various industries were conducted on risk occurrence mechanisms, risk probabilities, and baseline risk functions. In civil aviation flight safety, flight risk identification and assessment are critical to aviation risk management, which is a topic with significant theoretical and practical significance. Domestic and foreign scholars have conducted research on the theoretical model of risk management. The current flight safety risk management is mainly based on several existing theoretical models of accident causation [5–10], such as the Software, Hardware, Environment, Liveware model (SHEL), Reason's "Swiss cheese" model, the functional resonance analysis method (FRAM), the holographic modeling method and risk filtering, rating and management theory (HHM-RFRM), and Event Tree Analysis (ETA). Based on those theoretical models, scholars have researched the critical aspects of flight safety risk management. In the risk identification part, Shi et al. [11] used data mining methods to identify and classify risk factors in accident reports in the safety management system, which solved the cumbersome and subjective problems of manual identification. Still, there are limitations in the overall risk factor identification framework. Wu et al. [12] adopted the ETA method to identify single risk factors affecting flight safety and established a risk factor identification system. Paltrinieri et al. [13] proposed an atypical accident identification method, which showed promising results in identifying uncommon and complex coupled risk scenarios. In the risk assessment section, Gray et al. [14] utilized the 1% rule to assess the risk of aircrews with established medical problems, classifying them into risk classes with red/amber/green (RAG) colors. Tamasi et al. [15] proposed a methodology to determine risk qualitatively and quantitatively, using a risk assessment matrix combined with the ETA model. However, it still suffers from high uncertainty and lack of objectivity. Yong Gang et al. [16] used the N-K model to analyze the coupling effect of flight operation risk factors and systematically analyze the flight operation coupling while on the ground and in the air based on the coupled risk values.

The above research indicates the presence of two challenges in current flight training risk management. On the one hand, in the area of risk source identification, from the perspective of risk identification objects, some studies [17,18] have focused on the impact of single risk factors on the overall system risk, which is helpful for general system risk assessment. Still, for complex system risks [19,20], it is easy to ignore the impact of multi-factor coupling on flight training safety. For example, when the environment is poor and there is a human factor of pilot error, coupling these two risk factors increases the likelihood of an accident. Still, the risk of this multi-factor coupling has not been studied heavily. Relevant researchers have proposed a scenario-based risk response framework [21], but specific methods and measures for risk management are lacking. On the other hand, in the area of risk assessment, from the perspective of qualitative assessment, the risk assessment matrix [22] is an assessment method based on expert experience and cognitive level with natural uncertainties and is greatly influenced by assessors. From the perspective of quantitative evaluation, some studies [23–25] have only focused on the impact of coupling risk. Still, few have analyzed the specific coupling risk sub-scenarios under the coupling risk scenario.

When the flight instructor does not interfere as much as possible, and the flight student has a certain knowledge of risk management theory, this paper proposes a coupled risk scenario identification and assessment model based on HHM-RFRM theory. This model utilizes the advantages of the CM and N-K models to solve the above-mentioned issues in risk identification and assessment. First, in the risk identification section, the HHM method is used to find risk factors hierarchically and systematically, emphasizing the concept of coupled risk scenario and outputting flight training-related risk factors. Second, the risk assessment proposes the coupled risk scenario–hazard measurement model (CR-HM). The risk correction coefficient combines the multiple judgment criteria in RFRM with the risk assessment matrix–cloud model (PC-CM), which considers system resistance problems and human cognition's ambiguity and randomness to screen out the critical risk factors. With the IPC-CM model, the numerical characteristics of the risk factor cloud model (Ex, En,

He) are output. Then, a new set of evaluation ideas is formed using the N-K model and Bayesian theory to evaluate the coupled risk scenario quantitatively and output the final risk values. The flow of the research method is shown in Figure 1. Finally, taking the flight training of a domestic aviation school as an example, high-risk factors and key coupled risk scenarios are identified and evaluated.

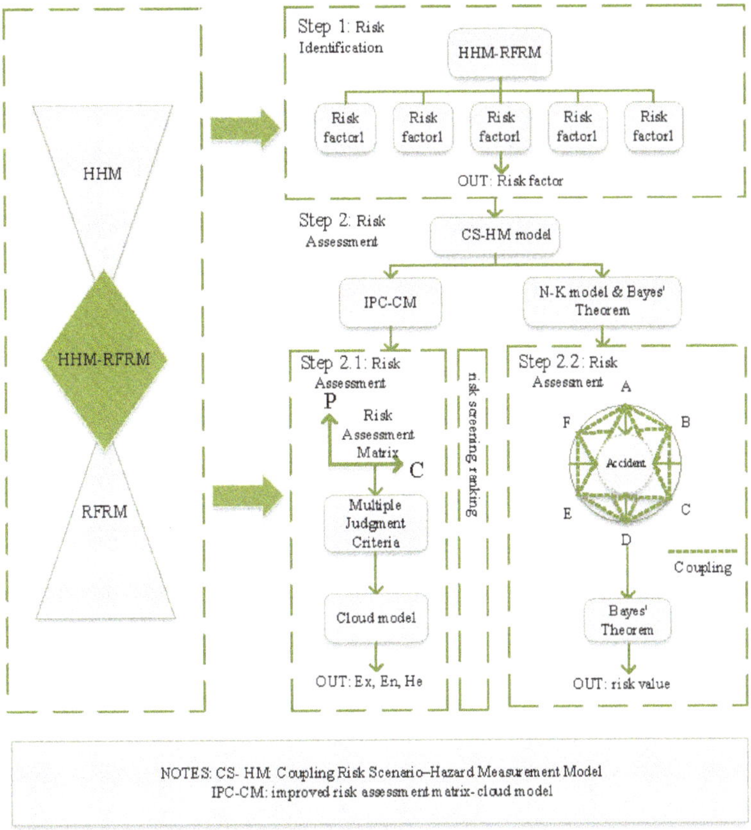

Figure 1. Schematic diagram of the proposed method.

2. Research Method

2.1. HHM-RFRM Method

The HHM-RFRM methodology [9] is a combination of hierarchical holographic modeling (HHM) [26] and risk filtering, rating and management (RFRM) [27,28] and embodies a philosophy of distinguishing "primary and secondary conflicts", filtering secondary risks through qualitative and quantitative assessment analysis and identifying primary risks. This paper focuses on the HHM approach and the five main stages of the RFRM approach, namely (1) scenario identification, (2) dual criteria filtering and rating, (3) multi-criteria assessment, (4) quantitative assessment, and (5) risk management. Although the classical HHM-RFRM method can help pilots better understand the possible risks in flight, it is difficult to achieve a scientific qualitative and quantitative risk assessment. Therefore, it is necessary to use the N-K model to filter out the key coupled risks by the probability of risk factors. Using the cloud model, a more accurate quantitative assessment is achieved by the numerical characteristics of the cloud model. In conclusion, the advantages of each model are utilized to improve the traditional HHM-RFRM to obtain better risk assessment results.

2.1.1. Risk Scenario Identification

Initially proposed by Kaplan and Garrick et al. [29], risk scenario identification is a critical step in HHM-RFRM and consists of three components: risk scenario, probability of occurrence and damage level. A comprehensive risk factor analysis is the starting point for risk identification.

The analysis of flight risk factors is the basis of risk identification. This paper uses the HHM model and the risk identification framework [30] (Pilot-in-Command, Aircraft, Environment, External Pressures—PAVE) to identify risk sources, which requires constructing a risk scenario framework for the risks encountered in flight. Based on the iterative idea of the hierarchical holographic modeling process and the Delphi method, this paper constructs the HHM-PAVE framework to identify the risk factors in flight training. The specific process is shown in Figure 2:

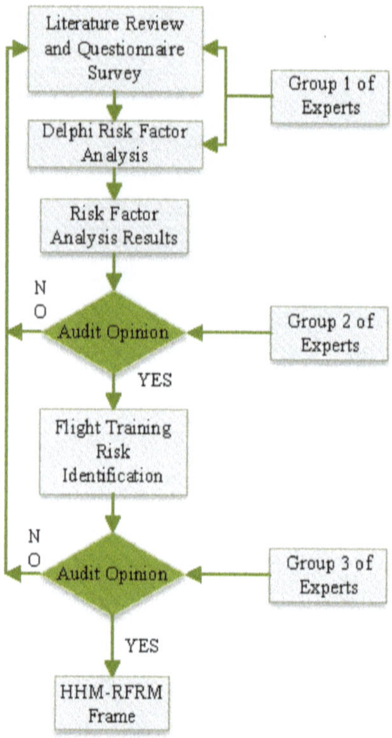

Figure 2. HHM-PAVE framework flow chart.

Based on the construction of the HHM-PAVE framework in Figure 2, individual risk factors are identified. However, in in-flight safety system risk, there is not only single-factor risk but also multi-factor coupled risk. This paper emphasizes the multi-dimensional risk factor coupling in flight training safety, as detailed in Section 2.3, the N-K model. Assume that $T_n(X_1, X_2 \ldots X_m)$ denotes an N-dimensional risk scenario consisting of M risk elements, which are defined as follows:

$$T_n(X_1, X_2 \ldots X_m) = X_1 \odot X_2 \odot \ldots \odot X_m \tag{1}$$

where $\odot$ represents the coupling effect, and the algorithm satisfies the commutative law $X_1 \odot X_2 = X_2 \odot X_1$, $T_2(a, b)$ represents a risk scenario where the risk factors within the two-dimensional risk subsystem a and b are coupled.

(1) Single-factor coupling risk

Single-factor coupling risk refers to the risk caused by the coupling effect and influence between the risk factors belonging to a single subsystem that affects flight training safety. For example: $T_1(a); T_1(b); T_1(c); T_1(d) \ldots$

(2) Two-factor coupling risk

Two-factor coupling risk refers to the risk caused by the coupling effect and influence between two subsystems that affect flight training safety. For example: $T_2(a,b); T_2(a,c); T_2(a,d); T_2(b,c); T_2(b,d) \ldots$

(3) Multi-factor coupling risk

Multi-factor coupling risk refers to the risk caused by the coupling effect and influence of three or more risk factors that affect flight training safety. For example: $T_3(a,b,c); T_3(a,b,d); T_3(a,c,d), T_3(b,c,d) \ldots$

2.1.2. Risk Scenario Assessment

Risk scenario assessment is the core part of the RFRM method. It systematically evaluates and screens risk scenarios to screen out high-risk factors and their coupled risk scenarios continuously. It mainly includes two assessment methods: double filtering criteria and multiple judgment criteria, and the assessment steps are as follows:

Step 1: Double Filtering Criteria—Risk Assessment Matrix (PC)

The dual filtering criteria make up the first filtering step in the RFRM method, which aims to initially screen and rank the risk factors according to the dual criteria. The double filtering criteria and the civil aviation risk assessment matrix assess the probability and severity of an accident. In this regard, this paper adopts a risk assessment matrix that is more applicable to civil aviation [15] to obtain the distribution of likelihood (P), consequence (C) and the corresponding five risk levels (R), as shown in Tables 1–3 below.

Table 1. Risk probability class distribution (P).

Possibility Description	**Almost Impossible**	**Rare**	**Occasional**	**Possible**	**Frequent**
Probability level	*A*	*B*	*C*	*D*	*E*
Qualitative description	Almost never happen	Rarely happen	Occurs by chance, infrequently	Very likely to happen	Occurs frequently

Table 2. Risk consequence degree distribution (C).

Consequence Description	**Ignorable**	**Slight**	**General**	**Serious**	**Catastrophic**
Consequence level	1	2	3	4	5
Consequence score	0–0.3	0.3–0.5	0.5–0.7	0.7–0.9	0.9–1.0

Table 3. Risk rating (R).

Level Description	**Ignorable**	**Slight**	**General**	**Serious**	**Catastrophic**
Rank	I	II	III	IV	V
Risk value	1	3	5	7	9
Risk level matrix					
$\begin{bmatrix} I & \text{II} & \text{II} & \text{II} & \text{II} \\ I & \text{II} & \text{III} & \text{III} & \text{IV} \\ I & \text{III} & \text{III} & \text{IV} & V \\ \text{II} & \text{III} & \text{IV} & \text{IV} & V \\ \text{II} & \text{IV} & \text{IV} & \text{IV} & V \end{bmatrix}$					

Step 2: Multiple Judgment Criteria

The above risk assessment matrix only assesses the possibility and severity of the consequences of risk factors from the perspective of the assessment object. However, it puts specific restrictions on the overall assessment. In this paper, the screened risk factors are further analyzed from the perspective of global systems thinking. From a systems theory perspective, the analysis focuses on the system's resistance and resilience to risk characteristics: stability, robustness and redundancy. Risks are further avoided by comparing the risk resistance nature of the system. This paper introduces the 11 criteria proposed by Matalas and Fiering et al. [31] revised on the defensive capability of risky scenario knockdown systems. Based on the content of the criteria, the judging rules [26], and the expert empirical determination, a multiple judgment matrix was obtained as shown in Table 4, where X_i is the risk factor (Rf); I, II, III..., and XI is the standard serial numbers (St) and A_a^x is the score of the risk factor x under the criteria.

Table 4. Multiple judgment matrix.

St\Rf	X_1	X_2	...	X_m	X_{m-1}
I	$A_1^{X_1}$	$A_1^{X_2}$	...	$A_1^{X_{m-1}}$	$A_1^{X_m}$
II	$A_2^{X_1}$	$A_2^{X_2}$	...	$A_2^{X_m-1}$	$A_2^{X_m}$
III	$A_3^{X_1}$	...	...	...	$A_3^{X_m}$
IV	$A_4^{X_1}$	...	...	...	$A_4^{X_1}$
...	...	...	...	...	...
X	$A_{10}^{X_1}$	...	...	...	$A_{10}^{X_m}$
XI	$A_{11}^{X_1}$	...	...	...	$A_{11}^{X_m}$

2.2. *Cloud Model*

In classical HHM-RFRM methods and risk assessment matrices, which often include qualitative risk assessment processes, there are inevitably two of the most critical uncertainties inherent to human cognition: randomness and ambiguity [32]. This paper applies a new cognitive model-cloud model (CM) proposed by Li et al. [33], which can synthetically describe the randomness and fuzziness of concepts, instantiate the uncertainty transformation between qualitative ideas and their quantitative concepts, and realize the uncertainty transformation between qualitative concepts and their quantitative ones.

Three values represent the overall characteristics of qualitative concepts in the CM: Expectation (E_x), Entropy (E_n), and Hyper Entropy (H_e). E_x represents a measure of the elemental certainty of a qualitative picture, which can best represent the characteristics of a qualitative concept. E_n represents a measure of the uncertainty range of the qualitative concept, determined by the vagueness and randomness of the qualitative concept, and reflects the degree of deviation of the actual affiliation E_x.H_e is a measure of E_n uncertainty, reflecting the degree of cohesion of cloud drops of tension in the discourse world, which is determined by the vagueness and randomness of E_n [33].

Improved P-C Cloud Model (IPC-CM)

Based on the above risk assessment matrix, multiple criteria, and the CM method, this paper proposes an improved risk assessment matrix-cloud model (IPC-CM), which aims to provide more accurate assessment results for quantitative risk assessment and obtain the cloud model of each risk factor after screening. The IPC-CM model is the core assessment model in the whole HHM-RFRM model. It mainly includes the above four steps, as shown in Figure 3. Steps 1 and 2, the P-C concept cloud and rule base, are described detailed in the literature [34,35]. This paper focuses on the uncertainty inference of the CM and the optimization of the CM, where the uncertainty inference steps are as follows:

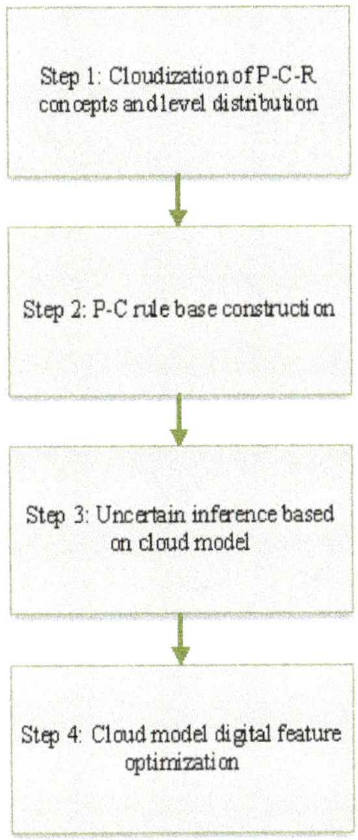

Figure 3. Steps to improve the cloud model.

(1) Generate two-dimensional random numbers

Equation (2) is used to generate a two-dimensional random value (X_p, X_c) with a two-dimensional normal distribution. At the same time, for each rule in the rule base, Equation (3) is used to generate a two-dimensional random value (X_{npi}, X_{nci})

$$(x_p, x_c) = NORMINC(Rand(), (E_{xp}, E_{xc}), (E_{np}, E_{nc})) \tag{2}$$

$$(E_{npi}, E_{nci}) = NORMINC(Rand(), (E_{np}, E_{nc}), (H_{ep}, H_{ec})) \tag{3}$$

(2) Calculate the activation strength μ matrix

Using (X_p, X_c) by Equation (2), the (E_{npi}, E_{nci}) corresponding rules caused by Equation (3) are substituted into Equation (4) to find the activation intensity when the conditional input of each direction in the rule base is (X_p, X_c). A total of 25 rules generated 25 μ values, which constitute the matrix μ.

$$\mu_i = \exp\left[-\frac{(X_p - e_{xpi})^2}{2(E_{npi}')^2} - \frac{(X_c - e_{xci})^2}{2(E_{nci}')^2}\right] \tag{4}$$

(3) Calculate cloud droplets (y, μ)

First, take the largest and second largest μ_i in the matrix. Then, use Equation (5) to generate the hierarchical cloud model's one-dimensional standard random value (E_{nR}'). Use Equation (6) to calculate the four y values for the μ_1 and μ_2 conditions to obtain four groups (y, μ)

$$E_{nR}' = NORMINC(Rand(), E_{nR}, H_{eR}) \tag{5}$$

$$\mu = \exp\left[-\frac{(y - E_x)^2}{2(E_x')^2}\right] \tag{6}$$

(4) Build virtual cloud

First, select the two closest cloud droplets (y_1, μ_1) and (y_2, μ_2) and construct a virtual concept with geometric methods. The three numerical characteristics of the virtual cloud are (E_x, E_n, H_e), where (E_x, E_n) are calculated by geometric forms using Equations (7) and (8). E_x can be designated as a critical parameter reflecting the risk value.

$$E_x = \frac{y_1\sqrt{-2\ln\mu_2} + y_2\sqrt{-2\ln\mu_1}}{\sqrt{-2\ln\mu_2} + \sqrt{-2\ln\mu_1}} \tag{7}$$

$$E_n = \frac{|y_1 - y_2|}{\sqrt{-2\ln\mu_2} + \sqrt{-2\ln\mu_1}} \tag{8}$$

where $x \in U$, x is the expectation of E_x, and $E_x{}'$ is a standard random variance number.

The CM obtained based on the risk assessment matrix is not quantitatively analyzed from the perspective of system stability. In this regard, In this paper, the new optimization method is proposed to use the correction coefficient P_i [36] combined with the multiple judgment matrix to correct the numerical characteristics of the cloud model under the risk assessment matrix to form the final IPC-CM, which can achieve the different scientific ranking of risk scenarios under the same risk level. The correction factor in Equations (9)–(12) is as follows:

$$P_i = \left(\frac{\alpha_i}{\beta_i}\right)^{\varepsilon_{ij}} \tag{9}$$

$$E_{xi}{}' = E_{xi} \times P_i \tag{10}$$

$$E_{ni}{}' = E_{ni} \times P_i{}^2 \tag{11}$$

$$H_{ni}{}' = H_{ni} \times P_i{}^2 \tag{12}$$

where P_i represents the correction coefficient under scenario i; a_i represents the safety and reliability of scenario i in the past period; β_i represents the safety and reliability of scenario i in the current period; ε_{ij} represents the risk coefficient ratio between factors i and j; E_{xi}, E_{ni}, and H_{ni} represent the original parameter values under scenarios i; $E_{xi}{}'$, and $E_{ni}{}'$, and $H_{ni}{}'$ represents the corrected value of the parameter.

2.3. N-K Model

Flight training is a complex system risk often involving multiple risk factors. Therefore, this paper introduces the concept of coupling. In physics, the phenomenon of two or more systems or two forms of motion interacting through various interactions to unite is called "coupling" [37]. Flight training risk coupling refers to the degree of mutual influence and dependence between or among various risk factors affecting aircraft flight during flight training. The coupling between or among risk factors changes the local or overall state of aircraft operation safety, resulting in flight accidents.

The N-K model consists of two parameters. N is the number of constituent factors in the system; and K is the number of inter-factor dependencies, reflecting the system's adaptability. If the system consists of N factors, and there are n states of factors, then there are n^N possible combinations of all the elements, the factors are combined in a certain way to form a network, and the range of K is [0, N−1]. Based on the evolutionary theory of biology, the interaction information between factors is calculated based on the N-K model to measure the coupling risk, and the coupling risk hazard is calculated according to Equations (13)–(15).

$$T_4 = T(A,B,C,D) = \sum_i\sum_j\sum_k\sum_m P_{ijkm}\log_2\left(P_{ijkm}/\left(P_{i\ldots} \times P_{.j..} \times P_{..k.} \times P_{\ldots m}\right)\right) \tag{13}$$

$$T_3 = T(A,B,C) = \sum_i\sum_j\sum_k P_{ijk}\log_2\left(P_{ijk}/\left(P_{i\ldots} \times P_{.j..} \times P_{..k.}\right)\right) \tag{14}$$

$$T_2 = T(A,B) = \sum_i \sum_j P_{ij} \log_2(P_{ij}/(P_{i...} \times P_{.j..})) \tag{15}$$

$$i \in \{0,1\}, j \in \{0,1\}, k \in \{0,1\}, m \in \{0,1\}$$

where i, j, k, m represent the status values of A, B, C, and D risk factors, respectively; status value 0 means that the risk factor has not broken through the defense system, and status value 1 means that the risk factor has broken through the defense system; P_{ijkm} represents the changing risk of the mutual coupling of $ABCD$ risk factors probability; T_x represents the coupling of X risk factors; $T(A,B,C)$ represents the risk of the mutual coupling of risk factors A, B, and C. A defense system is a complex system consisting of "human–machine–environment–management" subsystems that prevent unsafe events or accidents from occurring.

2.4. *Quantitative Model Based on Bayes' Theorem*

Bayes' theorem is a general form of the product rule for calculating the probability of two (or more) independent events [38].

Assuming that there is a risk coupling between the two risk factors A B, without considering the risk of B, the probability that risk factor A causes an accident is prior probability $P(A)$. The likelihood of occurrence of risk B with a known intelligence A risk factor is conditional probability $P(B|A)$. At the same time, considering the risk factor B, $P(A|B)$ is the posterior probability. The Bayesian Equations (16)–(18) are as follows:

$$P(AB) = P(A)P(B|A) = P(B)P(A|B) \tag{16}$$

$$P(A|B) = \frac{P(A)P(B|A)}{P(B)} \tag{17}$$

$$P(A) = P(A|B)P(B) + P(A|\overline{B})P(\overline{B}) \tag{18}$$

From the perspective of quantitative risk assessment, this paper introduces Bayes' theorem for quantitative calculation from the two dimensions of consequence and possibility, which are defined as follows:

$$D_{\text{risk}} = C_{\text{risk}} \times P_{\text{risk}} \tag{19}$$

Based on the modified consequence level C_{risk} of the IPC-CM, the posterior probability P_{risk} of each risk scenario is calculated by combining the coupling relationship between the risk factors in flight training. Associating Equation (19), the final coupled risk scenario's hazard values are calculated.

3. Case Research

In the next section, this paper analyzes the accident investigation report of China's civil aviation safety management system from 2018 through 2021 and the aviation safety briefing of an aviation school. Based on real data from actual scenarios, the flight school's risk focus is continuously adjusted in the event of unsafe events and accident experiences. We take a flight school as an example and start from risk identification and assessment to verify the risks in flight training.

3.1. *Risk Identification*

3.1.1. HHM Frame

Based on the accident report of China's civil aviation safety management system and the aviation safety briefing data, this paper completed the risk factor analysis through Figure 2. From the pilot's perspective, the PAVE framework [30] is adopted to cultivate the critical thinking of pilot trainees. All risk factors are divided into four subsystems of P, A, V, and E.

PAVE consists of four parts: P = Pilot-in-command (PIC); A = Aircraft; V = Environment; and E = External pressures.

(1) P = Pilot-in-Command (PIC)

The pilot in command is one of the risk factors in flight. A pilot must conduct a multi-faceted assessment of their risk profile as the controller of the aircraft. It mainly includes the pilot's physiological and psychological condition and provides comprehensive quality.

(2) A = Aircraft

As the carrier of the flight, the aircraft is also one of the risk factors in the flight. The pilot must fully understand the aircraft's performance, historical failures, and whether the corresponding airworthiness instructions have been completed, and it must check the maintenance of the aircraft.

(3) V = Environment

The flight environment is one of the flight risk factors, and the weather is a major environmental factor. Terrain assessment is another essential component in analyzing the flight environment, which is followed by airports, airspace, nighttime, and visual errors.

(4) E = External Pressures

External pressures are an effect outside of the flight, usually at the expense of safety, that creates a feeling of pressure to complete the flight.

Based on the analysis of the above risk factors, this paper establishes the flight training risk HHM-PAVE framework, as shown in Figure 4.

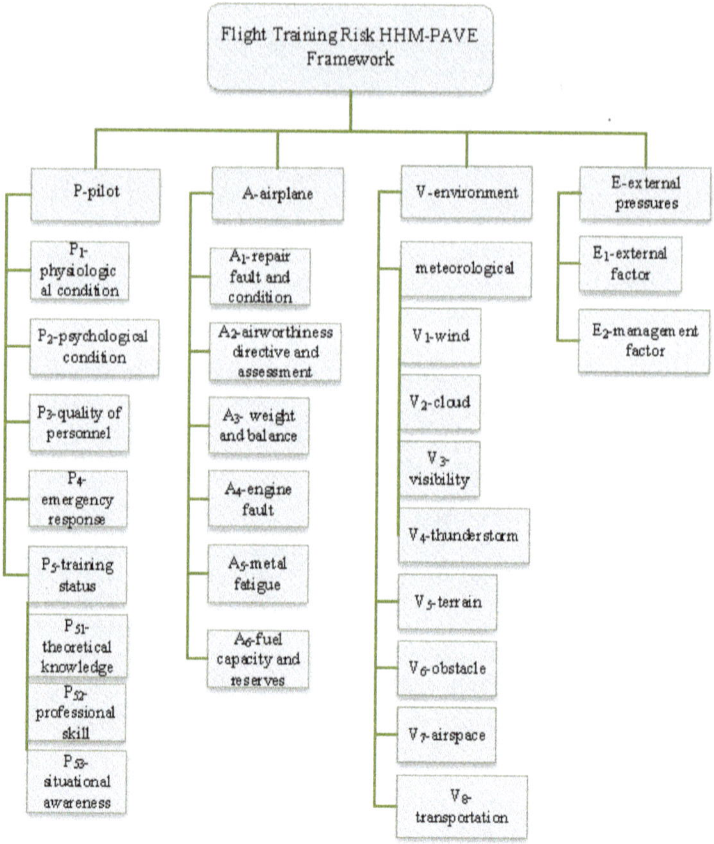

Figure 4. HHM-PAVE model block diagram.

3.1.2. Coupling Risk Scenario

According to the accident data of China's civil aviation safety management system from 2018 through 2021, the coupling theory is used to obtain the count and frequency of risk coupling in recent years, as shown in Table 5 below, where single-factor coupling risk means only one risk factor is involved, two-factor coupling risk means two risk factors are involved in risk coupling, and multi-factor coupling risk means three or more risks are involved in risk coupling; 1000 means P risk coupling effect; 0100 means A risk coupling effect; and 1110 represents PAV three-factor coupling effect.

Table 5. Number and frequency of risk coupling.

Risk Factor	Count and Frequency					
Single-factor coupling risk	0000	1000	0100	0010	0001	
Count	0	22	5	8	1	
Frequency	0.0000	0.4313	0.0980	0.1568	0.0196	
Two-factor coupling risk	1100	1010	1001	0110	0101	0011
Count	0	8	4	0	0	1
Frequency	0.0000	0.1568	0.0784	0.0000	0.0000	0.0196
Multi-factor coupling risk	1110	1101	1011	0111	1111	
Count	0	1	1	0	0	
Frequency	0.0000	0.0196	0.0196	0.0000	0.0000	

3.2. Risk Assessment–Coupling Risk Scenario–Hazard Measurement Model (CR-HM)

3.2.1. Risk Assessment Matrix Filtering

Through the identification of risk scenarios mentioned above, this paper identifies 23 risk factors and 16 main risk coupling scenarios, theoretically including 1630 risk coupling scenarios, from which key risk factors are identified, and the priority analysis of key risks is performed. First, the 23 risk factors are analyzed qualitatively, and the two criteria of likelihood and severity of consequences are filtered using a risk assessment matrix. This filtering is accomplished by interviewing experts and administering questionnaires to relevant people. Senior flight instructors made subjective judgments about the likelihood and consequences of each factor based on their own flight experience and then asked the opinions of 20 flight instructors based on a questionnaire asking for their judgments. The results are shown in Table 6 below.

Table 6. Risk assessment matrix.

Risk Assessment Matrix					
Possibility	Seriousness				
	Negligible 1	Slight 2	Normal 3	Serious 4	Catastrophic 5
Almost impossible A					A_2
Rare B			A_6	P_4, A_3, A_4, A_5	V_3, V_4
Occasional C			P_2	A_1, P_{53}	V_5, V_6, V_7, V_8
Possible D			E_2	$P_3, P_1, P_{51}, P_{52}, E_1$	V_1, V_2
Frequent E					

Here, green represents risk level I, blue represents level II, yellow represents level III, orange represents level IV, and red represents level V.

The risk assessment matrix gives an initial rating and filtering of each risk factor. The risk factors for grades I, II, and III were filtered out. There are 16 risk factors, P_1, P_3, P_{51}, P_{52}, P_{53}, A_1, V_1, V_2, V_3, V_4, V_5, V_6, V_7, V_8, E_1, and E_2, which were retained for further analysis.

3.2.2. Multi-Criteria Assessment of Flight Risk

According to the detailed scoring criteria and scoring rules of multiple criteria, the 16 risk factors mentioned above are further evaluated, and the evaluation criteria are divided into three levels: high (H), medium (M), and, low (L), which were expressed by the values of 1, 0.5, and 0.2, respectively. The final multiple judgment matrix was obtained as shown in Table 7 below, where St represents standard, Rf represents risk factor; and H, M and L represent the evaluation level respectively.

Table 7. Risk factor multiple judgment matrix.

St/Rf	V_3	V_4	A_1	P_{53}	V_5	V_6	V_7	V_8	E_1	P_3	P_1	P_{51}	P_{52}	E_2	V_1	V_2
I	L	L	L	H	M	L	M	M	H	H	L	H	H	H	L	L
II	L	M	L	H	M	M	M	M	H	L	L	M	H	H	M	L
III	L	L	M	M	M	M	M	M	M	M	M	H	H	H	M	H
IV	M	H	L	H	M	M	M	L	M	M	M	M	M	M	H	H
V	M	M	H	M	M	M	H	M	M	M	M	M	H	H	M	M
VI	M	M	M	M	M	M	M	M	H	H	H	H	H	H	H	H
VII	M	M	M	M	M	H	M	L	L	M	H	H	H	L	H	H
VIII	M	M	H	H	M	M	M	L	M	M	M	H	H	H	H	M
IX	M	M	M	L	M	M	M	L	L	L	L	H	H	H	H	M
X	M	H	M	H	L	M	H	M	M	M	M	M	H	M	H	L
XI	L	L	L	L	L	L	L	L	L	L	L	L	L	L	L	L

3.2.3. IPC-CM Assessment

The conventional risk assessment matrix, which assesses risk only qualitatively, has the problem of boundary uncertainty, and the rating process has no scientifically sound uncertainty reasoning mechanism. This paper adopts the IPC-CM model for risk grading. The method further evaluates and sorts the screened risk factors.

This paper uses the IPC-CM model to cloud R, P, and C to generate the expectation (E_x), entropy (E_n), and super-entropy (H_e) numerical features corresponding to each rank. The softened scores of the index levels were achieved. The clouding results are shown in Table 8, and the corresponding cloud model is shown in Figure 5.

Table 8. P, C, and R grade cloud model.

	Rank	A	B	C	D	E
P	E_x	1	3	5	7	9
	E_n	1/3	1/3	1/3	1/3	1/3
	H_e	0.02	0.05	0.05	0.05	0.02
	Rank	**1**	**2**	**3**	**4**	**5**
C	E_x	0.15	0.35	0.55	0.75	0.95
	E_n	0.1/3	0.1/3	0.1/3	0.1/3	0.1/3
	H_e	0.02	0.02	0.02	0.02	0.02
	Rank	**I**	**II**	**III**	**IV**	**V**
R	E_x	1	3	5	7	9
	E_n	1/3	1/3	1/3	1/3	1/3
	H_e	0.02	0.05	0.05	0.05	0.02

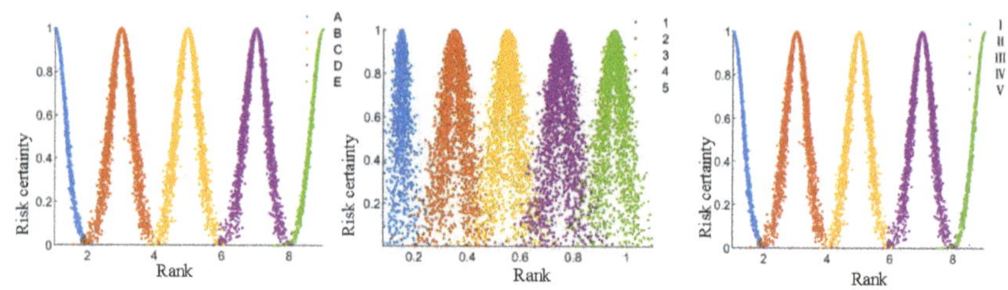

(a) P (b) C (c) R

Figure 5. P, C, and R cloud model diagrams.

Based on the multiple judgment data in Table 4 and the IPC-CM model, the new numerical features and rankings were obtained using Equations (2)–(13). The results are shown in Table 9:

Table 9. Numerical characteristics of risk factors.

Rf	E_x	E_n	H_e	Rank
V_7	6.9871	0.0335	0.015	9
V_1	8.9926	0.0410	0.014	1
V_5	5.9324	0.0563	0.015	12
P_3	7.0000	0.0104	0.014	8
A_1	7.7486	0.0574	0.014	6
V_8	2.890	0.0727	0.015	16
V_3	4.9558	0.0402	0.015	15
V_4	6.1573	0.0137	0.014	11
V_2	8.3141	0.0375	0.013	2
P_{53}	8.0109	0.0251	0.013	5
E_1	5.5060	0.0811	0.014	13
V_6	5.0004	0.0377	0.014	14
P_1	6.6985	0.0133	0.013	10
P_{51}	7.4925	0.0156	0.013	7
P_{52}	8.2033	0.0697	0.013	3
E_2	8.0865	0.0292	0.012	4

The cloud model of the above 16 risk factors is sorted and screened, and the standard cloud plots before and after filtering are shown in Figure 6.

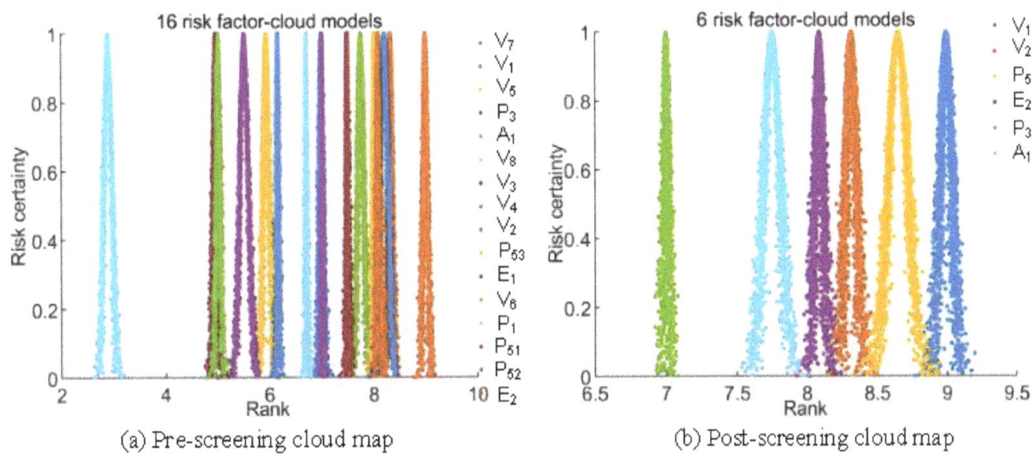

Figure 6. Before and after screening cloud map.

According to the sorting provided in Table 9 and the filtered data provided in Figure 6b, the six most critical risk factors in flight training are selected, namely P_3, P_5, A_1, V_1, V_2, and E_2. The other risk factors with low-risk values are screened out, which does not mean that pilots are not concerned about them, but compared with risk factors with high-risk values, pilots should follow the principle of attention distribution.

Based on this filtering, the coupled scenarios of key risk factors are further analyzed and evaluated based on the HHM framework and holographic theory. The critical flight risk HHM-PAVE sub-framework is shown in Figure 7.

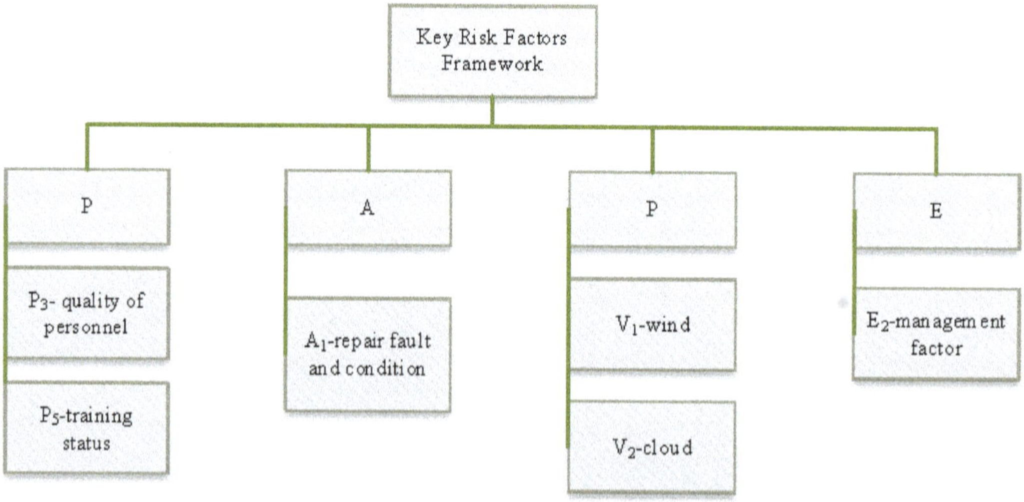

Figure 7. HHM-PAVE sub-framework.

3.2.4. N-K Coupling Risk Scenario Assessment

Risk coupling is performed according to the six key risk factors under the above HHM sub-framework. This paper considers the coupling of four subsystems and obtains 16 coupling scenarios. The coupling probabilities of risk factors are calculated by Table 5. The results are shown in Table 10.

Table 10. One-factor, two-factor, multi-factor coupling probability.

O-C	P_r	O-C	P_r	O-C	P_r	O-C	P_r
$P_{0\cdots}$	0.2941	$P_{1\cdots}$	0.7058	$P_{\cdot\cdot0\cdot}$	0.6471	$P_{\cdot\cdot1\cdot}$	0.3529
$P_{\cdot0\cdot\cdot}$	0.8824	$P_{\cdot1\cdot\cdot}$	0.1176	$P_{\cdots0}$	0.8431	$P_{\cdots1}$	0.1569
T-C	P_r	**T-C**	P_r	**T-C**	P_r	**T-C**	P_r
$P_{00\cdot\cdot}$	0.1961	$P_{01\cdot\cdot}$	0.0980	$P_{10\cdot\cdot}$	0.6862	$P_{11\cdot\cdot}$	0.0196
$P_{0\cdot0\cdot}$	0.1176	$P_{0\cdot1\cdot}$	0.1765	$P_{1\cdot0\cdot}$	0.5294	$P_{1\cdot1\cdot}$	0.1764
$P_{0\cdot\cdot0}$	0.2549	$P_{0\cdot\cdot1}$	0.0392	$P_{1\cdot\cdot0}$	0.5882	$P_{1\cdot\cdot1}$	0.1176
$P_{\cdot00\cdot}$	0.5294	$P_{\cdot01\cdot}$	0.3529	$P_{\cdot10\cdot}$	0.1176	$P_{\cdot11\cdot}$	0.0000
$P_{\cdot0\cdot0}$	0.7451	$P_{\cdot0\cdot1}$	0.1373	$P_{\cdot1\cdot0}$	0.0980	$P_{\cdot1\cdot1}$	0.0196
$P_{\cdot\cdot00}$	0.5294	$P_{\cdot\cdot01}$	0.1176	$P_{\cdot\cdot10}$	0.3137	$P_{\cdot\cdot11}$	0.0392
M-C	P_r	**M-C**	P_r	**M-C**	P_r	**M-C**	P_r
$P_{000\cdot}$	0.0196	$P_{001\cdot}$	0.1765	$P_{010\cdot}$	0.0980	$P_{011\cdot}$	0.0000
$P_{100\cdot}$	0.5098	$P_{101\cdot}$	0.1765	$P_{110\cdot}$	0.0196	$P_{111\cdot}$	0.0000
$P_{\cdot000}$	0.4314	$P_{\cdot001}$	0.0980	$P_{\cdot010}$	0.3137	$P_{\cdot011}$	0.0392
$P_{\cdot100}$	0.0980	$P_{\cdot101}$	0.0196	$P_{\cdot110}$	0.0000	$P_{\cdot111}$	0.0000
$P_{0\cdot00}$	0.0980	$P_{0\cdot01}$	0.0196	$P_{0\cdot10}$	0.1569	$P_{0\cdot11}$	0.0196
$P_{1\cdot00}$	0.4314	$P_{1\cdot01}$	0.0980	$P_{1\cdot10}$	0.1569	$P_{1\cdot11}$	0.0196
$P_{00\cdot0}$	0.1569	$P_{00\cdot1}$	0.0392	$P_{01\cdot0}$	0.0980	$P_{01\cdot1}$	0.0000
$P_{10\cdot0}$	0.5882	$P_{10\cdot1}$	0.0980	$P_{11\cdot0}$	0.0000	$P_{11\cdot1}$	0.0196

Here, O-C represents one-factor coupling, T-C represents two-factor coupling, M-C represents multi-factor coupling, Pr represents probability, and $P_{00..}$ represents the probability of occurrence when the pilot and aircraft are not involved in the coupling.

According to the risk coupling probability data in Table 10 and Equations (14)–(16), the risk values of each coupling scenario are calculated, respectively, as follows: T(PA) = 0.3635; T(PV) = 0.4173; T(PE) = 0.0395; T(AV) = 0.1953; T(AE) = 0.0067; T(VE) = 0.0914; T(PAV) = 0.6939;

T(PVE) = 0.4194; T(PAE) = 0.3572; T(AVE) = 0.2480. From the ranking result of risk coupling, T(PAVE) > T(PAV) > T(PVE) > T(PAE) > T(AVE) > T(PV) > T(PA) > T(AV) > T(VE) > T(AE) > T(VE), where the coupling risk value is the largest T_4, followed by T_3 and finally T_2.

3.2.5. Quantitative Evaluation of Bayesian Probabilities

Based on the above-identified risk coupling situation, an example analysis is carried out for a pilot of an aviation school to perform a specific flight mission. First, by collecting relevant historical data and consulting the flight safety accident statistical database, the frequency of various accidents and the influencing factors leading to them are analyzed to determine the prior probability of risk factors. For example, the priori probability of a flight accident occurring when a pilot is poorly trained is 0.80. Second, from the system theory perspective, combined with the PAVE hazard identification framework and decision makers, expert experience strengthens comprehensive judgment. When a pilot is well trained, the likelihood of a flight accident due to operational error or lack of knowledge is still higher, with a conditional probability of 0.25. According to Equations (17)–(19), the posterior probability is calculated as 0.5714, and the posterior probabilities of the other risk factors are obtained similarly, as shown in Table 11:

Table 11. Flight training risk probability.

Risk Factor	Priori Probability	Conditional Probability	Posterior Probability
P_5	0.80	0.25	0.5714
P_3	0.35	0.04	0.0219
A_1	0.80	0.15	0.4138
V_1	0.65	0.30	0.4432
V_2	0.55	0.25	0.2895
E_1	0.40	0.04	0.0270

The coupling effect of the six risk factors under the HHH sub-frame is analyzed through Table 11. This paper mainly evaluates the two-dimensional risk coupling scenario. According to Equation (20), the risk degree of the two-dimensional risk scenario is obtained, as shown in Table 12. Generally, a risk degree higher than 0.05 is considered high for two-dimensional risk scenarios.

Table 12. Risk of two-dimensional risk coupling scenarios.

Risk Scenario	Sub-Scene	Dangerous	Risk Scenario	Sub-Scene	Dangerous
$P\odot A$	$P_5\odot A_1$	0.23644	$A\odot V$	$A_1\odot V_1$	0.18339
	$P_3\odot A_1$	0.00906		$A_1\odot V_2$	0.11979
$P\odot V$	$P_5\odot V_1$	0.25324	$A\odot E$	$A_1\odot E_1$	0.01117
	$P_5\odot V_2$	0.16542	$V\odot E$	$V_1\odot E_1$	0.01196
	$P_3\odot V_1$	0.00970		$V_2\odot E_1$	0.00781
	$P_3\odot V_2$	0.00634	$P\odot E$	$P_5\odot E_1$	0.01542
				$P_3\odot E_1$	0.00059

From the above calculation, it can be seen that there are five risk scenarios with a risk degree exceeding 0.05, which, respectively, reflect the three main risk coupling scenarios of risk management in this flight mission, namely pilot human factors–environmental factors, human factors–aircraft factors, and aircraft factors–environmental factors. The main risk scenario includes a total of five risk coupling sub-scenarios, of which the top three key coupling sub-scenarios are $P_5\odot V_1$, $P_5\odot A_1$, and $A_1\odot V_1$, with risk degrees of 0.25324, 0.23644, and 0.18339, respectively. Pilot training, wind, and aircraft conditions are the critical risk factors for coupling, indicating that in flight training, the quality of pilot training will directly affect the risk value. In the case of poor flight training and other risks, the risk value in this scenario is high, and flight accidents are very likely to occur.

4. Results and Discussion

The proposed model firstly obtained all risk factors by risk identification, secondly ranked risk screening by the IPC-CM model, and finally output the final risk values by the N-K model and Bayesian formula. The following results were obtained and discussed.

(1) Regarding the research involving screening filtering and ranking in the RFRM method, the IPC-CM model is proposed, which abandons the traditional purely qualitative way of risk matrix assessment and takes advantage of the cloud model in terms of the uncertainty of subjective perception. Based on cloud theory, cloud vertices, ranges and thicknesses are used to show the risk value of risk factors visually. The cloud model obtained by this method is scientific, intuitive, and easy to understand. Figure 8 shows the results based on the IPC-CM, which achieves a further division of the same level in the traditional risk matrix assessment [15]. Figure 8a–f reflects the risk value of risk factors under different levels. As a result, a preliminary screening assessment algorithm for systemic risk is formed.

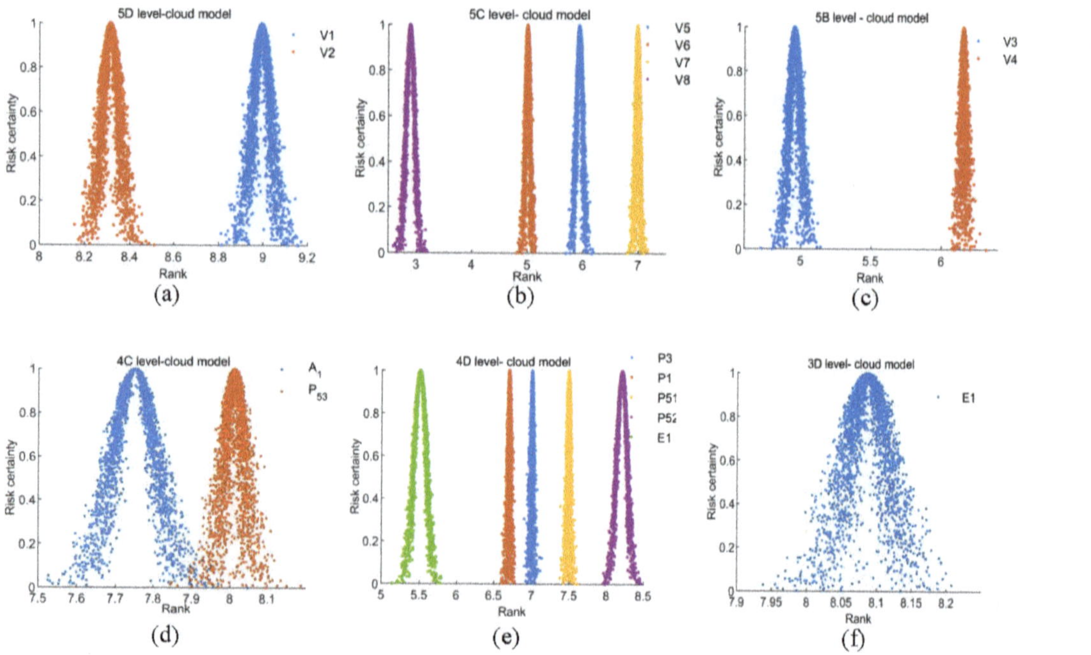

Figure 8. Risk factor cloud model.

(2) Regarding the quantitative assessment of coupling scenarios in RFRM, Table 13 shows that the risk value increases with the increase in coupling factors, T4 > T3 > T2. Table 12 reveals the key coupled risk scenarios and their hazard levels in flight training risk management. The results show that when $P_5 \odot V_1$, $P_5 \odot A_1$, and $A_1 \odot V_1$ factors are coupled, the risk values are large, 0.25324, 0.23644, and 0.18339, respectively—much higher than the high-risk level of 0.05. Among them, the pilot and environment coupling have the highest number and enormous risk value, which fully confirms that the pilot is still the primary cause of current flight training accidents [39]. At the same time, the findings show that the number of aircraft conditions involved is low, but the risk value is also high. Although the leading cause of flight accidents is no longer early mechanical failures, the degree of severe consequences caused by aviation equipment has not decayed in the slightest [40], so the risk value is still high. In addition, when pilot-related risk values are high, the quality of trainee flight training should be subsequently enhanced, and when environmental involvement risk values are high, the meteorological safety of training flights should be

strengthened. In conclusion, the assessment results guide the key direction of flight training risk management.

Table 13. Risk coupling value at risk.

Risk Coupling Scenario	Risk Value	Risk Coupling Scenario	Risk Value
T (a c)	0.4173	T (a b c d)	0.8257
T (a b)	0.3635	T (a b c)	0.6939
T (b c)	0.1953	T (a c d)	0.4194
T (c d)	0.0914	T (a b d)	0.3572
T (a d)	0.0395	T (b c d)	0.2480
T (b d)	0.0067		

In terms of overall flight training risk management, according to the final assessment results in Table 12, it is evident that the pilot training situation participates in a high number of couplings and has an increased risk of the coupled with other threats, implying that the management of the pilot training situation at the flight school is becoming more and more critical. In the study results, wind and aircraft condition factors also have higher risks of coupling with other threats. However, in the actual training process of domestic flight schools, the focus is still only on the operational skills of the aircraft, and most of the risks are often managed by the instructors on behalf of the pilots, although the risk values are significant. The perspective is prone to cause pilot dependency psychology [41] and to cause the Dunning–Kruger effect [42]. As China has entered the stage of high-quality development, reducing costs, increasing revenue, and improving efficiency will inevitably lead to the emergence of the adverse effects of risk overlap. In civil aviation flight safety, without a set of scientific risk identification and assessment methods, it is difficult to truly grasp the policy of moving forward the gate, controlling at source, and implementing prevention-oriented and comprehensive management to conduct scientific risk management. This paper fully demonstrates the existence of such critical risks from risk management identification and assessment. It also reflects the inadequacy of risk management in domestic flight schools. The aim is to systematically learn risk identification and assessment methods from the initial training theory stage, develop pilots' risk management capabilities, and enable them to autonomously identify risks, assess them, and eventually control them. This paper provides a new risk identification and assessment methodology to facilitate pilots' scientific risk management. More importantly, as risk management is one of a pilot's core competencies for flight school, the method can provide a positive reference for the development of risk management core competency of pilots by continuous identification, screening and assessment.

5. Conclusions

In this paper, a new HHM-RFRM risk identification and assessment method has been proposed. Based on the assessment results, the conclusions are as follows:

(1) Research on risk identification in HHM proposes the HHM-PAVE framework construction method. HHM iterative ideas address the holistic, logical aspects of system risk. The Delphi method reduces individual cognitive errors (randomness), while the PAVE framework enables pilots to reduce their workload and identify risk factors more clearly. The HHM-PAVE framework solves the fuzzy logic problem between risk factors in the existing text classification, making the identified risk factors more comprehensive and objective.

(2) Research on risk assessment, based on the uncertainty of qualitative evaluation and system resistance, proposed the CR-HM model, which uses the IPC-CM model to complete a more scientific ranking of risk factors and screening. The method based on risk factors can more objectively integrate system resistance. This method takes into account not only the likelihood of accidents caused by risk factors and the severity of the consequences but also the resistance of the overall system to the risk factors. The CM model obtained by this

method is significantly lower than the traditional CM algorithm in E_n and H_e, solving the uncertainty of human cognition in the qualitative risk assessment matrix and making the assessment results more scientifically segmented and intuitive. The introduction of the N-K model and Bayes' theorem in the coupled risk scenario is utilized to realize the quantitative assessment of the coupled scenario hazard degree.

(3) A new HHM-RFRM methodology is proposed for the overall risk identification and assessment. A case study including a flight training mission is conducted to identify key risk factors and coupled risk scenarios, assess their hazard levels, and identify weaknesses in risk management. The method can help pilots identify key risk factors; evaluate the degree of risk; help pilots establish a scientific approach to risk management; effectively improve the efficiency of risk prevention and control management; improve the development of core competency of pilots; and enhance risk management in domestic flight schools.

Future research will start with the risk identification of specific scenarios and further analyze the intrinsic mechanism of coupled risk scenarios and the impact on critical aspects of pilots. Based on the digital risk management platform, a pilot-oriented risk assessment and decision support model will be constructed to ensure flight training safety further.

Author Contributions: Conceptualization, F.Y. and H.S.; methodology, F.Y. and P.Z.; software, F.Y. and P.Z.; validation, F.Y. and Y.Z.; formal analysis, F.Y. and H.S.; writing—original draft preparation, F.Y.; writing—review and editing, F.Y. and P.Z.; supervision, H.S.; project administration, H.S. All authors have read and agreed to the published version of the manuscript.

Funding: This research was funded by the Key Project of the Civil Aviation Joint Fund of the National Natural Science Foundation of China (U2033213), the Special Project on Flight Technology (FZ2022ZX50) of the Key Laboratory of Civil Aviation Flight Technology and Flight Safety, and the Independent Project of the Research Base of Civil Aviation Flight Technology and Flight Safety (FZ2021ZZ01).

Institutional Review Board Statement: Not applicable.

Informed Consent Statement: Informed consent was obtained from all subjects involved in the study.

Data Availability Statement: All the data contained in this study can be obtained upon request to the corresponding author. Readers can also request part of the original data and the results of data processing in this paper.

Acknowledgments: The authors thank everyone who contributed to the article.

Conflicts of Interest: The authors declare that they have no conflict of interest to report regarding the present study.

References

1. ASN Safety Databases. Available online: https://aviation-safety.net/database/databases.php (accessed on 4 January 2022).
2. Zhang, W.-J.; Zhang, Y.-J.; Zhang, C. Intelligent ship navigation risk identification and screening based on HHM-RFRM method. *J. Saf. Environ.* **2022**, 1–10.
3. Oster, C.V., Jr.; Strong, J.S.; Zorn, C.K. Analyzing aviation safety: Problems, challenges, opportunities. *Res. Transp. Econ.* **2013**, *43*, 148–164. [CrossRef]
4. Shyur, H.-J. A quantitative model for aviation safety risk assessment. *Comput. Ind. Eng.* **2008**, *54*, 34–44. [CrossRef]
5. Sun, R.; Shi, R. Analysis of Civil Aviation Communications Error Events Based on SHEL Model. In Proceedings of the ICTIS 2011: Multimodal Approach to Sustained Transportation System Development: Information, Technology, Implementation, Wuhan, China, 26 April 2011; pp. 2115–2121.
6. Reason, J. Human error: Models and management. *BMJ* **2000**, *320*, 768–770. [CrossRef]
7. Hirose, T.; Sawaragi, T.; Horiguchi, Y.; Nakanishi, H. Safety analysis for resilient complex socio-technical systems with an extended functional resonance analysis method. *Int. J. Astronaut. Aeronaut. Eng.* **2017**, *2*, 1–18.
8. Walker, M.; Papadopoulos, Y. Qualitative temporal analysis: Towards a full implementation of the Fault Tree Handbook. *Control. Eng. Pract.* **2009**, *17*, 1115–1125. [CrossRef]
9. Zhu, J.; Liu, B.; Dong, P.; Zeng, L. Quality-Risk Identification and Screening of Aircraft Major Overhaul Project Based on HHM-RFRM. In Proceedings of the Journal of Physics: Conference Series, Harbin, China, 8–10 January 2021; p. 012194.
10. Liu, H.; Chen, G.; Lin, Z. Risk Assessment and Management of National Defense Engineering Construction Based on HHM-RFRM. *IOP Conf. Ser. Mater. Sci. Eng.* **2020**, *780*, 072026. [CrossRef]

11. Shi, D.; Guan, J.; Zurada, J.; Manikas, A. A data-mining approach to identification of risk factors in safety management systems. *J. Manag. Inform. Syst.* **2017**, *34*, 1054–1081. [CrossRef]
12. Wu, P.F.; Duanmu, J.S.; Du, J.Y.; Yang, J.; Zhang, C. Risk factor identification and classification of flight safety based on fault tree approach. In Proceedings of the International Conference on Mechatronics and Control Engineering (ICMCE 2012), Guangzhou, China, 29–30 November 2012; pp. 2232–2237.
13. Paltrinieri, N.; Dechy, N.; Salzano, E.; Wardman, M.; Cozzani, V. Towards a new approach for the identification of atypical accident scenarios. *J. Risk Res.* **2013**, *16*, 337–354. [CrossRef]
14. Gray, G.; Bron, D.; Davenport, E.D.; d'Arcy, J.; Guettler, N.; Manen, O.; Syburra, T.; Rienks, R.; Nicol, E.D. Assessing aeromedical risk: A three-dimensional risk matrix approach. *Heart* **2019**, *105*, s9–s16. [CrossRef]
15. Tamasi, G.; Demichela, M. Risk assessment techniques for civil aviation security. *Reliab. Eng. Syst. Saf.* **2011**, *96*, 892–899. [CrossRef]
16. Chen, Y.; Xiong, S.; He, Q.; Zhan, T. NK model based analysis of risk coupling in civil aviation maintenance unsafe events. *China Saf. Sci. J.* **2018**, *28*, 104.
17. Wu, F.; Mu, H.; Feng, S. Analysis of the Risk of Air Traffic Controllers' Fatigue Based on the SHEL Model. In Proceedings of the ICTE 2015, Dalian, China, 21 September 2015; pp. 2951–2958.
18. Wang, L.; Sun, R.; Yang, Z. Analysis and evaluation of human factors in aviation maintenance based on fuzzy and AHP method. In Proceedings of the 2009 IEEE International Conference on Industrial Engineering and Engineering Management, Hong Kong, China, 8–11 December 2009; pp. 876–880.
19. Guan, D.; Guo, P.; Hipel, K.W.; Fang, L. Risk reduction in a project portfolio. *J. Syst. Sci. Syst. Eng.* **2017**, *26*, 3–22. [CrossRef]
20. Guan, D.-J.; Guo, P. Project portfolio risk components and analysis from a complex systems perspective. *Modern. Manag.* **2015**, *35*, 90–92.
21. Ming-hong, Z.; Lian, S.; Bo, G. Research on the Similarity of the Structured Emergency Events Based on Scenario. *Chin. J. Manag. Sci.* **2017**, *25*, 151–159.
22. Alwi, A.; Ariffin, K.A.Z. Information Security Risk Assessment for the Malaysian Aeronautical Information Management System. In Proceedings of the 2018 Cyber Resilience Conference (CRC), Putrajaya, Malaysia, 13–15 November 2018; pp. 1–4.
23. Yantao, W.; Jianxun, T.; Yifei, Z. Coupling analysis of risk factors in flight operation. *China Saf. Sci. J.* **2017**, *27*, 77.
24. Dan, P.; Yong-zhou, L.; Fan, L. Aircraft operational safety risk coupling based on N-K model. *J. Saf. Environ.* **2022**, *22*, 606–614.
25. Luo, F.; Liu, T.-Q. Analysis of Coupled Risk of Air Traffic Safety Based on N-K Model. *J. Wuhan Univ. Technol. (Inform. Manag. Eng.)* **2011**, *33*, 267–270.
26. Kaplan, S.; Haimes, Y.Y.; Garrick, B.J. Fitting hierarchical holographic modeling into the theory of scenario structuring and a resulting refinement to the quantitative definition of risk. *Risk Anal.* **2001**, *21*, 807. [CrossRef] [PubMed]
27. Haimes, Y.Y.; Kaplan, S.; Lambert, J.H. Risk filtering, ranking, and management framework using hierarchical holographic modeling. *Risk Anal.* **2002**, *22*, 383–397. [CrossRef] [PubMed]
28. Lambert, J.H.; Ling, C.W.; Haimes, Y.Y. Remediation site prioritization by the risk ranking and filtering method. In Proceedings of the Risk-Based Decision Making in Water Resources V, Santa Barbara, CA, USA, 3–8 November 1991; pp. 311–321.
29. Kaplan, S.; Garrick, B.J. On the quantitative definition of risk. *Risk Anal.* **1981**, *1*, 11–27. [CrossRef]
30. Ruiz, L.E. Risk Analysis, Pilot Motivation, and Decision-Making: Application of the PAVE Personal Minimums Checklist to Pilot Decision-Making in Three General Aviation Accidents. *Colleg. Aviat. Rev. Int.* **2001**, *19*, 121–139. [CrossRef]
31. Matalas, N.C.; Fiering, M.B. Water-resource systems planning. In *Climate, Climatic Change, and Water Supply. Studies in Geophysics*; Bryson, R.A., Ed.; National Academy of Sciences: Washington, DC, USA, 1977; pp. 99–110.
32. Li, D.; Liu, C.; Gan, W. A new cognitive model: Cloud model. *Int. J. Intell. Syst.* **2009**, *24*, 357–375. [CrossRef]
33. Liu, Y.; Li, D. Statistics on atomized feature of normal cloud model. *J. Beijing Univ. Aeronaut. Astronaut.* **2010**, *36*, 1320.
34. Xian-guo, W.; Hai-zhou, Z.; Li-mao, Z.; Ya-wei, Q.; Yan-hong, W. Improved R = P × C Risk Evaluation Method Based on Cloud Model. *J. Civ. Eng. Manag.* **2014**, *31*, 37–42.
35. Zhang, L.; Wu, X.; Chen, Q.; Skibniewski, M.J.; Zhong, J. Developing a cloud model based risk assessment methodology for tunnel-induced damage to existing pipelines. *Stoch. Env. Res. Risk A* **2015**, *29*, 513–526. [CrossRef]
36. Yuan, Z. Study on Risk Quantitative Assessment Model for Civil Aviation Safety. *China Saf. Sci. J.* **2007**, *17*, 140–145.
37. Kob, W.; Andersen, H.C. Testing mode-coupling theory for a supercooled binary Lennard-Jones mixture I: The van Hove correlation function. *Phys. Rev. E* **1995**, *51*, 4626. [CrossRef]
38. Efron, B. Bayes' theorem in the 21st century. *Science* **2013**, *340*, 1177–1178. [CrossRef]
39. Kelly, D.; Efthymiou, M. An analysis of human factors in fifty controlled flight into terrain aviation accidents from 2007 to 2017. *J. Saf. Res.* **2019**, *69*, 155–165. [CrossRef]
40. Eliaz, N.; Latanision, R.M. Preventative maintenance and failure analysis of aircraft components. *Corros. Rev.* **2007**, *25*, 107–144. [CrossRef]

41. Wang, M. The Impact of Cultural Values on Chinese Students in American Higher Education. *Qualit. Rep.* **2016**, *21*, 611–628. [CrossRef]
42. Kruger, J.; Dunning, D. Unskilled and unaware of it: How difficulties in recognizing one's own incompetence lead to inflated self-assessments. *J. Pers. Soc. Psychol.* **1999**, *77*, 1121–1134. [CrossRef] [PubMed]

sustainability

MDPI

Article

Analyzing Health, Safety, and Environmental Risks of Construction Projects Using the Fuzzy Analytic Hierarchy Process: A Field Study Based on a Project Management Body of Knowledge

Ahmad Soltanzadeh [1], Mohsen Mahdinia [1], Alireza Omidi Oskouei [2], Ehsan Jafarinia [3], Esmaeil Zarei [4,5] and Mohsen Sadeghi-Yarandi [6,*]

1 Department of Occupational Health and Safety Engineering, Research Center for Environmental Pollutants, Faculty of Health, Qom University of Medical Sciences, Qom P.O. Box 3713649373, Iran
2 Department of Public Health, Faculty of Health, Qom University of Medical Sciences, Qom P.O. Box 3713649373, Iran
3 Department of Environmental Engineering, School of Engineering, Islamic Azad University, Tehran Central Branch, Tehran P.O. Box 3175719843, Iran
4 Department of Safety Science, College of Aviation, Embry-Riddle Aeronautical University, Prescott, AZ 86301, USA
5 Robertson Safety Institute (RSI), Embry-Riddle Aeronautical University, Prescott, AZ 86301, USA
6 Department of Occupational Health Engineering, School of Public Health, Tehran University of Medical Sciences, Tehran P.O. Box 3175719843, Iran
* Correspondence: m-sadeghiy@alumnus.tums.ac.ir

Citation: Soltanzadeh, A.; Mahdinia, M.; Omidi Oskouei, A.; Jafarinia, E.; Zarei, E.; Sadeghi-Yarandi, M. Analyzing Health, Safety, and Environmental Risks of Construction Projects Using the Fuzzy Analytic Hierarchy Process: A Field Study Based on a Project Management Body of Knowledge. *Sustainability* **2022**, *14*, 16555. https://doi.org/10.3390/su142416555

Academic Editor: Asterios Bakolas

Received: 7 November 2022
Accepted: 5 December 2022
Published: 9 December 2022

Publisher's Note: MDPI stays neutral with regard to jurisdictional claims in published maps and institutional affiliations.

Abstract: Due to their unique nature, construction projects are considered one of the world's most hazardous and incident-prone industrial sectors. The present study aimed to analyze health, safety and environmental (HSE) risks relating to construction projects based on the project management body of knowledge (PMBOK) and sustainability approach. This study was conducted with the participation of 30 experts, using the semi-quantitative risk assessment technique, in nine areas of the project management's body of knowledge, based on the fuzzy analytic hierarchy process. Risk, in this study, was estimated using a two-dimensional matrix of incident probability and severity, each of which has four sub-parameters. The HSE risks pertaining to each of the nine areas of PMBOK were identified. After that, the two dimensions of risk, including incident probability and severity, were measured. Thirty-seven risk sources associated with nine areas of the PMBOK were identified. Risk analysis revealed that 20 sources were at an unacceptable risk level, and 17 risks were at a tolerable risk level. Identifying HSE-related risk sources in accordance with the nine areas of PMBOK, and using FAHP to assess the risk of these hazards in construction projects, can lead to a more realistic estimate of risk in construction projects. The presented method in the current study can create a novel perspective in terms of the construction industry's risk management and assessment.

Keywords: construction project; risk; project management body of knowledge; safety; sustainability; fuzzy analytic hierarchy process

1. Introduction

Construction projects are generally complex and sometimes unsafe for workers and environments, thus affecting sustainable development [1]. They are one of the most hazardous workplaces because of the high number of accidents that occur. Consequently, construction safety can be regarded as one of the most severe problems in the construction industry worldwide, particularly when large construction projects are underway. This is because of the involvement of many workers, construction techniques, numerous large and heavy plants, the large amount of materials and equipment utilized, the complex construction operations, the multi-interfaces, and the different disciplinary aspects of the project's workforce. These measures eventually lead to higher accident rates during construction projects. Accidents that tend to occur during construction projects include falling from a height, collisions, collapsing, and electric shocks; of these, falling from height and collapsing are the most prevalent [2,3].

Due to their unique nature, construction projects are considered to be part of one of the most hazardous and incident-prone industrial sectors in the world [4–7]. The construction industry has always faced challenges in terms of risk factors and health, safety, and environmental (HSE) risks. The number of incidents and injuries in the construction industry has increased daily, making the construction industry one of the world's most hazardous industries [8]. Indeed, 25–50% of catastrophic and fatal incidents in industrialized countries are related to the construction industry [9]. A previous study has shown that the construction industry in the USA, South Korea, and China have consistently high fatal occupational injuries, and the most common accident types were "fell from a higher level" and "struck by an object". China recorded the highest average number of fatal occupational injuries in construction sites at 2328, followed by the U.S. at 881, and South Korea at 533; however, South Korea had the highest average mortality rate at 17.9, followed by the U.S. at 9.4, and China at 5.3 [10].

The presence of harmful occupational incidents in construction projects, such as falls and slips, thrown objects, abrasions, and collisions, are among the major incidents that tend to occur in this sector. These incidents have other consequences associated with them in addition to direct and indirect costs and adverse social consequences, such as legal prosecutions, damage to the organization's credibility, a reduction in the quality of the project, and so on. As such, paying due attention to these factors can play a very important role in the promotion and productivity of organizations [11].

Construction safety, as a result, continues to represent a severe problem, and it poses a challenge for researchers and practitioners. In Iran, society and the economy have suffered human and financial losses due to poor safety performance in the construction industry [2,3].

Today, one of the main reasons behind the economic development of any society is its success in advancing construction projects and creating the necessary infrastructure in that society. Realizing this requires the necessary technology and expertise during the management of these projects. There are several approaches and standards in this regard, one of which is the project management body of knowledge approach. The PMBOK approach emphasizes nine main areas of project management: project integration management, project scope management, project schedule management, project cost management, project quality management, project human resource management, project communication management, project risk management, and project procurement management [12].

In its PMBOK guide to the project risk management process, the Project Management Institute (PMI) defines six phases: risk management planning, risk identification, qualitative risk analysis, quantitative risk analysis, risk response planning, and risk monitoring and control [12]. Analyses of incidents in construction projects shows that improper risk management processes have caused many of these projects to encounter severe problems. Consequently, many organizations implementing construction projects have been removed from the competition cycle due to the lack of proper risk management of occupational safety and health (OS and H). Furthermore, supposing the root cause of the abovementioned issue is found, it then becomes apparent that the majority of problems are caused by the inadequate project management structure, chiefly, the occupational safety and health (OS and H) management of the project. In addition, there seems to be no coherent and appropriate method or algorithm to mitigate this issue.

In addition to management concepts that are appropriate to the nature of industrial risks, using accurate and reliable mathematical approaches, such as the Fuzzy Analytic Hierarchy Process, can be a practical step when assessing the risk factors in this industry [13,14].

Sustainable organizations persist in balancing the triple bottom line of people, planet, and profit to acquire long-term success and viability. This implies that organizations cannot be sustainable without protecting their human resources' safety, health, and welfare. Sustainability is not just about what is done but how it is done. It is a mindset that demands leadership, and not settling for second best in any aspect of the operation. Moreover, it requires setting and achieving goals beyond regulatory compliance measures [1,15].

Worldwide, organizations have assumed this mindset to showcase their values, to measure effects and consequences, and to increase their competitive benefit; however, workplace safety and health are often underemphasized or ignored entirely. Integrating safety and health into sustainability offers an opportunity to better protect employees and to create a sustainable organization. Although many worker points are embedded within the concept of sustainability, there is a unique chance to progress O and H using this framework. In this context, OS and H promotes workers' safety, health, and welfare. Employing a sustainability framework provides a way to reimagine approaches for protecting workers, it introduces new issues to analyze, and it offers opportunities that aid innovation [16].

Unexpectedly, this is not often the case, as little attention is given to safety concerns when a sustainable approach is being developed. Organizations' sustainability programs usually only focus on environmental and financial situations. Safety should be given suitable attention in order to create truly sustainable practices as it preserves human resources. Moreover, sustainability is about conserving resources such as the environment and measuring how socially responsible an organization has been when conducting its operations, including its ability to protect employees (human resources) from incidents and occupational injuries [15,17,18].

Experts argue that occupational safety and health fit squarely within the social responsibility component of sustainability [19].

One of the most important ways to decrease incidents and consequences in construction projects is to use risk assessment methods that are adapted to the working conditions.

One of the ways to achieve sustainability is to preserve the safety of employees, especially in high-risk work environments. This involves assessing the relevant risks associated with the dangers of work environments, and forming management plans with forward-looking and proactive approaches. To achieve this, all existing potential hazards must first be identified and assessed. Then, appropriate controls and corrective measures should be taken to obtain the following [11]: risk management and assessment, as an essential element to identify all HSE risks. Indeed, this can help the construction industry detect critical hazards [1,20,21]. All of the abovementioned issues call for the creation of an appropriate scientific and operational algorithm that is commensurate with the nature of HSE risks in the construction industry. As one of the highest-risk sectors in the occupational community, in both developed and developing countries, a scientific approach, such as a fuzzy analytic hierarchy process, could be beneficial. As such, this study was designed and conducted in order to analyze the HSE risks of construction projects, in accordance with the nine areas of the project management body of knowledge, using the fuzzy analytic hierarchy process approach.

2. Materials and Methods

2.1. Study Design

The current study was a descriptive–analytical, cross-sectional study that was conducted within one of the largest construction macro-projects in Iran, in 2020. This study used a semi-quantitative technique to assess HSE risks based on the sustainability approach and fuzzy analytic hierarchy process (FAHP) methodology. Risk, in this study, was estimated based on a two-dimensional matrix of incident probability and severity, each of which has four sub-parameters. In the present study, HSE risks related to the 9 area project management body of knowledge were identified and assessed in a large construction project with the participation of 30 experts in project management, health, safety, and environment (HSE), as well as construction.

All participants were male and employed in the largest construction project hub in Iran—Tehran. Among the participants, ten experts had a master's degree, and twenty experts had a bachelor's degree. The mean and standard deviation of the age and work experience of the participants were 41.6 ± 10.32 and 12.14 ± 8.10 years, respectively. Moreover, 50% of the participants had a degree in safety engineering, 35% had a degree in HSE engineering, and 15% had a degree in industrial management. In order to collect data, checklists to measure the two components of probability and severity, and eight parameters to determine the values of the mentioned components, were designed and given to experts for evaluation. In order to evaluate the reliability of the collected data, the most skilled and experienced experts in Iran were used. Additionally, at the beginning of the study, a training class was held to familiarize the participants with the evaluation model. During the study, the performance of participants was monitored by researchers. Implementation steps of the present study are presented in Figure 1.

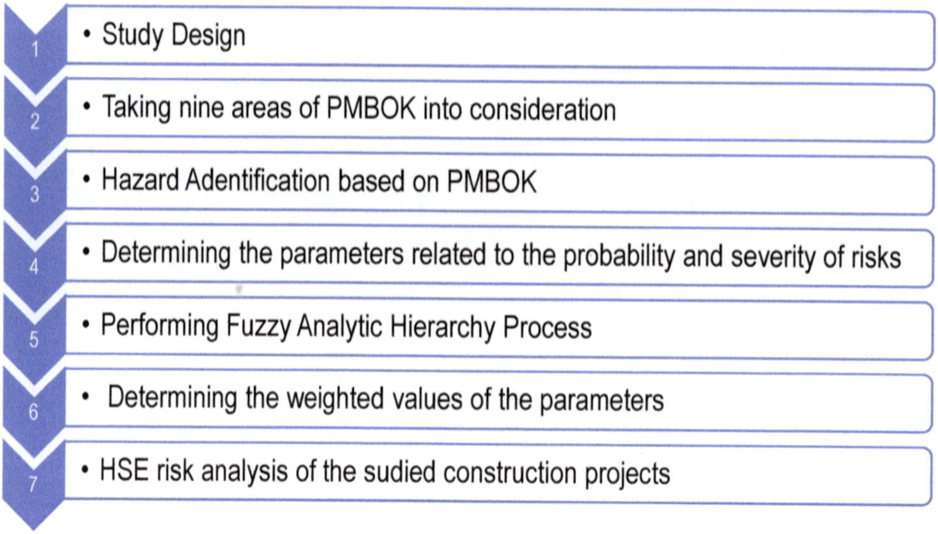

Figure 1. Implementation steps of the present study.

Project Management Body of Knowledge

PMBOK is the most well-known global standard in project management, and it is the most common benchmark for assessing project management systems (in other words, it is a familiar language in project management). The PMBOK guide is defined based on the following processes.

The three processes include:

Inputs (documents, maps, designs, etc.);

Tools and techniques (how to use inputs);

Outputs (documents, productions, etc.).

The nine areas of the PMBOK guide include the following items:

1. Integration;
2. Scope;
3. Time;
4. Cost;
5. Quality;
6. Human Resources;
7. Communications;
8. Risk;

9. Procurement.

Process groups categorize PMBOK processes according to their conceptual sequence. There is a process in five of the PMBOK groups:

The planning process group is in charge of project planning.

The implementation process group is responsible for the implementation of project plans.

The monitoring and control process group evaluates how well the project is being implemented and programmed.

The termination process group performs some of the final tasks for the project.

The PMBOK is a general term that describes a body of knowledge in the project management profession.

The PMBOK global standard is one of the best project management standards in the world, and it is revised every four years by the project management institute (PMI). The purpose of PMBOK is to provide its audience with an integrated approach to project management practices. In this study, we used the 6th edition of PMBOK.

This study was conducted in accordance with the following steps:

2.2. Identification of HSE-Related Risks

Identifying HSE risks in this study was based on the nine areas of PMBOK. These nine areas include: (1) project integration management, (2) project scope management, (3) project schedule management, (4) project cost management, (5) project quality management, (6) project human resource management, (7) project communication management, 8) project risk management, and (9) project procurement management [12]. Identifying the HSE risks in this large construction project was performed using a risk identification checklist that was related to construction projects. Moreover, a description and analysis of the various activities that were undertaken for this project were also used to identify HSE risks, as was a brainstorming approach that was employed by the panel of experts in the study.

2.3. Measurement of Sub-Parameters of Risk Dimensions

This study used the guide in order to perform the semi-quantitative risk assessment technique to calculate and estimate the sub-parameters of risk dimensions, including incident probability and severity. The dimension of risk repeatability in this study included the parameter of incident probability, which was measured based on four sub-parameters, including technical inspection, incident experience, detection probability, and human reliability (Table 1). The incident severity parameter was estimated as the dimension of risk outcome using the sub-parameters of human harm, cost imposition, damage to the organization's credibility, and impact on project time and operational interruption (Table 2) [3].

Table 1. Guide to the determination of incident probability [3].

Score	Technical Inspection	Incident Experience	Detection Probability	Human Reliability (HR)
1	Weekly	Incident data are available in similar projects and root analysis has been performed on them.	The risk is detected and revealed via the existing controls.	Regarding this risk, HR is assessed, BBS is implemented, a training program is implemented, and training outcomes are evaluated.
2	Monthly	Incident data are available through employer/contractor records, and root analysis has been performed on them.	The probability (>50%) is that the risk is detected and revealed via the existing controls.	Regarding this risk, a training program is implemented and training outcomes are evaluated.

Table 1. *Cont.*

Score	Technical Inspection	Incident Experience	Detection Probability	Human Reliability (HR)
3	Once in every three months	Incident data are available through employer/contractor records. Only a descriptive analysis has been performed on them.	The probability (<50%) is that the risk is detected and revealed via the existing controls.	Regarding this risk, a training program is implemented.
4	Once in every six months	Incident data are available through the employer/contractor records. No analysis has been performed on them.	It is unlikely (<10%) that the risk is detected and revealed via the existing controls.	Regarding this risk, compulsory and official training programs have been incompletely performed.
5	At least once during the project lifetime	No incident data is available.	There is no control, and in the case of any risk being present, it is not detectable	Regarding this risk, no measure is taken for HR assessment, BBS implementation, training, or evaluation.

Table 2. Guide to the determination of incident severity [3].

Score	Human Harm	Cost Imposition (Financial Damage, Legal Fine)	Damaging Organization Credibility	Impact on Project Time and Operational Interruption
1	Minor harm, injury, and trauma requiring first aid.	Less than USD 2500	Imperceptible repercussions	Operational interruption of less than 2 h
2	Moderate harm, lower trauma, and injury, leading to short-term hospitalization (up to three days).	USD 2500–5000	Repercussions among the stakeholders	Operational interruption of up to one day
3	Severe harm, and multiple traumas and injuries, leading to long-term hospitalization (more than three days).	USD 5000–10,000	Repercussions among the stakeholders and social networks	Operational interruption ranging from one day to one week
4	Harm leading to disability, amputation, and permanent disability.	USD 10,000–25,000	Repercussions among the stakeholders, social networks, and widely-circulated newspapers	Operational interruption ranging from one week to one month
5	Death of one person or more.	More than USD 25,000	Repercussions among the stakeholders, social networks, and widely-circulated newspapers, both at the national and international level	Operational interruption lasting more than one month

2.3.1. Probability of Occurrence

The probability feature that relates to the concept of risk is defined as the probability of an incident happening within a specific period, which, in this study, was determined using the following parameters.

- Detection probability;
- Human reliability;
- Technical inspection;
- Accident experience;
- Severity of occurrence.

2.3.2. Severity of Occurrence

The severity component that relates to the concept of risk is defined as the range of losses and injuries caused if the risk comes to fruition, and harm occurs. It is clear from this concept that this parameter can be calculated and specified through the following important factors:

- human injury;
- financial loss;
- operational interruption;
- reputational damage.

2.4. HSE Risk Analysis of Construction Projects

An analysis of the HSE risks that are related to the 9 PMBOK areas of this large construction project was conducted in accordance with FAHP. These risks were analyzed using a two-dimensional risk matrix (Figure 2). The weight factors presented in this figure were calculated and presented for each of the sub-parameters of the two dimensions of the risk matrix in accordance with FAHP.

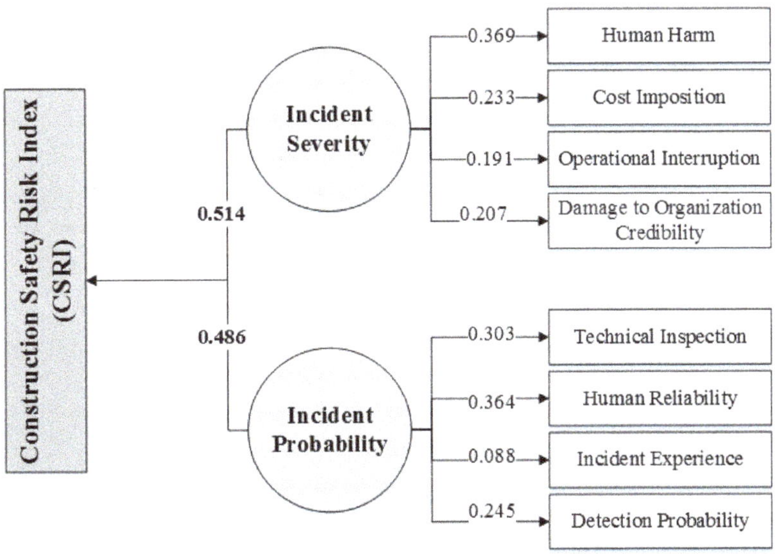

Figure 2. Algorithm of Construction Risk Assessment [3].

The current study was performed using the method proposed by Chang; this is because it is easier to perform and it yields accurate results [22,23]. As such, the construction risk index (CRI), and the incident probability and severity parameters, were calculated based on equations 1–3 and Figure 2 below. It should be noted that decision-making levels based on these calculations have been classified into acceptable risk (CRI < 1), ALARP (as low as reasonably practicable) (CRI = 1–3), and unacceptable risk (CRI > 3).

ALARP stands for "as low as reasonably practicable". "Reasonably practicable" means weighing a risk against the trouble, time, and money needed to control it; thus, ALARP describes the level to which we expect to see workplace risks controlled.

The ALARP concept can be used to define two sets of risk tolerance criteria: a minimum requirement and a target value. Between the two sets of criteria, a tolerable level of risk may be found. The residual risk should fall either in the acceptable region or close to the bottom of the tolerable region. The ALARP concept arises within a regulatory framework. Increasingly, it is used by companies around the world as it provides a reasonable basis for managing risks [24].

$$\boldsymbol{CRI} = [\boldsymbol{Probability} \times \mathbf{0.486}] \times [\boldsymbol{Severity} \times \mathbf{0.514}] \tag{1}$$

$$\boldsymbol{Probability} = \sum \boldsymbol{P_i PW_i} \tag{2}$$

$$\boldsymbol{Severity} = \sum \boldsymbol{S_i SW_i} \tag{3}$$

CRI: Construction risk index;
Probability: Incident probability;
Severity: Incident severity;
Pi: Numerical index of sub-parameters of incident probability (Table 1);
PWi: Normalized weight of each of sub-parameters of incident probability (Figure 2);
Si: Numerical index of sub-parameters of incident severity (Table 2);
SWi: Normalized weight of each of sub-parameters of incident severity (Figure 2) [3].

3. Results

The results of the HSE hazard identification, based on nine areas of the PMBOK, revealed that a total of 37 risks in this project threaten the safety and health of human, as well as economic and environmental investments (Tables 3–5). These risks include (1) integrated project management (four risks), (2) project scope management (three risks), (3) project scheduling (five risks), (4) project cost management (three risks), (5) project quality (five risks), (6) human resources (four risks), (7) project risk management (four risks), (8) project communication (three risks), and (9) project procurement (four risks). The factors leading to the incident probability for each of the 37 risk sources have been presented in Tables 3–5.

Table 3. The results of HSE risk analysis in three areas, including integrated project management, project scope, and project scheduling.

Risk Source	Causes of Incidence	Incidence Probability				Incidence Severity				CRI
		Technical Inspection	Incidence Experience	Detection Probability	Human Reliability	Human Resources	Cost Imposition	Damage to Organization's Credibility	Project Interruption	
Integrated project management										
Lack of policy definition in the field of HSE.	Lack of commitment from the manager and staff.	N.A †	N.A	3	3	5	5	5	5	2.282 ‡
Lack of definition concerning the HSE needs of stakeholders	Lack of detection of the HSE needs of stakeholders.	N.A	N.A	4	4	5	3	4	4	2.517
Lack of developing HSE programs during project lifetime.	Lack of an index for evaluating HSE performance.	N.A	N.A	3	2	5	4	4	4	1.597
Absence of/HSE unit and defects in organization chart.	Improper status of HSE in projects, wrong decisions, and vulnerability to incidences occurring.	N.A	N.A	1	3	5	5	5	5	1.670 ‡
Project scope management										
Insufficient knowledge of organization's managers concerning the hazards and risks of the construction industry.	Lack of obedience to managers to resolve notified discrepancies.	N.A	N.A	4	1	5	4	5	5	1.600
Lack of attention to HSE issues when developing work breakdown structure.	Lack of attention to HSE in the breakdown structure, and a violation of these guidelines in the execution phase.	N.A	N.A	3	4	5	3	3	5	2.255

Table 3. *Cont.*

Risk Source	Causes of Incidence	Incidence Probability				Incidence Severity				CRI
		Technical Inspection	Incidence Experience	Detection Probability	Human Reliability	Human Resources	Cost Imposition	Damage to Organization's Credibility	Project Interruption	
Failure to comply with the prioritization of activities based on HSE requirements.	Lack of proper fire alarm and extinguisher systems, and a lack of fall prevention guard installations.	N.A	N.A	2	4	5	3	3	5	2.003
Project scheduling										
Failure to specify the time for receiving Machinery Health Certificate at WBS.	Defects in machinery and equipment.	4	3	3	4	5	4	5	5	4.367
Failure to allocate time for HSE training at WBS.	Lack of proper training and occurrence of unsafe practices.	N.A	4	4	4	5	3	4	5	3.014
Failure to allocate time to make the workplace safe before the start of work (installing lifeline).	Failure to make the workplace safe.	1	3	3	3	5	5	5	5	2.990 ‡
Failure to allocate time for HSE-focused meetings (incident analysis).	Few HSE problems are presented and there is a lack of commitment at the managerial level to solve problems.	N.A	N.A	4	2	5	3	3	4	1.676
Failure to determine the time of clinical and para clinical tests (high risk occupations).	Physical incompetence of the individual at time of employment and occurrence of occupational diseases and incidents.	N.A	4	5	3	5	3	4	4	2.758

† No index has been defined for this sub-parameter which concerns the identified risk sources. ‡ Based on the construction of the semi-quantitative risk assessment technique, if the index of any of the incidence probability or severity parameters equals five, regardless of the final construction risk index, the risk level is estimated to be unacceptable.

Table 4. The results of HSE risk analysis in three areas: project costs, quality, and human resource management.

Risk Source	Causes of Incidence	Incidence Probability				Incidence Severity				CRI
		Technical Inspection	Incidence Experience	Detection Probability	Human Reliability	Human Resources	Cost Imposition	Damage to Organization's Credibility	Project Interruption	
Project cost management										
Failure to pay employees' salaries on time, thus creating job stress and lack of focus on the assigned tasks.	Creating job stress and setting the stage for incidents to occur.	N.A †	N.A	4	5	4	4	4	5	2.785
Lack of finances to provide standard equipment and machinery.	Defects in equipment and machinery and the occurrence of incidents.	N.A	3	5	5	3	2	5	5	2.306 ‡
No cost allocation for hiring an HSE supervisor, expert, and officer, proportional to the project phases.	Failure to set up an integrated HSE monitoring system.	N.A	N.A	5	5	2	2	5	5	1.521 ‡
Non-allocation of civil liability insurance policies to third parties and machinery.	Failure to receive insurance with full coverage.	N.A	4	5	1	2	3	5	5	1.598
No funds allocated to the implementation of improvement projects (such as lifeline systems and fuse supplies).	Failure to implement improvement projects and resolve HSE inconsistencies.	2	3	5	5	2	2	5	5	2.608 ‡
Project quality										
Non-compliance with QC concepts regarding concrete and structural welding.	Poor execution of the project and occurrence of incidents in the implementation phase.	3	4	3	3	5	5	4	4	3.550

Table 4. *Cont.*

Risk Source	Causes of Incidence	Incidence Probability				Incidence Severity				CRI
		Technical Inspection	Incidence Experience	Detection Probability	Human Reliability	Human Resources	Cost Imposition	Damage to Organization's Credibility	Project Interruption	
Lack of quality materials being supplied.	Failure to properly implement the project.	2	5	3	4	5	5	5	3	3.801
Lack of quality personal protective equipment.	Inability to prevent incidents.	2	3	2	2	5	5	4	3	2.301
Absence of a maintenance system for the machinery.	Inability to properly repair and maintain machinery.	3	4	4	3	5	5	4	4	3.898
Supplying consumer equipment without the necessary quality.	Inability to supply quality consumer equipment.	4	3	3	3	5	4	4	3	3.387
Human resources										
Lack of training at the outset of employment.	Lack of familiarity with workplace hazards.	N.A	3	2	1	5	3	4	4	1.155
Failure to determine the plans and responsibilities of individuals.	Lack of familiarity with HSE tasks and performing unsafe practices.	5	N.A	3	4	5	3	3	3	3.461
Failure to determine the qualification requirements for HSE to employ all the staff in the project.	Employing unqualified individuals and the occurrence of unsafe acts.	5	5	3	4	5	4	5	4	4.739
Failure to develop a special training program for all occupations.	Lack of awareness of specialized HSE information when conducting activities.	4	4	4	3	5	5	4	5	4.353

† No index has been defined for this sub-parameter which concerns the identified risk sources. ‡ Based on the construction of the semi-quantitative risk assessment technique, if the index of any of the incidence probability or severity parameters equals five, regardless of the final construction risk index, the risk level is estimated to be unacceptable.

Table 5. Results of HSE risk analysis in three areas, including project risk, project communication, and project procurement management.

Risk Source	Causes of Incidence	Incidence Probability				Incidence Severity				CRI
		Technical Inspection	Incidence Experience	Detection Probability	Human Reliability	Human Resources	Cost Imposition	Damage to Organization's Credibility	Project Interruption	
Project risk management										
Lack of proper risk management planning (identifying stakeholders, legal requirements and risk assessment).	Lack of readiness to use the full capacity of risk management techniques to enter the identification stage.	N.A †	N.A	3	4	5	3	4	2	2.055
Inability to identify HSE risks.	Lack of incident data collection, and lack of risk identification meetings.	N.A	N.A	2	3	5	4	4	3	1.651
Lack of proper assessments to identify risks.	Failure to prioritize eliminating the identified risks.	N.A	N.A	2	3	5	4	4	3	1.651
Failure to formulate the necessary controls based on the risk control pyramid and failure to follow up on the implementation of control measures.	Lack of development of all, or parts of, the control methods, and a lack of readiness of the organization to deal with risk.	N.A	N.A	3	3	5	5	5	5	2.306 ‡
Project communication										
Lack of using experienced HSE advisors.	Lack of specialized inspections of equipment and lack of determination of improvement methods.	N.A	N.A	4	4	5	4	3	3	2.416
Lack of executive methods and operation controls.	Failure to determine the risks associated with each executive operation.	N.A	N.A	3	2	5	5	5	5	1.827 ‡

Table 5. *Cont.*

Risk Source	Causes of Incidence	Incidence Probability				Incidence Severity				CRI
		Technical Inspection	Incidence Experience	Detection Probability	Human Reliability	Human Resources	Cost Imposition	Damage to Organization's Credibility	Project Interruption	
Underuse of the HSE experiences of other projects.	Failure to register HSE records and use them in future projects.	N.A	3	2	3	5	5	5	5	2.306 ‡
Project procurement management										
Failure to explicitly state HSE provisions in contracts.	Defects in contracts and lack of calculation of HSE provisions in the contract by the contractor, and lack of obligation to fulfill the requirements during execution.	N.A	N.A	3	2	5	5	5	5	1.827 ‡
Failure to announce and continuously monitor HSE executive regulations.	Lack of HSE disciplinary criteria for dealing with contractors.	N.A	N.A	2	3	5	4	4	5	1.802
Failure to timely supply personal protective equipment and other items related to HSE.	Inability of the project to reduce the risk of incidents.	N.A	2	2	2	5	5	5	5	1.741
Lack of contractor evaluations.	Lack of managing contractors.	N.A	N.A	4	4	5	4	5	5	2.901

† No index has been defined for this sub-parameter which concerns the identified risk sources. ‡ Based on the construction of the semi-quantitative risk assessment technique, if the index of any of the incidence probability or severity parameters equals five, regardless of the final construction risk index, the risk level is estimated to be unacceptable.

The results of the HSE risk assessment for this construction project showed that the risk index for 20 risk sources was estimated to be at an unacceptable level, and it was at the ALARP level for 17 risk sources.

None of the risk sources had a risk index at an acceptable level. It is worth mentioning that ten risk sources, despite the estimated risk index of three, were placed at the unacceptable risk level, thus requiring immediate corrective measures.

Based on the results, out of the four identified risks in the area of integrated project management, the HSE risk level related to two risk sources was found to be at the ALARP level, and it was found to be unacceptable for the other two risk sources. Table 3 also revealed that the risk level of three identified risks under the project's management was at the ALARP level. In addition, the risk level of four risk sources which pertained to project scheduling was at the ALARP level.

Moreover, the risk level regarding 'not allocating time to make the environment safe before starting work' (such as installing a lifeline) was assessed to be at unacceptable level (Table 3).

According to the results presented in Table 4, the level of risk of three risk sources related to project cost management was unacceptable, whereas for two risk sources, it was estimated to be at the ALARP level. The risk level of the five risk sources related to project quality was estimated to be at an unacceptable level. The results in this table also showed that the HSE risk of three sources regarding human resources in project management knowledge was unacceptable (Table 4).

As is shown in Table 5, one risk source in the area of project risk management was at an unacceptable level, whereas the risk level for the other three risks was estimated to be within the ALARP range. The results of the risk assessment related to project communication showed that the risk level of the two risk sources was assessed to be at an unacceptable level. Furthermore, the risk level of two risk sources in the project procurement area was found to be within the ALARP range, and it was assessed to be unacceptable for two other sources.

4. Discussion

Construction projects are one of the most hazardous and incident-prone industries due to their unique and dynamic nature [20,25,26]. A construction site is a dynamic, continuously evolving workplace that accommodates multiple groups and suppliers working in parallel. In addition, the impulsive nature of weather, deliveries, and unexpected events put pressure on stakeholders to manage tight deadlines and limit costs.

Effective safety management in construction projects is a core consideration for all types of organizations that are responsible for protecting and optimizing the efficiency of human resources. Concerning construction, ensuring workplace safety is not an easy task. Occupational accidents in the construction industry will have an impact on economic and social issues in organizations, as well as countries. The growth of the construction industry has been mitigated by accidents or injuries, which occur frequently. It has been calculated that around 60,000 construction fatalities occur worldwide annually, equaling one accident every nine minutes. Among all industries in the world, construction has the highest accident rate, including deaths and disabling damages [2,3].

The severity of the damage caused by construction projects is so great that creating a suitable platform for risk management and reducing incidents has become a national priority in many developed and developing countries [27,28]. Despite the very favorable turnover of construction projects, many construction worksites worldwide still do not provide good, safe conditions. Studies have shown that construction projects have a wide array of risk factors that lead to reduced safety levels and increased incident rates in the industry [1,29,30]. In these projects, each worker is directly exposed to a high volume of risk factors which contribute to incidents. In addition to causing human damage, the hazards associated with these projects can impact various aspects of the industry, such as a project's existing costs, the quality of work, time scheduling, and organizational credibility [31].

Various studies revealed that the construction industry is one of the most dangerous, due to its exceptional and dynamic nature [32–35]. The severity of losses and damages caused by construction projects is such that creating a suitable platform for the risk management process, in order to reduce accidents in many developed and developing countries, has become a national priority [36]. In these projects, each person was directly exposed to a high volume of risk factors which can cause accidents. In addition to generating harm to human resources, the risks associated with these projects can affect various aspects of the industry, such as current project costs, quality of work, time management, the credibility of the organization, and so on. [37]. Other studies show that the mentioned risk factors include personal risks, occupational risks, environmental risks (unsafe conditions), unsafe acts, and managerial–organizational factors [38,39].

The study performed by O. Sanni-Anibire et al. revealed that the type of accident with the highest risk score involved "falling objects", whereas the most significant cause was excessive winds on the project site. Their results showed that slips, trips, and falls had the best safety performance. Furthermore, using a six sigma evaluation, the average project safety performance was 2.33-sigma, which implies that 228,739 accidents may occur in every million opportunities [40].

The results of the current study showed that the existing risks, based on nine areas of the PMBOK, consist of integrated project management, project scope management, project scheduling, project cost management, project quality, human resources, project risk management, project communication, and project procurement. The results of the HSE risk assessment that were related to the Project Management Body of Knowledge also revealed that the risk index was estimated to be at an unacceptable level for 20 risk sources, and it was at an ALARP level for 17 risk sources; however, none of the identified risk sources were assessed as having an acceptable risk level, thus indicating the presence of high-risk levels in this industry.

The results of the HSE risk analysis of the construction project in this study were based on three areas of project management knowledge, including: integrated project management, project scope management, and project scheduling. Moreover, the results pointed to a lack of attention to the status and principles of HSE in all stages of the construction project. This is the most important principle for controlling the risk factors on the worksite, and one of the major reasons for the high frequency of incidents in the construction industry. This issue implies that most construction employers pay the least attention to the subject of HSE risk management in the initial and time scheduling phases of the project. One consequence that can increase the risk index is the impact on the scheduling of construction projects, which can create fundamental challenges for defined project scheduling. Some of these challenges include the deaths of key members of the project, and operational interruptions which occur until their replacements are found. The death or disability of employees, equipment damage, and project interruptions until these issues are resolved can cause a loss of employee morale. Moreover, long-term interruptions of the project may also occur due to governmental organization intervention as a result of non-compliance with HSE rules and regulations [27,28,41,42].

With regard to the importance of human resources, there are times when the incident involving the worker is simple and non-technical; sometimes, the same happens to the project manager or CEO. Obviously, the consequences of the incidents in the two cases are different. As such, in risk assessment, it should be made clear which of these two groups are exposed to incidents and how many human resources are exposed to them.

In addition, disregarding the issues related to risk management in financial management and project procurement can impose direct and indirect costs on the project. Failure to comply with safety precautions can cause high costs because of non-compliance with HSE regulations, monetary compensation for death, medical expenses, an increase in insurance rates, indirect costs due to reduced work efficiency, damage to equipment, and so on [5,29].

The impact of incidents on the direct and indirect costs of the project is considered to be only a minor part of the consequences related to HSE risks and project costs; therefore, one of the most tangible ramifications of HSE risks lies with project costs. Failure to observe safety measures and the improper management of existing risks can lead to heavy financial costs for the project [42,43]. The results of this study also showed that paying attention to costs is of paramount importance in risk assessment and prioritization.

According to several studies of this nature, various factors affect the levels of HSE risks. The effect of HSE risks on quality in the study by Husin et al. [44], the effect of HSE risks on cost in the study by Ikpe et al. [45], and the effect of HSE risks on human resources in JW Garrett and Teizer [46], are examples of such studies.

In a study conducted by Debasish Majumder et al., the results revealed that FRA and FAHP approaches could evaluate the worksite's actual status, and important hazards can be identified to motivate proprietors to invest in safety in their industry. With this technique, all the input parameters are measured in terms of fuzzy numerals (accident percentage, accident severity, and expenses of safety measures). The overall risk is calculated as the sum of the products of the RS and the weight of each body part in terms of damage sustained in an accident [13]; therefore, the use of management principles, as well as different and reliable mathematical methods, such as the fuzzy analytic hierarchy process, can lead to a more accurate estimation of the risk levels in construction projects.

In today's world, sustainability is attracting considerable attention as many governments have integrated it into their economic development strategies. According to the World Health Organization (WHO), sustainable development is defined as a strategy to "meet the requirements of the present world population without generating an adverse impact on health and the environment, and without consuming or endangering the global resource, therefore without compromising the ability of future generations to meet their needs." Sustainable development depends on several regulations for preparing its actions, many of which can be involved in occupational health and safety. These principles include the necessity for attention to people's health and quality of life, the prevention of known risks, and the application of precautions when there is uncertainty concerning certain dangers [16].

One of the ways with which to achieve sustainability is to preserve the safety of workers, especially in high-risk work environments, by assessing the relevant risks in accordance with the risks of work environments and in the form of management plans using a prospective and preventive approach. In this study, practical steps were taken to promote sustainable safety; a forward-looking approach was adopted by using project management concepts that addressed the types of risk in the construction industry.

The results of the study performed by Jilcha et al. revealed that innovations in workplace safety and health bring sustainable development via healthy people, a safer workplace, decreased costs associated with accidents, a controlled environment, managed workplace accidents, and improved workplace safety knowledge [47].

Hui Zhou et al. indicated that safety accidents cause significant losses of life and property, which expose the problems in construction management and hinder the sustainable development of society [1]. This issue reveals the need for innovation in the field of safety assessment and management in this industry.

However, the authors of the current study, in their literature review, found that no study deals with the various aspects of risk assessment, namely, the impact of risk on cost, quality, project scheduling, damage to the credibility of the organization, legal and criminal penalties, the importance of human resources, and the impact on human resources. It should be mentioned that since the present study was conducted in Iran (a developing country) and the safety levels observed in the construction industry in developed countries are much higher, it is suggested that the method used in these countries should be used with caution. It is also recommended that researchers in developed countries conduct studies in the future using a similar algorithm.

Considering that the issue of risk assessment is at the heart of the risk management concept, it is suggested that future studies consider the present method when making management decisions during construction projects.

The present study was performed in order to introduce and implement a unique approach that assesses construction projects' safety risks, in accordance with the dynamic and specific characteristics of construction projects and activities that are based on the PMBOK and FAHP. This study determined the most critical factors affecting construction projects' occupational accident frequency and severity. The present technique could be a practical step toward decreasing occupational accident risk levels in the construction industry and developing control plans, especially in developing countries, where there exists lower risk management performance.

One of the limitations of the current approach is that there is no quantitative method to calculate and evaluate the effective parameters pertaining to the construction industry's probability and severity of risks; thus, it is suggested that researchers in the future develop and apply quantitative methods with the same algorithm as the present study. They should ensure that such methods are developed in line with safety management systems via international management guidelines.

5. Conclusions

The results of the current study indicate that the integration of HSE and PMBOK can improve the effectiveness of the risk assessment and management process. Identifying HSE-related risk sources in accordance with the nine areas of PMBOK, as well as using a fuzzy analytic hierarchy process to assess the risk of these hazards in a construction project, can help provide a more realistic estimation of the risk index in construction projects. Using the existing guidelines in various areas of project management knowledge, and integrating it with practical methods such as FAHP, can be an effective step toward creating a suitable and specialized operational algorithm. Using the developed model in the present study can be a practical step in evaluating risk sources and implementing effective control measures.

Author Contributions: Conceptualization, A.S.; Methodology, E.J., E.Z. and M.S.-Y.; Software, M.M. and E.J.; Investigation, A.O.O.; Resources, E.J.; Data curation, M.M., A.O.O. and E.Z.; Writing—original draft, M.S.-Y.; Writing—review & editing, A.S., E.Z. and M.S.-Y.; Visualization, A.O.O.; Project administration, A.S. All authors have read and agreed to the published version of the manuscript.

Funding: This research received no external funding.

Institutional Review Board Statement: Not applicable.

Informed Consent Statement: Not applicable.

Data Availability Statement: Not applicable.

Conflicts of Interest: The authors declared no potential conflicts of interest with respect to the research, authorship, and/or publication of this article.

References

1. Zhou, X.-H.; Shen, S.-L.; Xu, Y.-S.; Zhou, A.-N. Analysis of Production Safety in the Construction Industry of China in 2018. *Sustainability* **2019**, *11*, 4537. [CrossRef]
2. Othman, I.; Shafiq, N.; Nuruddin, M. Effective safety management in construction project. In Proceedings of the International Conference on Architecture and Civil Engineering (ICACE 2017), Petaling Jaya, Malaysia, 8–9 May 2017.
3. Mahdinia, M.; Yarandi, M.S.; Jafarinia, E.; Soltanzadeh, A. Development of a New Technique for Safety Risk Assessment in Construction Projects Based on Fuzzy Analytic Hierarchy Process. *Eng. Struct.* **2021**, *7*, 04021037. [CrossRef]
4. Kang, Y.; Siddiqui, S.; Suk, S.J.; Chi, S.; Kim, C. Trends of Fall Accidents in the U.S. Construction Industry. *J. Constr. Eng. Manag.* **2017**, *143*, 04017043. [CrossRef]
5. Soltanzadeh, A.; Mohammadfam, I.; Moghimbeigi, A.; Ghiasvand, R. Key factors contributing to accident severity rate in construction industry in Iran: A regression modelling approach/Primjena regresijskog modela u analizi ključnih čimbenika koji pridonose težini nesreća u građevinskoj industriji u Iranu. *Arch. Ind. Hyg. Toxicol.* **2016**, *67*, 47–53. [CrossRef]
6. Shafique, M.; Rafiq, M. An Overview of Construction Occupational Accidents in Hong Kong: A Recent Trend and Future Perspectives. *Appl. Sci.* **2019**, *9*, 2069. [CrossRef]

'. Zhou, Z.; Li, C.; Mi, C.; Qian, L. Exploring the Potential Use of Near-Miss Information to Improve Construction Safety Performance. *Sustainability* **2019**, *11*, 1264. [CrossRef]
3. Winge, S.; Albrechtsen, E.; Mostue, B.A. Causal factors and connections in construction accidents. *Saf. Sci.* **2018**, *112*, 130–141. [CrossRef]
9. Mohandes, S.R.; Sadeghi, H.; Mahdiyar, A.; Durdyev, S.; Banaitis, A.; Yahya, K.; Ismail, S. Assessing Construction Labours' Safety Level: A Fuzzy MCDM Approach. *J. Civ. Eng. Manag.* **2020**, *26*, 175–188. [CrossRef]
10. Choi, S.D.; Guo, L.; Kim, J.; Xiong, S. Comparison of fatal occupational injuries in construction industry in the United States, South Korea, and China. *Int. J. Ind. Ergon.* **2019**, *71*, 64–74. [CrossRef]
11. Mohammadi, A.; Tavakolan, M.; Khosravi, Y. Factors influencing safety performance on construction projects: A review. *Saf. Sci.* **2018**, *109*, 382–397. [CrossRef]
12. Project Management Institute. *Guide to the Project Management Body of Knowledge*; Project Management Institute: Newtown Square, PA, USA, 2018.
13. Majumder, D.; Debnath, J.; Biswas, A. Risk Analysis in Construction Sites Using Fuzzy Reasoning and Fuzzy Analytic Hierarchy Process. *Procedia Technol.* **2013**, *10*, 604–614. [CrossRef]
14. Zarei, E.; Khan, F.; Abbassi, R. An advanced approach to the system safety in sociotechnical systems. *Saf. Sci.* **2023**, *158*, 105961. [CrossRef]
15. Liu, B.; Lu, Q. Creating a Sustainable Workplace Environment: Influence of Workplace Safety Climate on Chinese Healthcare Employees' Presenteeism from the Perspective of Affect and Cognition. *Sustainability* **2020**, *12*, 2414. [CrossRef]
16. Boileau, P.-E. Sustainability and prevention in occupational health and safety. *Ind. Health* **2016**, *54*, 293–295. [CrossRef]
17. Zarei, E.; Khan, F.; Abbassi, R. A dynamic human-factor risk model to analyze safety in sociotechnical systems. *Process Saf. Environ. Prot.* **2022**, *164*, 479–498. [CrossRef]
18. Zarei, E.; Karimi, A.; Habibi, E.; Barkhordari, A.; Reniers, G. Dynamic occupational accidents modeling using dynamic hybrid Bayesian confirmatory factor analysis: An in-depth psychometrics study. *Saf. Sci.* **2021**, *136*, 105146. [CrossRef]
19. Nawaz, W.; Linke, P.; Koç, M. Safety and sustainability nexus: A review and appraisal. *J. Clean. Prod.* **2019**, *216*, 74–87. [CrossRef]
20. Alkaissy, M.; Arashpour, M.; Ashuri, B.; Bai, Y.; Hosseini, R. Safety management in construction: 20 years of risk modeling. *Saf. Sci.* **2020**, *129*, 104805. [CrossRef]
21. Mohammadfam, I.; Soltanzadeh, A.; Moghimbeigi, A.; Akbarzadeh, M. Modeling of Individual and Organizational Factors Affecting Traumatic Occupational Injuries Based on the Structural Equation Modeling: A Case Study in Large Construction Industries. *Arch. Trauma Res.* **2016**, *5*, e33595. [CrossRef]
22. Kramar, U.; Dragan, D.; Topolšek, D. The Holistic Approach to Urban Mobility Planning with a Modified Focus Group, SWOT, and Fuzzy Analytical Hierarchical Process. *Sustainability* **2019**, *11*, 6599. [CrossRef]
23. Hatefi, S.M.; Tamošaitienė, J. Construction projects assessment based on the sustainable development criteria by an integrated fuzzy AHP and improved GRA model. *Sustainability* **2018**, *10*, 991.
24. Nesticò, A.; He, S.; De Mare, G.; Benintendi, R.; Maselli, G. The ALARP Principle in the Cost-Benefit Analysis for the Acceptability of Investment Risk. *Sustainability* **2018**, *10*, 4668. [CrossRef]
25. Mohammadfam, I.; Soltanzadeh, A.; Moghimbeigi, A.; Akbarzadeh, M. Confirmatory Factor Analysis of Occupational Injuries: Presenting an Analytical Tool. *Trauma Mon.* **2016**, *22*. [CrossRef]
26. Jo, B.W.; Lee, Y.S.; Kim, J.H.; Khan, R.M.A. Trend Analysis of Construction Industrial Accidents in Korea from 2011 to 2015. *Sustainability* **2017**, *9*, 1297. [CrossRef]
27. Seker, S.; Zavadskas, E.K. Application of Fuzzy DEMATEL Method for Analyzing Occupational Risks on Construction Sites. *Sustainability* **2017**, *9*, 2083. [CrossRef]
28. Yang, Y.; Tang, W.; Shen, W.; Wang, T. Enhancing Risk Management by Partnering in International EPC Projects: Perspective from Evolutionary Game in Chinese Construction Companies. *Sustainability* **2019**, *11*, 5332. [CrossRef]
29. Kim, J.Y.; Lee, D.S.; Kim, J.D.; Kim, G.H. Priority of Accident Cause Based on Tower Crane Type for the Realization of Sustainable Management at Korean Construction Sites. *Sustainability* **2021**, *13*, 242. [CrossRef]
30. Soltanzadeh, A.; Mahdinia, M. Path analysis of occupational injuries based on the structural equation modeling approach: A retrospective study in the construction industry. *Iran Occup. Health* **2019**, *16*, 47–57.
31. Woolley, M.J.; Goode, N.; Read, G.J.M.; Salmon, P.M. Have we reached the organisational ceiling? A review of applied accident causation models, methods and contributing factors in construction. *Theor. Issues Ergon. Sci.* **2019**, *20*, 533–555. [CrossRef]
32. Falahati, M.; Karimi, A.; Mohammadfam, I.; Mazloumi, A.; Khanteymoori, A.R.; Yaseri, M. Multi-dimensional model for determining the leading performance indicators of safety management systems. *Work* **2020**, *67*, 959–969. [CrossRef]
33. Qi, H.; Zhou, Z.; Li, N.; Zhang, C. Construction safety performance evaluation based on data envelopment analysis (DEA) from a hybrid perspective of cross-sectional and longitudinal. *Saf. Sci.* **2021**, *146*, 105532. [CrossRef]
34. Lin, F.; Wu, P.; Xu, Y. Investigation of Factors Influencing the Construction Safety of High-Speed Railway Stations Based on DEMATEL and ISM. *Adv. Civ. Eng.* **2021**, *2021*, 9954018. [CrossRef]
35. Johansson, J.; Berglund, L.; Johansson, M.; Nygren, M.; Rask, K.; Samuelson, B.; Stenberg, M. Occupational safety in the construction industry. *Work* **2019**, *64*, 21–32. [CrossRef] [PubMed]
36. Sousa, V.; Almeida, N.M.; Dias, L.A. Risk-based management of occupational safety and health in the construction industry—Part 1: Background knowledge. *Saf. Sci.* **2014**, *66*, 75–86.

37. Pereira, E.; Taghaddos, H.; Hermann, R.; Han, S.; Abourizk, S. A conceptual accident causation model based on the incident root causes. In Proceedings of the5th International/11th Construction Specialty Conference, Vancouver, British Columbia, Canada, 8–10 June 2018.
38. Rowlinson, S.; Jia, Y.A. Construction accident causality: An institutional analysis of heat illness incidents on site. *Saf. Sci.* **2015**, *78*, 179–189. [CrossRef]
39. Kim, J.-M.; Kim, T.; Bae, J.; Son, K.; Ahn, S. Analysis of plant construction accidents and loss estimation using insurance loss records. *J. Asian Arch. Build. Eng.* **2019**, *18*, 507–516. [CrossRef]
40. Sanni-Anibire, M.O.; Mahmoud, A.S.; Hassanain, M.A.; Salami, B.A. A risk assessment approach for enhancing construction safety performance. *Saf. Sci.* **2019**, *121*, 15–29. [CrossRef]
41. Kim, M.; Lee, I.; Jung, Y. International Project Risk Management for Nuclear Power Plant (NPP) Construction: Featuring Comparative Analysis with Fossil and Gas Power Plants. *Sustainability* **2017**, *9*, 469. [CrossRef]
42. Yang, K.; Kim, K.; Go, S. Towards Effective Safety Cost Budgeting for Apartment Construction: A Case Study of Occupational Safety and Health Expenses in South Korea. *Sustainability* **2021**, *13*, 1335.
43. Nguyen, L.H.; Watanabe, T. The Impact of Project Organizational Culture on the Performance of Construction Projects. *Sustainability* **2017**, *9*, 781. [CrossRef]
44. Husin, H.N.; Adnan, H.; Jusoff, K. Management of Safety for Quality Construction. *J. Sustain. Dev.* **2008**, *1*, 41–47. [CrossRef]
45. Ikpe, E.; Hammon, F.; Oloke, D. Cost-Benefit Analysis for Accident Prevention in Construction Projects. *J. Constr. Eng. Manag.* **2012**, *138*, 991–998. [CrossRef]
46. Garrett, J.; Teizer, J. Human factors analysis classification system relating to human error awareness taxonomy in construction safety. *J. Constr. Eng. Manag.* **2009**, *135*, 754–763.
47. Jilcha, K.; Kitaw, D. Industrial occupational safety and health innovation for sustainable development. *Eng. Sci. Technol. Int. J.* **2017**, *20*, 372–380. [CrossRef]

sustainability MDPI

Article

New Approaches to Project Risk Assessment Utilizing the Monte Carlo Method

Andrea Senova [1], Alica Tobisova [2,*] and Robert Rozenberg [2]

[1] Faculty of Mining, Ecology, Process Control and Geotechnologies, Technical University of Kosice, Letna 9, 042 00 Kosice, Slovakia
[2] Faculty of Aeronautics, Technical University of Kosice, Rampova 7, 041 21 Kosice, Slovakia
* Correspondence: alica.tobisova@tuke.sk; Tel.: +421-556026184

Abstract: An environment of turbulence in the market in recent years and increasing inflation, mainly as a result of the post-COVID period and the ongoing military operation in Ukraine, represents a significant financial risk factor for many companies, which has a negative impact on managerial decisions. A lot of enterprises are forced to look for ways to effectively assess the riskiness of the projects that they would like to implement in the future. The aim of the article is to present a new approach for companies with which to assess the riskiness of projects. The basis of this is the use of the new Crystal Ball software tool and the effective application of the Monte Carlo method. The article deals with the current issues of investment and financial planning, which are the basic pillars for effective management decisions with the goal of sustainability. The article has verified a methodology that allows companies to make effective investment decisions based on assessing the level of risk. For practical application, the Monte Carlo method was chosen, as it uses sensitivity analysis and simulations, which were evaluated for two types of projects. Both simulations were primarily carried out based on a deterministic approach through traditional mathematical models. Subsequently, stochastic modeling was performed using the Crystal Ball software tool. As a result of the sensitivity analysis, two tornado graphs were created, which display risk factors according to the degree of their influence on the criterion value. The output of this article is the presentation of these new approaches for financial decision-making within companies.

Keywords: financial risk; sustainability; Monte Carlo method; sensitivity analysis; investment planning

Citation: Senova, A.; Tobisova, A.; Rozenberg, R. New Approaches to Project Risk Assessment Utilizing the Monte Carlo Method. *Sustainability* **2023**, *15*, 1006. https://doi.org/10.3390/su15021006

Academic Editors: Esmaeil Zarei, Samuel Yousefi and Mohsen Omidvar

Received: 5 December 2022
Revised: 30 December 2022
Accepted: 2 January 2023
Published: 5 January 2023

1. Introduction

Creating the conditions for correct investment decisions is a key factor leading to the sustainability of businesses in the future. A systemic approach focused on the sustainability of businesses in the field of financial and investment planning can create a comprehensive view of the issue of effective managerial decision-making.

Currently, several authors are interested in and draw relationships between financialization and technological innovations, as well as analyzing the behavior of nonfinancial enterprises in financing from both a macro and micro perspective [1–3].

Risk is generally perceived as the uncertainty of future development, the uncertainty of whether the projects that the company invests in will be profitable or will make a loss. The success or, on the contrary, the failure of business projects can significantly affect the economic result of the company and, in the worst case, even the very existence of the company [4–6]. For this reason, companies should pay attention to the assessment of the risks of individual business projects before their implementation. Currently, risk management is very neglected in practice, but globalization forces our entrepreneurs to apply new methods of risk management to their businesses in order to be competitive [7,8]. Risk analysis is usually understood as a process of defining threats, the probability of their occurrence, and the impact on assets, i.e., determining risks and their severity [9–11]. Other authors [12–15] describe risk analysis as part of five basic phases of risk management: the determination of project risk factors, the determination of the significance of risk factors, the determination of project risk, the assessment of project risk and the adoption of measures to reduce it, and the preparation of a corrective action plan.

We currently know several risk management methods for each business activity or strategy. In general, we distinguish between deterministic and stochastic (probabilistic) approaches to risk measurement. Deterministic approaches assume that a certain value of one variable is assigned a certain value of the second variable. In stochastic approaches, it is assumed that a certain value of one variable corresponds with certain probabilities of different values of the other variable. Stochastic approaches incorporate variability into the risk measurement model itself by specifying a probability distribution for the random variables. In particular, the following types of probability models can be used to measure risk: models based on an expert determination of subjective probability distributions, analytical models, and simulation models [4,8,9,16].

For new business plans, the greater part of the required probability distributions of risk factors must be determined by subjective estimation based on expert evaluation. It is usually easier to determine them in the form of a discrete probability distribution for three decision variants: pessimistic, most probable, and optimistic. In the second type of probabilistic model, an analytical approach is used using standard theoretical probability distributions for the continuous and discrete variables. The result of the solution is the determination of the consequences of risk variants in the sense of determining the probability distribution of the values of the evaluation criteria for individual risk variants. The third type of probabilistic model—simulation models–is used when the problem is too complex for the use of the previous methods. The main phases of simulation studies are the definition of the problem, the creation of a simulation model, the specification of input variable parameters and their mutual relations, and the simulation and design of new experiments. Currently, the use of simulation models is associated with the application of Monte Carlo computer simulations [4].

Large portfolios of financial assets or commodities with high variability, which can significantly affect the financial stability of the company, will require more sophisticated techniques, including statistical analyses based on the value at risk and cash flow at risk models. VaR models make it possible to estimate the value of the risk in the portfolio as a maximum loss in the event that the portfolio had to be held for a fixed period with a predetermined level of significance—usually with a probability of 95% or 99% based on past experience [17].

The categorization of individual methods for risk analysis is presented in Table 1.

Table 1. Overview of risk analysis methods [16,18–20].

Group of Risk Analysis Methods	Types of Methods
Qualitatively	What-if method, scenario analyses, failure mode and consequence questionnaires, criticality analyzes (FMEA / FMECA), hazard and operability analysis (HAZOP), human error analysis (HEA), block reliability scheme, fault tree analysis (FTA), event tree analysis (ETA), probability risk analysis and safety assessment (PRA &PSA), survey questionnaires
Quantitatively	Statistical, cost and efficiency analysis, expert systems, analysis of the relative value of risk, sensitivity analyses, Monte Carlo simulations, critical point analysis; reduced standard methods, cost–benefit analysis, the Delphi method
Combined (qualitative and quantitative approaches)	Fault tree analysis, the Delphi method, value chain analysis

Table 1. *Cont.*

Group of Risk Analysis Methods	Types of Methods
Qualitative methodologies used in nuclear and chemical processing plants	Preliminary hazard analysis (PHA), hazard and operability analysis (HAZOP), failure mode and consequence analysis (FMEA/FMECA)
Tree techniques used to quantify the probability of occurrence of accidents and other adverse events leading to loss of life or economy	Fault tree analysis (FTA), event tree analysis (ETA), cause and effect analysis (CCA), fault tree risk management (MORT), organizational safety management by assessment technique (SMORT)
Techniques for a dynamic system	Dynamic event logic analytical method (DYLAM), dynamic event tree analytical method (DETAM), Markov model, transition method
Updated (positive) risk	Market research, prospecting, test marketing, research and development, business impact analysis
Downside risk (negative)	Approach analysis, fault tree analysis (FTA), failure mode and consequence analysis (FMEA)
Both	Dependency modeling, swot analysis (strengths, weaknesses, opportunities, threats), tree and event analysis (ETA), business continuity planning, bpest analysis (business, political, economic, social, technological), real option modeling, decision making under conditions of risk and uncertainty, statistical inference, measures of central tendency and dispersion PESTLE (political, economic, social, technological, and legal environment)
Intuitive technique	Guided discussion (brainstorming)
Inductive technique (What if?)	Preliminary hazard analysis (PHA), checklists, human error analysis (HEA), hazard and operability analysis (HAZOP), criticality failure mode and consequence analysis (FMECA)
Deductive method (so how?)	Events and fault trees

The basis of risk management is a certain systematic procedure for working with risk and uncertainty aimed at increasing the quality of project preparation and evaluation. The first three phases of risk management include determining the risk factors, determining their significance, and determining project risk [21–23]. These three phases are collectively referred to as project risk analysis. The next two phases are referred to as the project's own risk management [10,24,25]:

- The 1st stage of risk management is the determination of risk factors. The content of this phase is the determination of risk factors as quantities whose possible future development could affect the economic results, the criteria of the economic efficiency of the project (profit, return on capital, and net present value), and its financial stability;
- The 2nd stage of risk management is the determination of project risk. The importance of the risk factors can basically be determined in two ways, namely expertly or by using sensitivity analysis;
- The 3rd phase determines the risk of investment projects. Project risk can be determined numerically or indirectly. In numerical form, the risk is determined using statistical characteristics (dispersion, standard deviation, coefficient of variation), which serve as a measure of risk in financial management. Project risk is indirectly determined using certain managerial characteristics, which, in their summary, provide information on a greater or lesser degree of risk.

Hertz and Thomas [26] prescribe the content of risk analysis, which includes the analysis of input variables (resulting in the determination of the risk factors and their distribution functions), Monte Carlo simulation (generation of risk situations), and the evaluation of outputs based on the obtained probability distributions. Berkowitz [27] divides the risk analysis into two basic parts: the identification of risk factors and their impact on the value of the portfolio and a model that connects the risk factors with the observed output quantity.

Savvides [28] presents a risk analysis model, which consists of a sequence of seven basic steps, ensuring the processing of a certain number of inputs (random variables, i.e., risk factors, deterministic variables, and parameters) for the calculation of the outputs (selected criteria for evaluating business projects).

Several authors [29–31] discuss the procedure for determining the significance of risk factors in two ways, namely, the expert assessment of risk factors or sensitivity analysis.

The expert assessment of the significance of risk consists of a professional evaluation by managers who have the necessary knowledge and experience in the areas where the individual risk factors fall. The significance of the risk is assessed from two points of view. The first is the probability of the occurrence of the risk factor, and the second is the intensity of the negative impact that the occurrence of the risk factor has on the results of the project [32].

The purpose of the sensitivity analysis is to determine the sensitivity of the project's economic criterion, such as its net present value, profit, and profitability of invested funds, depending on the factors that influence this criterion. So, it means determining how certain changes in these factors, for example, changes in the volume of production, or utilization of production capacity, reflect changes in the selling prices of products, the prices of the basic raw materials, the materials and energy, the investment costs, the interest and tax rates, the exchange rates, the project lifetime, and the discount rates that affect the chosen economic criterion of the project [18,33]. For those factors in which certain changes, e.g., a deviation in the size of 10% from the most probable value, cause only a small change in this criterion, we then can consider them to have little importance because the sensitivity of the chosen criterion to changes in these factors is small.

On the contrary, those factors in which the same change causes significant changes in the chosen criterion will certainly be significant for us. The given criterion is highly sensitive to changes in these factors. However, in the case of risk factors with smaller impacts on the project's profit, it is necessary to remember that the percentage changes in profit refer to an increase in these factors by a specified percentage. However, if possible, changes in some risk factors with a small impact on profit can be significantly greater (e.g., in the case of energy prices); it is also necessary to consider such a factor as a significant risk factor. Therefore, not only the results of the sensitivity analysis but also the possible range of these factors are essential to define unimportant risk factors that can be neglected and work only with their most probable estimates [34].

The main goal of these methods is to allow those managers who are responsible for risk management to have more transparent access to information about threats and to ensure integrated risk management throughout the enterprise at the level of strategic management. In the current conditions of business uncertainty, simple deterministic models are not sufficient; we need to focus more on the use of probabilistic methods for measuring risk, which provide greater possibilities in terms of information security of decision-making processes.

In our opinion, these methods most accurately determine the extent of risks and allow investors to more easily decide on which investment project to invest in, as well as help them decide on reducing or transferring risk to another entity.

The basic shortcoming of the traditional methods for evaluating investment projects is a single-scenario approach based on an optimistic assumption of the development of the business environment. An increase in the quality of investment decision-making, in terms of respect for risk and uncertainty, can be brought about by probabilistic approaches, a significant representative of which is the Monte Carlo simulation [35].

This tool requires the identification of risk factors affecting investment projects and, thus, their evaluation criteria. The result of the application of Monte Carlo simulation is the distribution of the probability of these quantities and, subsequently, an easier decision for the investor to accept or reject investment projects based on valuable information about the size of the project's risk obtained by this method [19,36].

The Monte Carlo method originated in the 20th century. Even so, this method is currently considered one of the most advanced methods today. The wide application of this method results from its simple modification to current conditions and the usability of modern software tools. For this very reason, this method has become a multidisciplinary method used in various scientific branches, such as the field of physics and electrical engineering [37–40], chemistry [41,42], safety assessment [43], industry [44,45], the public sector [46], economics [36,47–49], and many other fields. Practice has shown that the use of the Monte Carlo method leads to a significant reduction in variance but, above all, to a reduction in computing time [50,51].

The goal of our contribution is to apply Crystal Ball software tools and Monte Carlo simulation in the evaluation of investment projects, which creates prerequisites for expanding the applied use of simulation software tools in risk management in practice. The article is aimed at solving the issue of financing the investment activities of companies in order to decide on a more effective project. The modeling process was based on the evaluation of the economic efficiency of the investment and a decision about which of the two projects is more advantageous and less risky in terms of future sustainability.

The secondary goal was to integrate the use of new classical and modern economic-statistical methods, which are an effective tools for the sustainability of businesses [1,3,52]. The application verification was based on the methodology presented by us in our published article [19]. The methodology shows two approaches to eliminating risk in enterprises in Slovakia. The first approach represents the modeling of financial risks using the principles of financial mathematics in order to optimize them. The second approach is stochastic modeling, which is based on the use of simulations.

The purpose of the article is to present new approaches to assessing the riskiness of projects and investment decisions. At the same time, the aim of the article is to verify, using a practical example, the methodology created by us aimed at achieving the sustainability of businesses in the territory of the Slovak Republic. The problem is primarily that businesses in the territory of the Slovak Republic use traditional and outdated methods that do not take risks and the factor of time into account in decision-making processes and in the processes of assessing projects and investments. The purpose of this contribution is to provide guidance for these companies on how to integrate new modern approaches into decision-making processes. The article applies the methodology of assessing project and investment decisions to the environment of a real company with the aim of introducing new software tools to companies that will facilitate the decision-making processes of the company's management and, thus, make the decision-making about the future investments of these companies more efficient.

Despite the wide applicability of the Monte Carlo method in published studies, there is no guide for the simple integration of this method into decision-making processes in companies. A methodology was therefore created for the conditions of companies in the territory of the Slovak Republic, which provides simple instructions for companies on how to integrate new approaches in the form of the Monte Carlo method into their internal processes.

The use of the Monte Carlo method through the software environment creates space for companies to implement simulations that integrate risk assessment, especially when taking time into account. The businesses will obtain a realistic idea of the future development of their investments. The main advantage of the methodology is the fact that the introduction of such an approach for companies in the conditions of the Slovak Republic does not represent high initial investments and will contribute to their sustainability.

2. Materials and Methods

The article deals with the issue of investment decision-making in enterprises in the territory of the Slovak Republic. The basic principle of the article is the verification of the methodology that was presented in the authors' previous publications [19]. The methodology is aimed at solving the investment decisions of the company when implementing modern software tools. Several companies operating in the territory of the Slovak Republic were chosen to verify the methodology. To fulfil the objective of the presented article, the article presents the outputs obtained from the methodology verification process within the company, which acts as a partner company ensuring security in transport sector companies, such as airports and transport companies. We will not name the company due to GDPR. Among other things, the analyzed company provides a number of products for companies in the transport sector that are essential as part of a security solution. The list of products is shown in Figure 1.

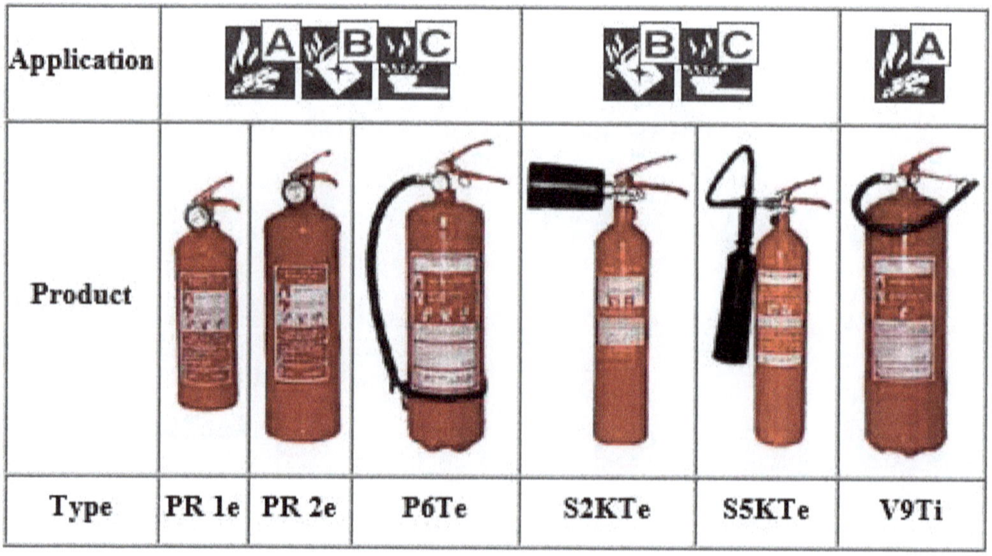

Figure 1. Products of the analyzed company.

The analyzed company was forced to make a decision in 2022 to modernize their technological procedures in production manufacturing. The company considered purchasing two types of lines:

- A project: the purchase of a new sheet metal ringer SIHR 6/3, 2050 × 6 mm. The amount of this investment is EUR 47,422.08;
- B project: the purchase of a new welding machine, amounting to EUR 88,000.

For research purposes, the lifetime of both devices was 12 years in the company's accounting records. The introduction of full automation brings with it an increase in production in direct proportion to the requested quantity, a reduction in labor costs, and a reduction in nondelivery. However, an increase in the variable costs associated with energy consumption is also expected.

It is focused on the use of the Monte Carlo method applied through the Crystal Ball software tool in the MS Excel environment. The sequence of steps is shown in Figure 2. As the algorithm of the methodology shows, the first step is to develop mathematical apparatus, which was processed in the MS Excel environment. The mathematical apparatus represents the modeling of deterministic variables that do not take into account changes in time. The basic monitored value was the profit.

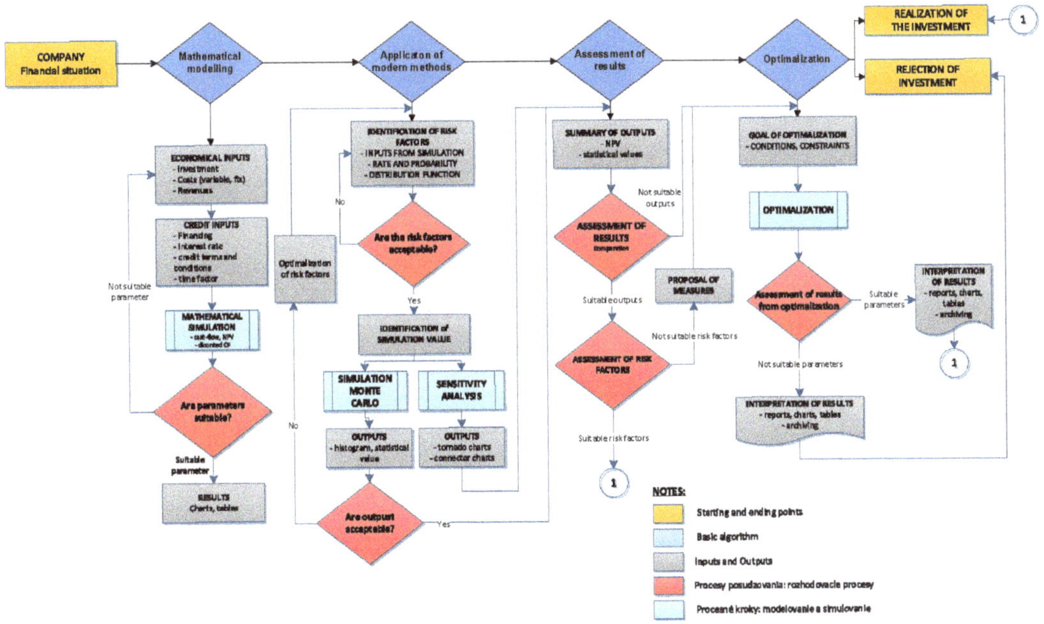

Figure 2. The assessment methodology algorithm for the investment decisions [19].

The following relations have been used in the calculation:

1. Depreciation: The company primarily uses linear depreciation, and this has also been modeled for the purpose of verifying the methodology, while the value of such depreciation is expressed by the following relationship:

$$Depreciation = \frac{\text{Asset entry price}}{\text{Period of depriciation}} \quad (1)$$

2. The value of operating costs has been calculated according to the following relationship:

$$Operating\ cost = \sum(\text{DC} + \text{IC} + \text{D} + \text{OC}) \quad (2)$$

where DC Direct cost; IC Indirect cost; D Depreciations; OC Other costs.

3. Revenues are calculated using the following relationship:

$$R = \sum_{i=0}^{n}(P + S) \quad (3)$$

where R Revenue; P Price; S Sale (quantity of sales).

4. The financial risk assessment model also took into account the tax burden in the form of income tax calculation. According to § 15 letter (b) of the Income Tax Act, the corporate income tax rate in Slovakia is 21% and is calculated from the tax base after the deduction of the tax loss [53]. The tax base is calculated according to this relationship:

$$Taxbase = \sum earlytaxbase - partof\,\text{non} - \text{taxable tax base}, \quad (4)$$

5. Profit after tax is calculated according to the relationship:

$$EAT = EBT - incometaxforordinaryactivity - incometaxforextraordinaryactivity, \quad (5)$$

where EAT earnings after taxes; EBT earnings before taxes.

In order to perform the necessary analyses, defining the basic parameters of the Monte Carlo simulation was required. The criterion value that has been assessed is profit before tax (EBT). Fixed costs, variable costs, sales price, and production are considered to be risk values (given that risk mapping has shown that they are the riskiest financial risks).

3. Results

3.1. Risk Mapping

As part of the risk mapping, a risk factor assessment matrix has been created. The matrix is based on an expert risk assessment. The essence of the expert assessment of a risk's significance when using risk assessment matrices is that this significance is assessed by two aspects. First of all, the probability of the occurrence of the risk was defined, and then the intensity of the negative impact that the occurrence of the risk had on the company was assessed.

The significance of the risk was assessed on the basis of a higher probability of occurrence and the higher intensity of the negative impact of this risk on the company. The output is a semiquantitative assessment of the significance of the company's risks based on the risk assessment matrix or its graphic display. The resulting risk assessment matrix is shown in Figure 3.

Probability	Intensity				
	Very small	Small	Medium	High	Very high
Very high		Human factor	Legal risks	Technological risks	Financial, exchange rate and market risks
High		Political risks		Economical and social risk	Production risks
Medium			Technical risks		Business risks
Small					
Very small					

Figure 3. Matrix of risk.

The risk matrix interprets a graphical representation of the probability of occurrence of a risk and its intensity. The significance of the impact of the risk is shown by a color scale: red, orange, and green. The risks that are the highest for the company are marked in red. On the contrary, the least risks are those marked in green. The yellow color indicates the risks with a medium level of riskiness. From the risk matrix, it can be stated that red risks are unacceptable for the company, and the company must immediately minimize them. The orange risks are temporarily acceptable risks, which require the clean implementation of measures within the company. The green risks are acceptable risks and do not require immediate action. It is clear from the elaborated risk matrix that financial risks are considered the riskiest for the company. For this reason, a profit was set for the criterion value in the simulations.

3.2. Sensitivity Analysis in the Simulation Model

The software tool Crystal ball, which was used for the Monte Carlo simulation, enables a sensitivity analysis to be performed through a tornado plot and a spider plot. The goal of this analysis was to get a basic idea of the impact of individual risk factors on the criterion value: profit and cash flow, and thus also a kind of control, whether the impact makes sense and whether there is, by chance, an error in the model. The principle of this analysis is that the resulting values of the criterion value are calculated based on the selection of the values from the predefined intervals of the possible values of the risk factors.

The output of the analysis is a tornado graph, which displays the individual risk factors in descending order according to the degree of their influence on the criterion value. The degree of influence is calculated according to the resulting values that the criterion variable achieves in the values of the considered risk factor intervals. For the needs of the sensitivity analysis in the simulation environment, the quantiles of 10% and 90% were chosen. Even in this case, the influence of only one risk factor is always considered without taking into account the simultaneous influence of other risk factors. The tornado graphs for both monitored projects—the A project and the B project—are shown in Figures 4 and 5.

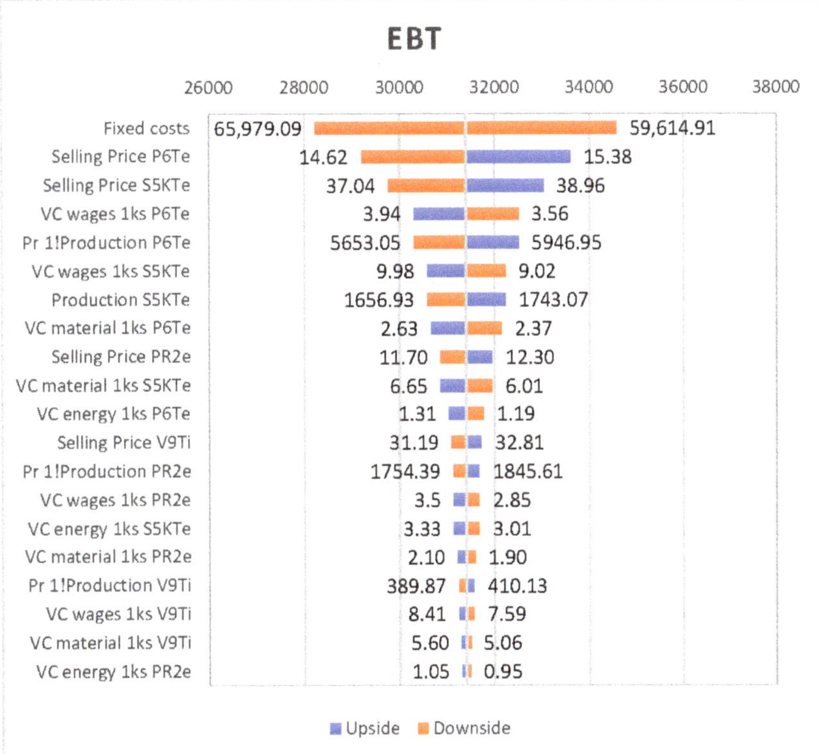

Figure 4. Tornado graph of the A project.

As can be seen from both graphs, the main risk factors are the fixed costs and the selling price of the P6Te product. The figures show that the 10% quantile of the risk factor in the form of the fixed costs in project A has a value of EUR 59,614.91, and in the B project, EUR 177,866.91. Subsequently, the 90% quantile reaches a value of EUR 65,979.09 and a value of EUR 196,855.09 in the B project for the fixed costs in the A project. It follows from the above that the range of values of the criterion value is the highest between the 10% and 90% quantile of the considered fixed costs. This means that if the fixed costs of the A project are only 10%, the value of the profit will be EUR 34,585.09. This can interpret the other values from the tornado charts of both projects in the same way.

The spider chart is also part of this analysis. The principle of this graph is practically identical to that of the tornado graph, with the difference that the resulting values of the criterion value are monitored not only in the interval values of the risk factors, but also between them. The spider charts of both projects are shown in Figures 6 and 7.

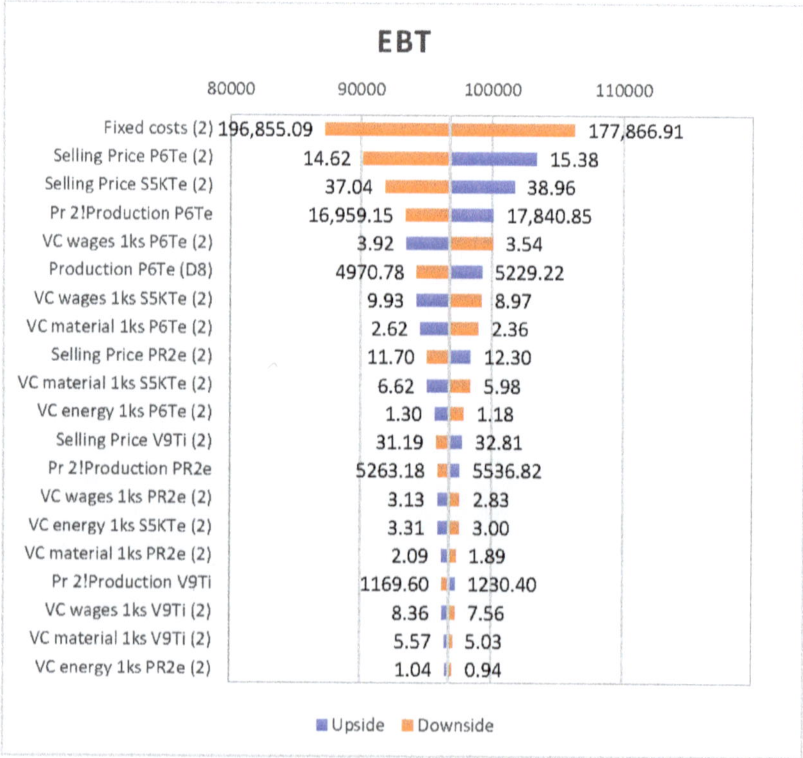

Figure 5. Tornado graph of the B project.

The spider chart shows the degree of influence of the risk factors using the slope of the lines. The advantage of this graph compared to the tornado graph is that it can also capture the possible nonlinear influence of the risk factor in the observed quantile interface precisely because the recalculation of the criterion value is carried out at several points from the interval of the possible values of the risk factor and not just from two. Additionally, in this case, the results of both charts confirmed the results obtained from the tornado charts.

3.3. Monte Carlo Simulation

If the behavior of the model seems "reasonable", it is possible to proceed to the Monte Carlo simulation itself in the Crystal Ball software environment. Setting the number of simulation steps is important when starting the simulation. For the needs of the simulation in the analyzed company, the number of simulation steps was set to 10,000, which means that a total of 10,000 values were generated within the simulation for each of the risk factors, for which, of course, 10,000 values were also obtained for each criterion quantity.

The primary result of the Monte Carlo simulation is the frequency histogram of the criterion variable and its automatic recalculation—normalization of the probability distribution. This fact enables the calculation of a whole range of statistical data. The main meaning of the number/probability distribution from the point of view of risk analysis is the overall view of the possible values of the criterion quantity and their number/probability. The results of the Monte Carlo simulation and the statistical analysis of the selected company for the A project are shown in Figure 8, and for the B project, in Figure 9.

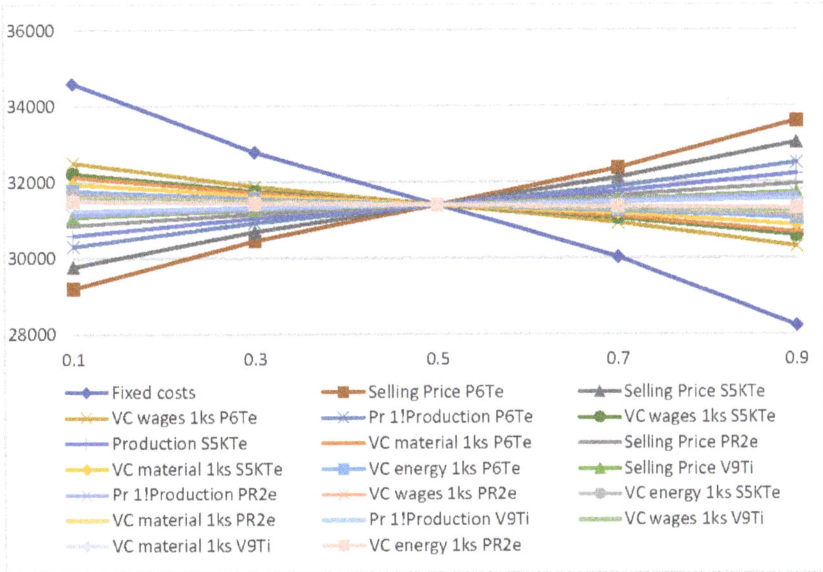

Figure 6. Spider chart of the A project.

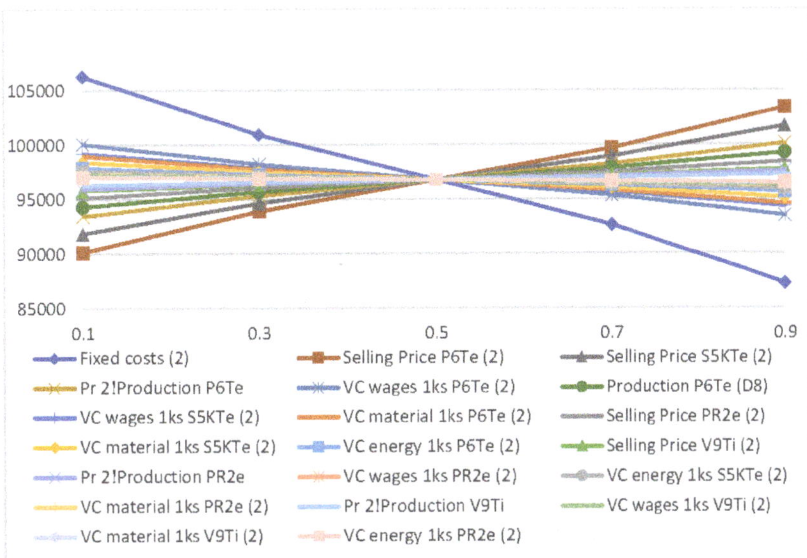

Figure 7. Spider chart of the B project.

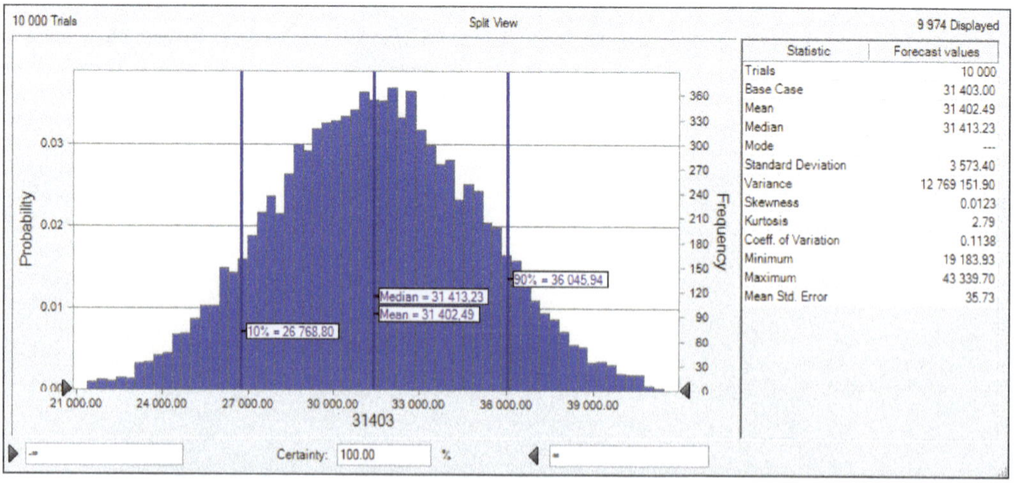

Figure 8. Probability/numerical distribution of profit for the A project.

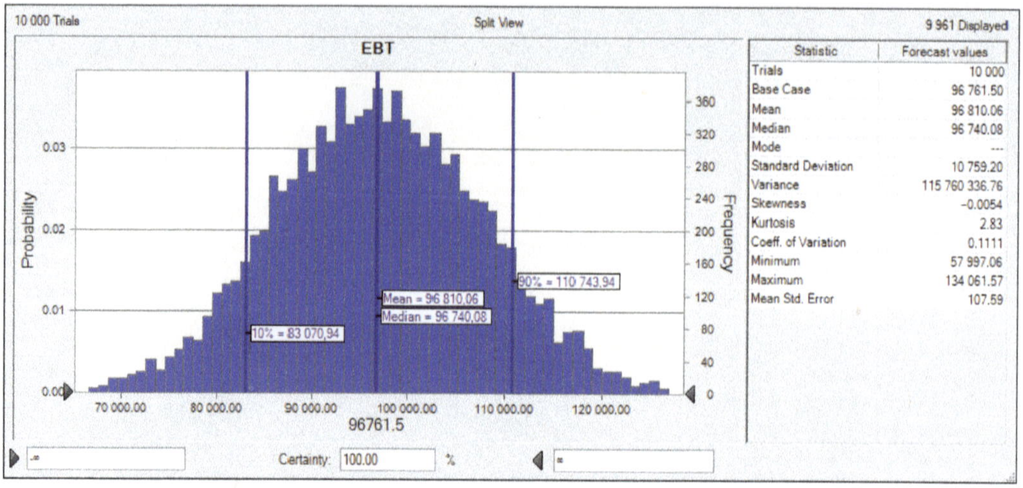

Figure 9. Probability/numerical distribution of profit for the B project.

Both graphs show that the distribution for both projects is symmetrical according to the mean value and the probability. At the same time, it follows from both graphs that in the case of the A project and the B project, the company will achieve a positive value for the criterion value with a 100% probability, i.e., profit.

Another important analysis was obtained using the Monte Carlo simulation: the Monte Carlo sensitivity analyses. It should be noted that although these results are similarly interpreted as per the classic sensitivity analyses mentioned above, the sensitivity analysis using Monte Carlo simulation is based on a completely different principle. This means that individual risk factors are analyzed from the point of view of their contribution to the total variance of the distribution of the criterion quantity. The graphic outputs of these analyses for the A project are shown in Figure 10, and for the B project, in Figure 11.

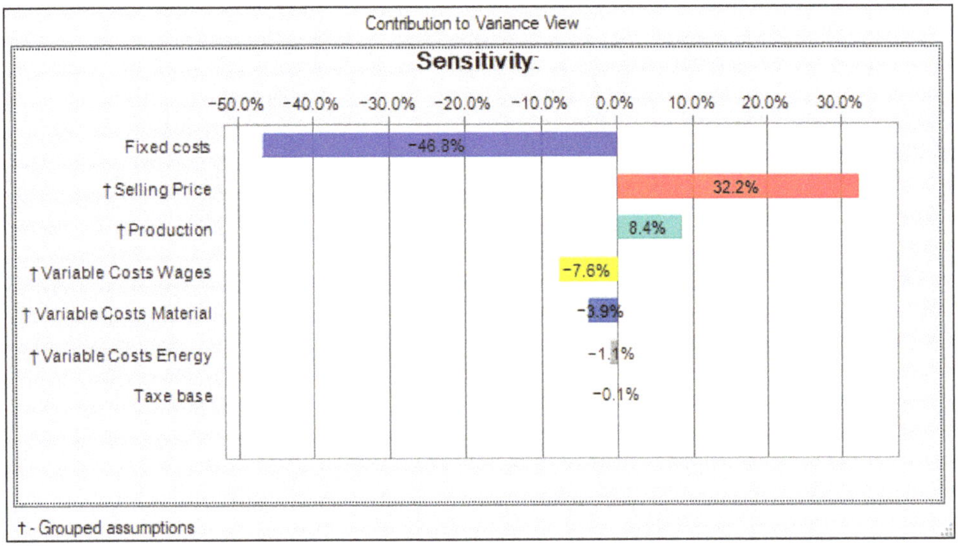

Figure 10. Profit sensitivity analysis—Monte Carlo Simulation in the A project.

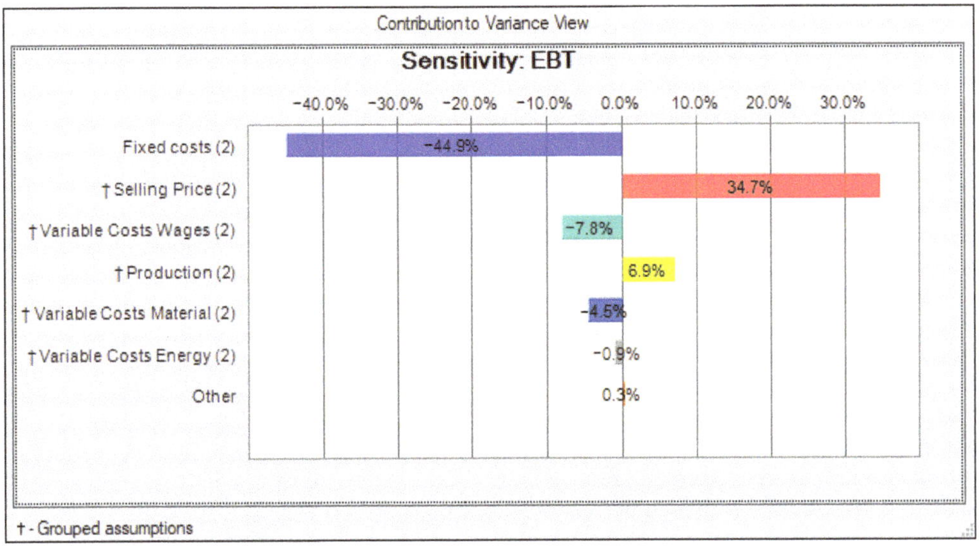

Figure 11. Profit sensitivity analysis—Monte Carlo Simulation in the B project.

Crystal Ball calculates the sensitivity by computing the rank correlation coefficients between every assumption and every forecast while the simulation is running. Correlation coefficients provide a meaningful measure of the degree to which assumptions and forecasts change together. If an assumption and a forecast have a high correlation coefficient, it means that the assumption has a significant impact on the forecast (both through its uncertainty and its model sensitivity). Positive coefficients indicate that an increase in the assumption is associated with an increase in the forecast. Negative coefficients imply the opposite situation. The larger the absolute value of the correlation coefficient, the stronger the relationship. It is important to note that the "Contribution To Variance" method is only an approximation and is not precisely a variance decomposition. Crystal

Ball calculates Contribution To Variance by squaring the rank correlation coefficients and normalizing them to 100%. Both the alternate "Rank Correlation View" and the Contribution To Variance view display the direction of each assumption's relationship to the target forecast. Assumptions with a positive relationship have bars on the right side of the zero line; assumptions with a negative relationship have bars on the left side of the zero line [54].

The influence of risk factors on the criterion value described in this way is very illustrative and can be shared mainly by laymen. However, from an analytical point of view, it is necessary to bear in mind that this is a derived and not completely accurate calculation. The principle of this sensitivity analysis is a rank correlation, within which the values of individual risk factors are generated, and the resulting criterion values are calculated. This is a kind of contribution to the variance based on squaring the rank correlation values and normalizing them to 100%. Subsequently, all the generated values are ranked, and the degree of rank correlation between the risk factors and criterion variables is calculated. In this way, the influence of individual risk factors on the criterion value is proven through the correlation value while simultaneously including the influence of all the other variables.

Despite the fact that a problem may arise when comparing both sensitivity analyses, in the case of the A project and the B project, the results are uniform in the identification of the riskiest factors, i.e., the fixed costs and selling price.

4. Discussion

Applying risk analysis to financial and investment decision-making is not easy due to the fundamental differences between deterministic and probabilistic approaches. Important barriers to successful implementation include, above all, the fact that it requires a change in thinking and a change in the traditional, long-established system processes for decision-making, and it is necessary to overcome resistance to changes.

An important limiting factor within sensitivity analysis in a simulation environment is that it analyzes the impact of individual risk factors in isolation, i.e., without including the dependencies between risk factors. Therefore, there is a danger arising from the exclusion of one of the risk factors, which, based on this sensitivity analysis, appears to be insignificant due to the neglect of its influence in connection with another risk factor. However, if we summarize the conclusions from the sensitivity analysis in the simulation environment, whether in the form of a tornado or spider web graph, it is significant mainly because of the following reasons:

1. A certain first visual check of the consistency of the relationships between the risk factors and the criterion value;
2. Evaluation of the significance of individual assumed risk factors in relation to the criterion value and a compilation of a certain possible list of risk factors that are unlikely to be important for further analyses;
3. Detection of the possible nonlinear relationships between risk factors and the criterion value.

The sensitivity analysis is a relatively complex method, which is the result of two influences:

1. The sensitivity of the model—in general, the sensitivity of the criterion quantity is to the risk factor, which results from the relationships defined in the mathematical model, e.g., how the criterion value changes when the value of the risk factor changes by 1%;
2. Uncertainty of risk factor values—possible values the risk factor can reach.

If the sensitivity of the model is high, even small changes in the values of the risk factors will lead to significant changes in the resulting criterion value. On the contrary, if the sensitivity of the model is relatively small, even with larger deviations in the values of the risk factors, significant changes in the criterion value may not occur.

As the sensitivity analysis showed, fixed costs and selling prices can be considered the riskiest factors. The correctness of the methodology was also confirmed by the fact that both sensitivity analyses—classical (in the simulation environment) and sensitivity analysis (in the Monte Carlo method)—demonstrated the significance of the same risk factors for the criterion variable EBT.

The core of the presented methodology is the Monte Carlo method. Monte Carlo simulation requires much more complex analysis than traditional deterministic models. The objective of the verifiability of the methodology was the assessment of the profitability of the projects in the selected company. The probability of project implementation within the given time limit is determined after completing the total number of cycles. The statistical metrics derived from these iterations are useful for determining the resulting decision for the success of the project [55,56]. Monte Carlo simulation involves choosing a statistical distribution representing the risk factor, which, in our case, is the duration of each activity, and then running a large number of iterations, creating the same number of different schedules for the project and calculating its total duration [57].

In order to assess the profitability of the projects, the profit output values and statistical indicators were obtained through Monte Carlo simulation. A comparison of the outputted statistical indicators is presented in Table 2.

Table 2. Comparison of the A project and the B project statistics.

Statistic	A Project	B Project
Base Case	31,403.00	96,761.50
Mean	31,402.49	96,810.06
Median	31,413.23	96,740.08
Standard Deviation	3573.40	10,759.20
Variance	12,769,151.90	115,760,336.76
Skewness	0.0123	−0.0054
Kurosis	2.79	2.83
Coefficient of Variability	0.1138	0.1111
Minimum	19,183.93	57,997.06
Maximum	43,339.7	139,061.57

The most interesting value is the difference between the mean value and the median, which is given by the skewness of the distribution. The distribution of the B project is skewed to the disadvantage of the company to the left (skewness is negative), i.e., the probability of significant negative profit values is greater than the analogous probability of positive values. In the case of the B project, the difference between the minimum and maximum values generated by the simulation is significant.

When deciding on two projects, the following characteristics were applied:

- If two projects have the same average value of expected revenues, the project with a lower standard deviation is preferred;
- If two projects have the same standard deviation, the project with a higher average value of expected revenues is preferred;
- In each project, a higher mean value and a lower standard deviation are preferred;
- If the project has a higher mean value and a lower deviation than all the other projects, it is optimal;
- If the projects have a different mean value and a different deviation, the project with a lower coefficient of variation is preferred.

On the basis of the above-mentioned findings, it can be concluded that the A project is a more advantageous and less risky project for the analyzed company.

Many companies today rely on well-known traditional methods for decision-making processes. However, in order for the decisions of the company's management to be effective, it is necessary that they take into account individual risks and provide management with information about developments over time. For this reason, it is necessary to imply new approaches not only in decision-making processes but also in the system procedures of individual companies.

In professional contributions, it is possible to find studies dealing with the application of the Monte Carlo method in partial calculations or in the solution of partial problems. Despite the multidisciplinary nature and wide applicability of the Monte Carlo method, there is no study that could provide guidance to companies on how to imply this method in decision-making processes. The research carried out enabled the creation of a methodology that integrates this method into decision-making processes in companies in the transport sector in the territory of the Slovak Republic. At the same time, the article demonstrated the applicability of such an approach in practice. The application of this approach in enterprises in the territory of the Slovak Republic, thus, becomes unique.

However, the methodology is limited by the conditions of the market environment of the companies in the territory of the Slovak Republic. It is primarily about the legislative conditions or the financial and educational possibilities of individual companies. However, with sufficient knowledge of the Monte Carlo method, its wide applicability provides scope for use in other types of businesses as well. However, the feasibility of such an approach needs to be subjected to future research.

5. Conclusions

Our methodology for evaluating investment projects was focused on solving the financing of investment activities in transport companies, where simulations and calculations in the MS Excel software environment were chosen as a tool to achieve this goal. The simulation tool used was the Crystal Ball simulation software, which is based on the Monte Carlo method. As part of the verification of the methodology, two approaches that focused on the analysis and evaluation of financial risks of investment projects were implemented. In order to fulfill the goal of the article, deterministic calculations were used to assess the riskiness of two projects using mathematical apparatus based on the principle of financial mathematics. The resulting ranking was used to assign an uncertainty to activity duration and estimate the probability of a project being completed on time, employing the Monte Carlo simulation approach. The main contribution of this article is the development of an innovative framework that co-ordinates an established qualitative and quantitative risk classification approach with a powerful simulation approach to effectively predict time deviations while executing complex projects under uncertainty [55,56]. The integration of new software tools into investment decisions is represented by the simulations of the Monte Carlo method based on the stochastic approach in the Crystal Ball software environment. The simulation is based on the modeling of the criterion value in the form of profit, taking into account risk factors defined as the distribution functions of input variables. The application of such an approach to managerial decision-making when assessing investment projects is unknown in Slovak companies and thus becomes unique. The uniqueness of the project assessment lies in the integration of various multicriteria approaches. The outputs of the article form part of the research into the VEGA project, which verifies the methodology on a sample of 100 enterprises in the transport sector in the Slovak Republic. The transport industry is an investment-intensive industry, and the question of how to mitigate risks in this sector is currently being discussed intensively. This article presents the verification of the effective assessment of the investment projects of enterprises. The goal is to ensure the sustainability of businesses based on the integration of new approaches to managerial decision-making. The application of probabilistic approaches in financial decision-making is negatively affected, mainly by a lack of the necessary knowledge or the weak support of sophisticated computer methods in the practice of companies. It is, therefore, necessary for companies, in their future research, to focus attention on the education of managers and the use of sophisticated modern tools for managing the risk of business projects. The result of such an effort should be a gradual change in corporate culture that supports expert work with risk. The possibility of applying the procedure in specific Slovak companies can be considered a practical contribution of the article. The proposals presented in the thesis form a system of solutions and are applicable under certain conditions in the practice of other industrial enterprises through the selected selection of individual methods and models by supplementing, replacing, or expanding with other specific characteristics and processes, according to a specific type of industry [52,58]. The basis of this will be the ability of colleges, universities, and scientific and research institutions to transmit the widest possible spectrum of the latest knowledge and findings in the field of risk management, with the aim of creating a platform for business practice for further development in this area. It is possible to state that, even at present, many of the methods that are defined have shortcomings and errors, which are pointed out by several authors dealing with this issue. These shortcomings often limit the application of these models in the practice of the companies themselves [2,3]. Therefore, it is advisable for every expert, evaluator, and risk manager to use not only the results of a risk analysis but to use several methods for such an evaluation at the same time and draw conclusions from their results that will bring them objective, more correct results. The implementation of the methods and models built in this way will enable Slovak companies, as well as other companies in the European region, to create space for the further rationalization and streamlining of business processes, increasing economic efficiency and performance and establishing their own business strategies for the future. At the same time, such methods of risk analysis could be an impetus (mainly for medium-sized enterprises) for the application of not only traditional, already proven methods but also modern researched methods and approaches, which will bring them a new perspective on the field of risk management and the possible complete elimination of risks, from which they will start their business development potential.

Author Contributions: Conceptualization, A.S., A.T. and R.R.; methodology, A.S., A.T. and R.R.; software, A.S., A.T. and R.R.; validation, A.S., A.T. and R.R.; formal analysis, A.S., A.T. and R.R.; investigation, A.S., A.T. and R.R.; resources, A.S., A.T. and R.R.; data curation, A.S., A.T. and R.R.; writing—original draft preparation, A.S., A.T. and R.R.; writing—review and editing, A.S., A.T. and R.R.; visualization, A.S., A.T. and R.R.; supervision, A.S., A.T. and R.R.; project administration, A.S., A.T. and R.R. All authors have read and agreed to the published version of the manuscript.

Funding: This research received no external funding.

Institutional Review Board Statement: Not applicable.

Informed Consent Statement: Not applicable.

Data Availability Statement: Not applicable.

Acknowledgments: The article was solved within the project VEGA, grant number 1/0770/22.

Conflicts of Interest: The authors declare no conflict of interest.

References

1. Liu, Y.; Failler, P.; Ding, Y. Enterprise financialization and technological innovation: Mechanism and heterogeneity. *PLoS ONE* **2022**, *17*, e0275461. [CrossRef] [PubMed]
2. Li, Z.; Yang, C.; Huang, Z. How does the fintech sector react to signals from central bank digital currencies? *Finance Res. Lett.* **2022**, *50*, 103308. [CrossRef]
3. Tanaka, H.; Tanaka, C. Sustainable investment strategies and a theoretical approach of multi-stakeholder communities. *Green Financ.* **2022**, *4*, 329–346. [CrossRef]
4. Fabianova, J.; Kacmary, P.; Janekova, J. Operative production planning utilising quantitative forecasting and Monte Carlo simulations. *Open Eng.* **2019**, *9*, 613–622. [CrossRef]
5. Vochozka, M.; Rowland, Z.; Vrbka, J. Financial analysis of an average transport company in the Czech Republic. *Our Sea* **2016**, *63*, 227–236. [CrossRef]
6. Feng, Z. Financial risks from three dimensions and risk identification model of enterprise. *Int. J. Manag. Sci. Eng. Manag.* **2011**, *6*, 71–80. [CrossRef]
7. Scholleova, H. *Investment Controlling: How to Evaluate Investment Intentions and Manage Company Investments*, 1st ed.; Grada: Praha, Czech Republic, 2009.
8. Janekova, J.; Fabianova, J.; Rosova, A. Environmental and economic aspects in decision making of the investment project "wind park". *Pol. J. Manag. Stud.* **2016**, *13*, 90–100. [CrossRef]
9. Smejkal, V.; Rais, K. *Řízení rizik ve firmách a jiných organizacích*; Grada: Praha, Czech Republic, 2010.

10. Fotr, J.; Souček, I. *Podnikatelský Zámer a Investiční Rozhodování*; Grada Publishing: Praha, Czech Republic, 2005; 356p.
11. Bartošová, V.; Majerčák, P.; Hrašková, D. Taking Risk into Account in the Evaluation of Economic Efficiency of Investment Projects: Traditional Methods. *Procedia Econ. Financ.* **2015**, *24*, 68–75. [CrossRef]
12. Kim, K.-J.; Lee, K.; Ahn, H. Predicting Corporate Financial Sustainability Using Novel Business Analytics. *Sustainability* **2018**, *11*, 64. [CrossRef]
13. Nunes, M.; Abreu, A.; Saraiva, C. A Model to Manage Cooperative Project Risks to Create Knowledge and Drive Sustainable Business. *Sustainability* **2021**, *13*, 5798. [CrossRef]
14. Larionov, A.; Nezhnikova, E.; Smirnova, E. Risk Assessment Models to Improve Environmental Safety in the Field of the Economy and Organization of Construction: A Case Study of Russia. *Sustainability* **2021**, *13*, 13539. [CrossRef]
15. Chapman, C.; Ward, S. *Project Risk Management. Processes, Techniques and Insights*; John Wiley & Sons: Hoboken, NJ, USA, 1997.
16. Bedford, T.; Cooke, R.M. *Probabilistic Risk Analysis: Foundations and Methods*; Cambridge University Press: Cambridge, UK, 2001.
17. Value at Risk (VAR). Available online: https://www.risk.net/definition/value-at-risk-var (accessed on 29 December 2022).
18. Deloach, J.W.; Temple, N. *Enterprise-Wide Risk Management: Strategies for Linking Risk & Opportunity*; Financial Times Management: London, UK, 2000.
19. Tobisova, A.; Senova, A.; Rozenberg, R. Model for Sustainable Financial Planning and Investment Financing Using Monte Carlo Method. *Sustainability* **2022**, *14*, 8785. [CrossRef]
20. Carbone, T.A.; Tippett, D.D. Project Risk Management Using the Project Risk FMEA. *Eng. Manag. J.* **2004**, *16*, 28–35. [CrossRef]
21. Qin, H.; Stewart, M.G. Risk perceptions and economic incentives for mitigating windstorm damage to housing. *Civ. Eng. Environ. Syst.* **2020**, *38*, 1–19. [CrossRef]
22. Valaskova, K.; Kliestik, T.; Svabova, L.; Adamko, P. Financial Risk Measurement and Prediction Modelling for Sustainable Development of Business Entities Using Regression Analysis. *Sustainability* **2018**, *10*, 2144. [CrossRef]
23. Vicig, P. Financial risk measurement with imprecise probabilities. *Int. J. Approx. Reason.* **2008**, *49*, 159–174. [CrossRef]
24. Sidorova, E.; Kostyukhin, Y.; Korshunova, L.; Ulyanova, S.; Shinkevich, A.; Ershova, I.; Dyrdonova, A. Forming a Risk Management System Based on the Process Approach in the Conditions of Economic Transformation. *Risks* **2022**, *10*, 95. [CrossRef]
25. Bahamid, A.R.; Doh, I.S. A review of risk management process in construction projects of developing countries. *IOP Conf. Series: Mater. Sci. Eng.* **2017**, *271*, 012042. [CrossRef]
26. Hertz, D.B.; Thomas, H. *Risk Analysis and Its Applications*; John Wiley & Sons: Hoboken, NJ, USA, 1983.
27. Berkowitz, J.A. Coherent Framenwork for Stress–Testing. *J. Risk* **2000**, *2*, 1–15. [CrossRef]
28. Savvides, S. Risk analysis in investment appraisal. *Proj. Apprais.* **1994**, *9*, 3–18. [CrossRef]
29. Razavi, S.; Jakeman, A.; Saltelli, A.; Prieur, C.; Iooss, B.; Borgonovo, E.; Plischke, E.; Piano, S.L.; Iwanaga, T.; Becker, W.; et al. The Future of Sensitivity Analysis: An essential discipline for systems modeling and policy support. *Environ. Model. Softw.* **2020**, *137*, 104954. [CrossRef]
30. Pesenti, S.M. Reverse Sensitivity Analysis for Risk Modelling. *Risks* **2022**, *10*, 141. [CrossRef]
31. Tsanakas, A.; Millossovich, P. Sensitivity Analysis Using Risk Measures. *Risk Anal.* **2015**, *36*, 30–48. [CrossRef] [PubMed]
32. Cont, R.; Deguest, R.; Scandolo, G. Robustness and sensitivity analysis of risk measurement procedures. *Quant. Financ.* **2010**, *10*, 593–606. [CrossRef]
33. Borgonovo, E.; Plischke, E. Sensitivity analysis: A review of recent advances. *Eur. J. Oper. Res.* **2016**, *248*, 869–887. [CrossRef]
34. Janekova, J.; Fabiánová, J.; Onofrejová, D.; Puškáš, E.; Buša, M. Implementation of Deviation Analysis Method in the Utilisation Phase of the Investment Project: A Case Study. *Pol. J. Manag. Stud.* **2017**, *15*, 99–109. [CrossRef]
35. Hammersley, J. *Monte Carlo Methods*; Springer Science & Business Media: Berlin/Heidelberg, Germany, 2013.
36. Züst, S.; Huonder, M.; West, S.; Stoll, O. Life-Cycle Oriented Risk Assessment Using a Monte Carlo Simulation. *Appl. Sci.* **2021**, *12*, 8. [CrossRef]
37. Di Nitto, A.; Davide, F.; Vardaci, E.; Bianco, D.; La Rana, G.; Mercogliano, D. The New Physics in LILITA_N21: An Improved Description of the Reaction 190 MeV ^{40}Ar + ^{27}Al. *Appl. Sci.* **2022**, *12*, 4107. [CrossRef]
38. Loffredo, F.; Vardaci, E.; Bianco, D.; Di Nitto, A.; Quarto, M. Protons Interaction with Nomex Target: Secondary Radiation from a Monte Carlo Simulation with Geant4. *Appl. Sci.* **2022**, *12*, 2643. [CrossRef]
39. Yang, Y.; Zhuo, W.; Zhao, Y.; Xie, T.; Wang, C.; Liu, H. Estimating Specific Patient Organ Dose for Chest CT Examinations with Monte Carlo Method. *Appl. Sci.* **2021**, *11*, 8961. [CrossRef]
40. Kaiser, W.; Popp, J.; Rinderle, M.; Albes, T.; Gagliardi, A. Generalized Kinetic Monte Carlo Framework for Organic Electronics. *Algorithms* **2018**, *11*, 37. [CrossRef]
41. Radosz, W.; Pawlik, G.; Mituś, A.C. Characterization of Monte Carlo Dynamic/Kinetic Properties of Local Structure in Bond Fluctuation Model of Polymer System. *Materials* **2021**, *14*, 4962. [CrossRef] [PubMed]
42. Pawlik, G.; Mitus, A. Photoinduced Mass Transport in Azo-Polymers in 2D: Monte Carlo Study of Polarization Effects. *Materials* **2020**, *13*, 4724. [CrossRef] [PubMed]
43. Lu, Z.; Zhuang, L.; Dong, L.; Liang, X. Model-Based Safety Analysis for the Fly-by-Wire System by Using Monte Carlo Simulation. *Processes* **2020**, *8*, 90. [CrossRef]
44. Koroteev, M.; Romanova, E.; Korovin, D.; Shevtsov, V.; Feklin, V.; Nikitin, P.; Makrushin, S.; Bublikov, K.V. Optimization of Food Industry Production Using the Monte Carlo Simulation Method: A Case Study of a Meat Processing Plant. *Informatics* **2022**, *9*, 5. [CrossRef]

45. Jones, H.G.; Lee, A.M. Monte Carlo Methods in Heavy Industry. *J. Oper. Res. Soc.* **1955**, *6*, 108–116. [CrossRef]
46. Foa, R.; Howard, M. Use of Monte Carlo Simulation for the Public Sector: An Evidence-Based Approach to Scenario Planning. *Int. J. Mark. Res.* **2006**, *48*, 27–48. [CrossRef]
47. Zhang, Y. The value of Monte Carlo model-based variance reduction technology in the pricing of financial derivatives. *PLoS ONE* **2020**, *15*, e0229737. [CrossRef]
48. Mucha, V.; Páleš, M.; Sakálová, K. Calculation of the Capital Requirement Using the Monte Carlo Simulation for Non-life Insurance. *J. Econ.* **2016**, *64*, 878–893.
49. Arenas, D.J.; Lett, L.A.; Klusaritz, H.; Teitelman, A.M. A Monte Carlo simulation approach for estimating the health and economic impact of interventions provided at a student-run clinic. *PLoS ONE* **2017**, *12*, e0189718. [CrossRef] [PubMed]
50. Ross, S.M.; Ghamami, S. Efficient Monte Carlo Barrier Option Pricing When the Underlying Security Price Follows a Jump-Diffusion Process. *J. Deriv.* **2010**, *17*, 45–52. [CrossRef]
51. Fleming, P.J.; Schaeffer, D.; Simon, B. Efficient Monte-Carlo simulation of a product-form model for a cellular system with dynamic resource sharing. *ACM Trans. Model. Comput. Simul.* **1995**, *5*, 3–21. [CrossRef]
52. Liao, G.; Hou, P.; Shen, X.; Albitar, K. The impact of economic policy uncertainty on stock returns: The role of corporate environmental responsibility engagement. *Int. J. Financ. Econ.* **2020**, *26*, 4386–4392. [CrossRef]
53. Act, No. 461/2003 Coll. on Social Insurance as Amended, Act No. 580/2004 Coll. on Health Insurance as Amended. Available online: https://www.podnikajte.sk/dan-z-prijmov/sadzby-dane-z-prijmov-fyzickych-pravnickych-osob-2019-2020 (accessed on 13 April 2022).
54. How Crystal Ball Calculates Sensitivity. Available online: https://docs.oracle.com/cd/E12825_01/epm.111/cb_user/frameset.htm?ch07s04s03.html (accessed on 29 December 2022).
55. Koulinas, G.K.; Demesouka, O.E.; Sidas, K.A.; Koulouriotis, D.E. A TOPSIS—Risk Matrix and Monte Carlo Expert System for Risk Assessment in Engineering Projects. *Sustainability* **2021**, *13*, 11277. [CrossRef]
56. An, J.; Kim, D.-K.; Lee, J.; Joo, S.-K. Least Squares Monte Carlo Simulation-Based Decision-Making Method for Photovoltaic Investment in Korea. *Sustainability* **2021**, *13*, 10613. [CrossRef]
57. Rees, M. *Business Risk ans Simulation Modelling in Practice Using Excel, VBA and @RISK*; John Wiley & Sons Ltd.: Hoboken, NJ, USA, 2015.
58. Kanupriya Indian textile sector, competitiveness, gender and the digital circular economy: A critical perspective. *Natl. Account. Rev.* **2022**, *4*, 237–250. [CrossRef]

sustainability

Article

Sustainable Food Production: An Intelligent Fault Diagnosis Framework for Analyzing the Risk of Critical Processes

Hamzeh Soltanali [1,2], Mehdi Khojastehpour [1,*], José Edmundo de Almeida e Pais [3,4] and José Torres Farinha [2,5]

1 Department of Biosystems Engineering, Ferdowsi University of Mashhad, Mashhad 9177948974, Iran; ha.soltanali@mail.um.ac.ir
2 Centre for Mechanical Engineering, Materials, and Processes (CEMMPRE), 3030-199 Coimbra, Portugal; torres.farinha@dem.uc.pt
3 CISE—Electromechatronic Systems Research Centre, University of Beira Interior, 6201-001 Covilhã, Portugal; edmundo.pais@ubi.pt
4 EIGeS—Research Centre in Industrial Engineering, Management and Sustainability, Lusófona University, Campo Grande 376, 1749-024 Lisboa, Portugal
5 Coimbra Institute of Engineering, Polytechnic of Coimbra, 3030-199 Coimbra, Portugal
* Correspondence: mkhpour@um.ac.ir; Tel.: +98-935-238-3031

Citation: Soltanali, H.; Khojastehpour, M.; Pais, J.E.d.A.e.; Farinha, J.T. Sustainable Food Production: An Intelligent Fault Diagnosis Framework for Analyzing the Risk of Critical Processes. *Sustainability* **2022**, *14*, 1083. https://doi.org/10.3390/su14031083

Academic Editors: Esmaeil Zarei, Samuel Yousefi and Mohsen Omidvar

Received: 20 December 2021
Accepted: 13 January 2022
Published: 18 January 2022

Abstract: Fault diagnosis and prognosis methods are the most useful tools for risk and reliability analysis in food processing systems. Proactive diagnosis techniques such as failure mode and effect analysis (FMEA) are important for detecting all probable failures and facilitating the risk analysis process. However, significant uncertainties exist in the classical-FMEA when it comes to ranking the risk priority numbers (RPNs) of failure modes. Such uncertainties may have an impact on the food sector's operational safety and maintenance decisions. To address these issues, this research provides a unique FMEA framework for risk analysis within an edible oil purification facility that is based on certain well-known intelligent models. Fuzzy inference systems (FIS), adaptive neuro-fuzzy inference systems (ANFIS), and support vector machine (SVM) models are among those used. The findings of the comparison of the proposed FMEA framework with the classical model revealed that intelligent strategies were more effective in ranking the RPNs of failure modes. Based on the performance criteria, it was discovered that the SVM algorithm classifies the failure modes more accurately and with fewer errors., e.g., RMSE = 7.30 and MAPE = 13.19 with that of other intelligent techniques. Hence, a sensitivity FMEA analysis based on the SVM algorithm was performed to put forward suitable maintenance actions to upgrade the reliability and safety within food processing lines.

Keywords: fault diagnosis; risk analysis; risk priority number; support vector machine; food industry; maintenance; sustainability; uncertainty

1. Introduction

With the increasing automation and development of smart technologies in modern food industries, the higher guarantee of functional safety and reliability is poised to be the major challenge towards sustainable food production [1–3]. In this context, the intelligent platforms provide the hardware and software solutions for process control and safety management within many food manufacturing systems [4,5]. They attempt to represent the novel fault diagnostic and prognostic methods for risk predicting and analysis processes [6,7]. One of the most essential parts of risk in analyzing system reliability and safety is the risk analysis procedure [8–10]. In general, the novel methods are mainly classified into the knowledge-based and data-driven approaches for risk and reliability analysis and prediction under various situations [11–13].

In such circumstances, there are many types of knowledge-based approaches that refer to fault diagnosis and risk analysis, such as fault tree analysis (FTA), hazard analysis, critical control points (HACCP), root cause analysis (RCA), etc. [14–16]. Among them, the failure mode and effect analysis (FMEA) technique is widely used in numerous industries to assess and mitigate the risk of unexpected failures [17]. Besides, it has been a well-established procedure for upgrading the production quality and reducing the severity and occurrence of failure using corrective tasks [18]. A complete FMEA dominated by experts' knowledge includes the following four main steps: identifying the failure modes, determining their causes and effects, ranking the risk of failure modes, and finally suggesting the maintenance activities for the high-risk failures [19]. A risk priority number (RPN) is frequently inserted in a traditional FMEA to evaluate the risk level of a process, rank failures, and prioritize maintenance operations [20]. The RPN value is calculated by multiplying the following three risk parameters: occurrence (O), severity (S), and detection (D). They are ranked from 1 to 10 on a discrete ordinal scale. Ultimately, by arranging the RPNs in a descending order, the most critical failures can be identified [21].

The classical-FMEA has been particularly effective in detecting system bottlenecks and assessing the risk of failure modes in food production systems. They include the possibility of having the same RPN values, failing to assess the relative importance of risk parameters, and estimating the precise value of risk parameters incorrectly. Such major fluctuations in the real situation may not only affect the accuracy of estimated risks, but also the proposed maintenance and safety functions within food processing systems [21–23]. The main objective of this study is to take such uncertainties into account, particularly when ranking the RPNs of failure modes to supplement the current classical-FMEA in the food sector. The key contribution is a new systematic FMEA framework for risk analysis procedure based on certain well-known intelligent models to overcome RPN issue classification within an edible oil purification plant. The intelligent techniques include the fuzzy inference systems (FIS), adaptive neuro-fuzzy inference systems (ANFIS), and support vector machine (SVM) models. The findings of the current study could help managers to establish practical functional safety and maintenance programs in the edible oil industry.

The remainder of this research is organized as follows: A description of the literature linked to various types of FMEA in the food sector and its associated uncertainties in the risk analysis process is included in the part "Literature review." The "Research methodology" section compares the traditional and intelligent-FMEA risk analysis methodologies to come up with an upgraded fault diagnosis framework. The "Results and Discussion" section contains the key comparison data of traditional and intelligent-FMEA risk analysis approaches, as well as how to use the results to propose appropriate maintenance tasks. Finally, the "conclusion" section is provided, along with further remarks and perspectives.

2. Literature Review

Over the years, various types of FMEA, such as process-FMEA (PFMEA), design-FMEA (DFMEA), and total-FMEA (TFMEA) have been conducted within a wide range of applications in food processing industries. Table 1 presents a summary review of the applied FMEAs in the food sector. The PFMEA is known as the main practical solution tool for analyzing various risks in food processing. For example, a PFMEA framework was performed to recognize the main critical points and analyze the risk by determining the RPN in the processing of potato chips. The results revealed that packaging, storage, potato receiving, frying, and distribution were the main critical points with the highest RPN, respectively [23]. In another study, a combined structure of PFMEA and ISO22000 was carried out on poultry slaughtering and manufacturing. In their work, the critical failure modes with high risks were identified by determining the RPN. [24]. Following this study, analyzing the risk of salmon processing has been conducted using PFMEA and its conjunction with the ISO 22000. The research findings could be beneficial for the manufacturers and their customers [25]. One of the FMEA applications is to control the quality and safety of food products. For example, the high quality of products has been a major challenge in the tea manufacturing industry. In this direction, a TFMEA model combined with the total quality management (TQM) technique was theoretically explored [26]. Following this, a FMEA structure for risk management in the confectionery industry has been designed to control system safety and quality [22]. In another work, a practical safety improvement plan for dairy product manufacturing under PFMEA analysis was suggested [27]. The results could be used by the manufacturers to produce safer dairy products. Another practical aspect of FMEA methods is its application to fault detection and optimization in food industries. For instance, the FMEA model was dedicated to allowing precise identification of food safety in verified HACCP systems. The incorporation of FMEA was verified to the procedure of the HACCP system in the bakery industry for better food safety assurance and fault detection [28]. Furthermore, a general structure of FMEA was suggested to detect the potential faults and their effects in primary food processing [29].

Table 1. A summary of literature review for FMEA applications in food industries.

Ref.	Year	Plant/ Process	Fault Diagnosis-Based Model			Maintenance Activity
			FMEA Model	Computational/ Intelligent Model	Sensitivity Analysis	
[23]	2007	Chips manufacturing plant	Classical PFMEA	-	-	-
[30]	2007	Corn curl manufacturing	Classical PFMEA	-	-	-
[25]	2008	Salmon processing and packing	Classical PFMEA	-	-	-
[24]	2009	Poultry product processing	Classical PFMEA	-	-	-
[26]	2011	Tea processing plant	Classical TFMEA	-	-	-
[22]	2012	Confectionery manufacturing	Classical PFMEA	-	-	-
[27]	2013	Dairy products manufacturing	Classical PFMEA	-	-	-
[28]	2014	Bakery critical equipment	Classical PFMEA	-	-	-
[29]	2016	General study	PFMEA	Fuzzy set theory	-	-
[31]	2017	Vegetable processing	PFMEA	Fuzzy set theory	-	-
[32]	2018	Meat production and processing	PFMEA	Fuzzy inference system	-	-
[33]	2019	General study	Classical PFMEA	-	-	-
Current study		Edible oil industry	PFMEA	Fuzzy inference system, ANFIS & SVM	✓	✓

A summary of the literature, the application of FMEAs in the food sector can be divided into several topics such as analyzing the risks, finding the critical points, improving the quality and safety, and selecting the maintenance activities. Despite the advantages of classical-FMEAs in the food industry, they have been criticized for several flaws and limitations that may affect proposed maintenance and safety decisions. The majority of epistemic uncertainties are included in the new systematic FMEA framework to improve the prior classical-FMEA in the food business. Intelligent approaches, on the other hand, have been deemed a very valuable alternative to enhance the accuracy of classical-FMEA for risk analysis under various uncertainties [34,35].

During the last few years, intelligent techniques such as support vector machine (SVM), fuzzy inference systems (FIS) and, adaptive neuro-fuzzy inference systems (ANFIS) have given great attention to modeling the FMEA and risk analysis processes. The FIS model, for example, has been used in the field of FMEA due to its software programming-based approach and its capacity to avoid cumbersome computations [19,36,37]. Currently, a comprehensive survey on the FIS-FMEA model was conducted with various rules and membership functions [MFs]. Based on the results, the combined MFs and model with a 10-class of fuzzy numbers have a higher possibility to create the larger risk cluster of failure modes [17]. Simşek and Ic [38] conducted an FMEA using a FIS model to evaluate and eliminate potential failure modes in a ready-mixed concrete plant. Their findings revealed that the fuzzy-rule-based system was effective in identifying and eliminating potential failure modes. Yucesan et al. [39] proposed a holistic FMEA approach based on a fuzzy-based Bayesian network and the best–worst method to deal with uncertain failure data. The proposed model might resolve the uncertainty in failure data and give a strong probabilistic risk analysis logic to represent the dependency between failure events in a manufacturing plant. The FUCOM and CoCoSo approaches were considered by Yousefi et al. [40] to improve the classical-FMEA technique in an unpredictable setting. Furthermore, Z-number theory was used to combine the ideas of reliability and uncertainty in evaluating the weight of risk variables. In an actual case study, the Z-FUCOM-CoCoSo approach was compared to the Fuzzy FMEA technique and a fuzzy variant of this approach. It was found that the Z-FUCOM-CoCoSo approach could provide the most feasible separation among failure modes when compared to traditional techniques. Rezaee et al. [41] presented a hybrid approach based on the Linguistic FMEA, FIS, and fuzzy data envelopment model to calculate a score for covering some RPN shortcomings and the prioritization of risks within the chemical industry. The results demonstrated that the proposed approach was very effective in prioritizing risks by taking uncertainty into account. In addition, to handle the uncertainties of classical-FMEA in other literature, the hybrid perception of fuzzy rule-based theories has been given a lot of attention [42–44].

On the other hand, the ANFIS model, with the benefits of both neural networks (NNs) and FIS principles in a single framework, has been used to reinforce the FMEA capabilities and manage the uncertainties in risk analysis [45–47]. For instance, an ANFIS model was developed to improve risk management and manage the uncertainties in risk variables. The proposed model was more convenient and efficient concerning risk management for single and clustered station facilities in transportation systems [48]. Moreover, the SVM algorithms constitute powerful regression and classification capabilities with that of FIS, neural networks (NNs), or genetic algorithms (GAs). They generally suffer from the presence of multiple local minima, structure selection problems, and overfitting issues [49–51]. Meanwhile, the SVMs have been approved as validation methods for failure mode analyses, fault detection as well as risk assessment in industrial fields [52–55].

Based on the literature, the performance comparison of such intelligent models in risk analysis, especially in food processing systems has not been previously evaluated. Hence, as the main motivation and innovation, we have contributed to proposing a new FMEA framework by intelligent techniques and comparing their outcomes with the classical model within food processing systems. In addition, given the need for monitoring the complex processes in the food sector, the proposed framework was implemented in the edible oil purification process. The outcomes were used to help the engineers to establish convenient safety and maintenance programs. Therefore, the main objective of this study is to propose a novel FMEA framework under intelligent techniques for analyzing the risks of the edible oil purification process.

3. Materials and Methods

An improved fault diagnosis framework for risk analysis with three main steps is shown in Figure 1. The first step includes process description such as main functions, potential failure modes as well as failure effects for the edible oil purification process. The main risk factors are defined in the second step using a knowledge-based approach, and the factors are then used as the main inputs of diagnostic models such as classical and intelligent-FMEAs. The multiplication and rule-based methods are used to determine the interaction of risk factors. The final step is to estimate RPN and use sensitivity analysis to investigate the impact of risk factors on RPN as well as suggest the convenient maintenance activities for the failure modes with the highest RPN. The details of the first and second steps are provided in the next sub-sections. The last one will be discussed in Section 4.

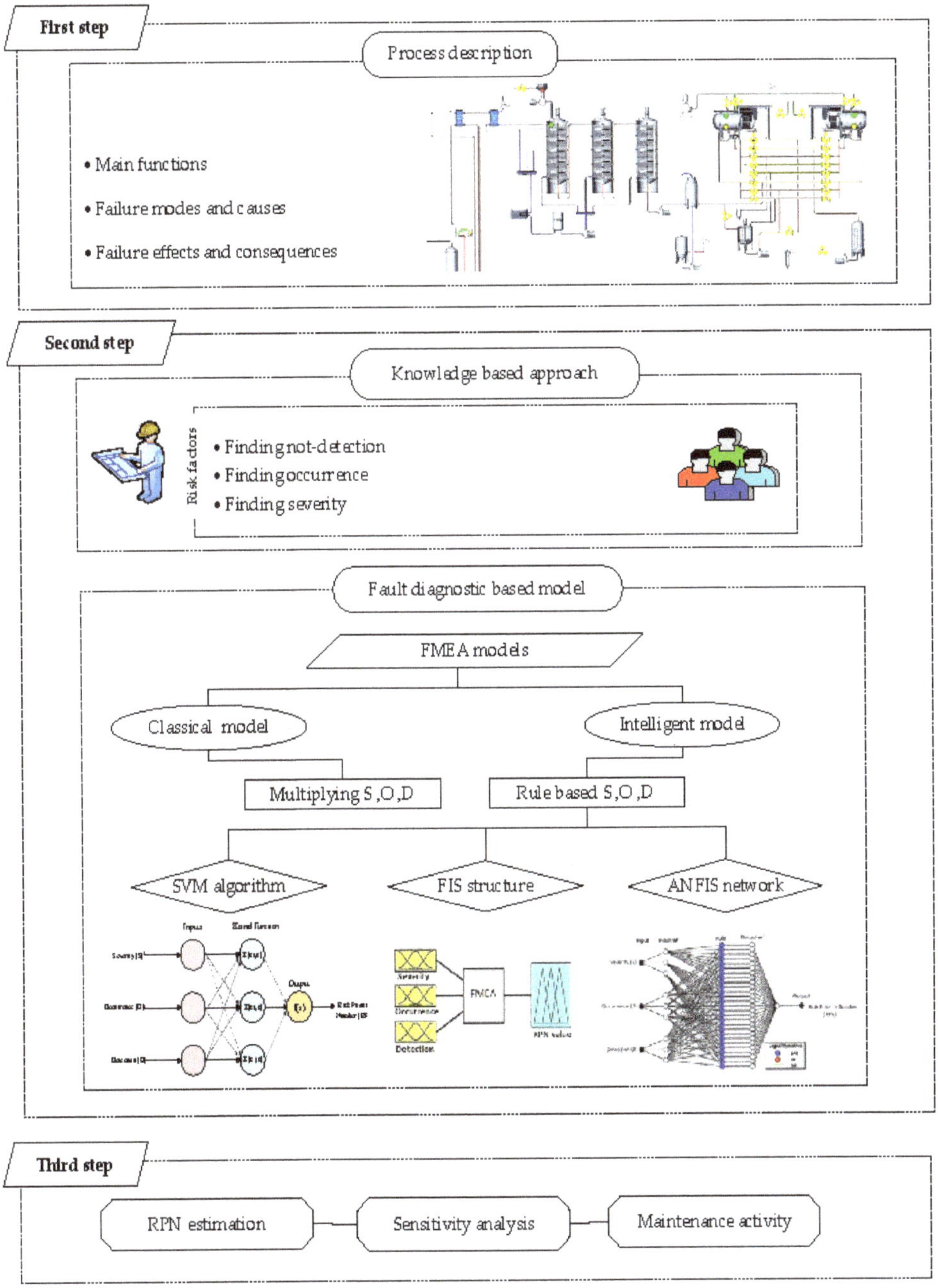

Figure 1. The proposed framework of the FMEA model.

3.1. First Step: Process Description

This study focuses on an edible oil purification facility and its processes in Iran to apply the proposed intelligent framework. Investigating the operational risk of such a process would provide a great opportunity to achieve higher reliability and safety guarantee. Figure 2 depicts the fundamental procedure for purifying the following two types of edible oils: liquid and solid. The basic stages of edible oil purification are neutralization, decolorization, winterization, deodorization, hydrogenation, and bleaching, as illustrated in Figure 2.

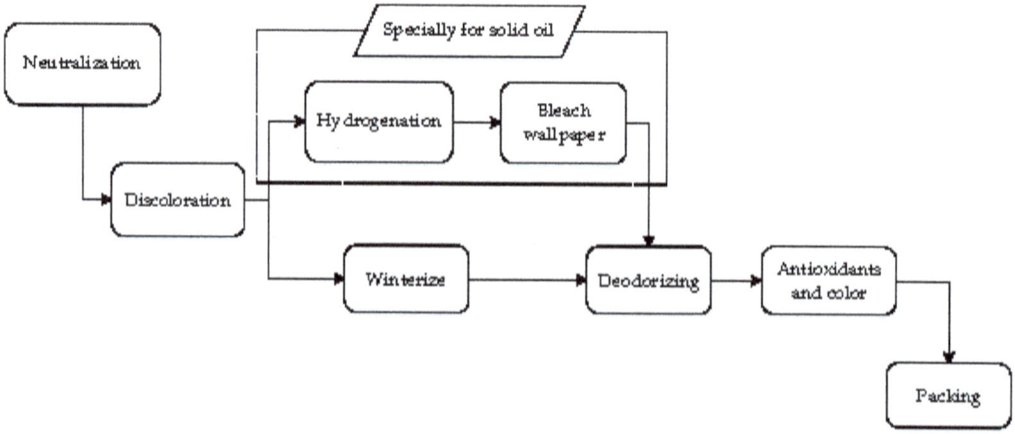

Figure 2. Production and processing of edible solid and liquid oils.

3.1.1. Neutralization Process

A process of neutralization or alkaline purification is shown in Figure 3 in which sodium hydroxide is used to react with free fatty acid to produce soap. To ensure the removal of soap and liquids, the outlet oil enters into the exchanger and is heated to 80 °C and then enters into a mixer where water is added to allow the soap to be completely discharged. Finally, the oil is inserted into a dryer to completely remove its moisture contents. The most important equipment for neutralization operations include separators, centrifugal pumps, heaters, mixers, hydraulic-pneumatic valves, vacuum dryers, and their attached pipes and fittings. So, the importance of their proper maintenance program for safe operation and high reliability is inevitable.

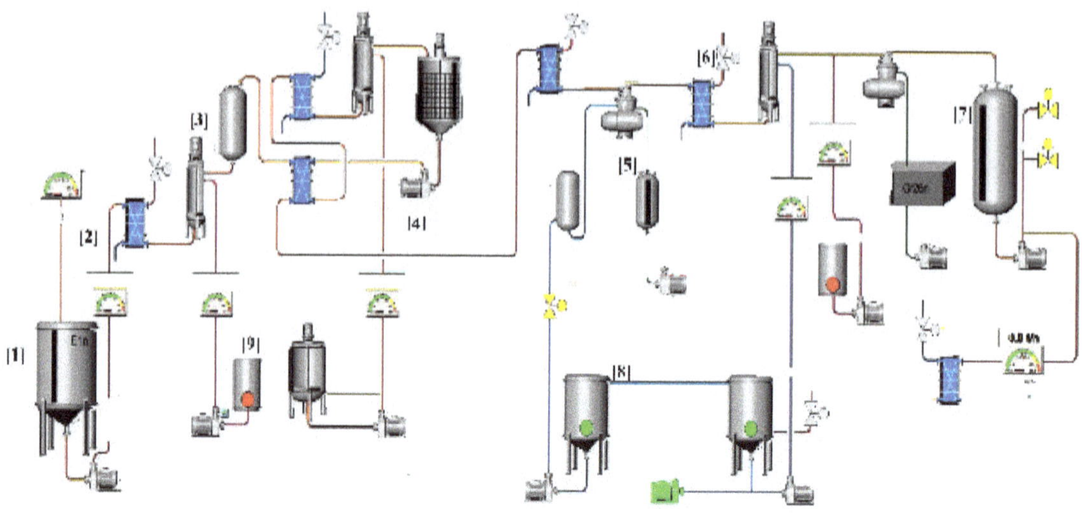

Figure 3. A schematic of the neutralization process: storage oil tank (1), exchanger (2), mixer (3), pump (4), separator (5), valve (6), dryer (7), water tank (8), sulfuric acid tank (9).

3.1.2. Discoloration Process

According to Figure 4, the color of the oil is reduced through decolorized soil and some particles of oil-based paint are removed. The decolorized soil is dissolved in oil or colloidal, and attractive colored particles are separated from the oil by a press. In general, decolorization is a physical absorption activity that removes pigments and impurities from the oil by absorption. The non-continuous [batch] system is used to decolorize the liquid oil, which has a larger volume and stronger stirrers than the solid oil tank, and the shelf life of the oil is much longer. After this step, the oil is transferred into the winterization process. The most important equipment in this phase includes hydraulic-pneumatic valves, pumps, mixing tanks, electrical systems, etc.

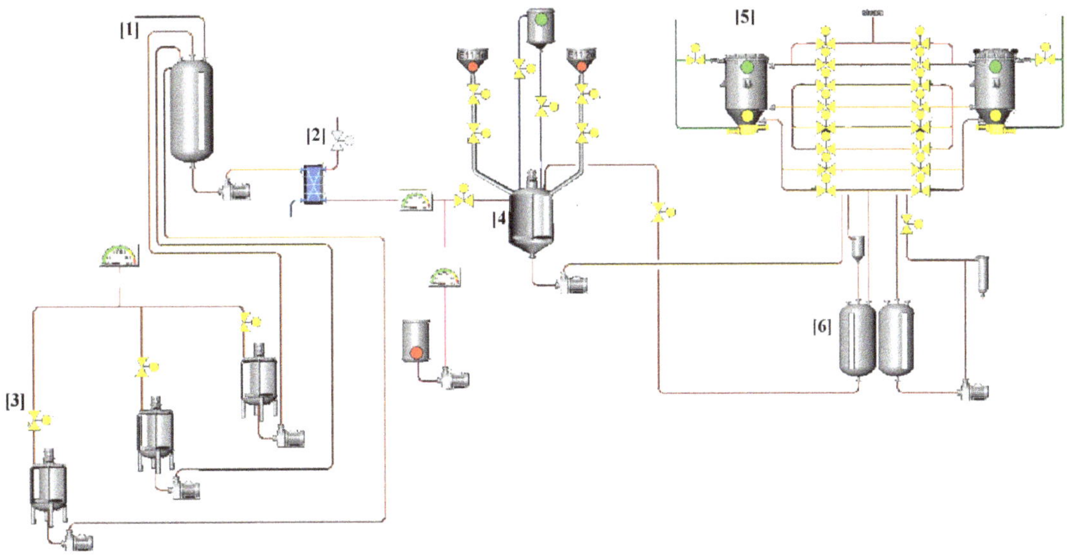

Figure 4. A schematic of the discoloration process: storage oil tank (1), exchanger (2), mixer (3), pulverizing tank (4), filtering tank (5), discoloration oil tank (6).

3.1.3. Winterize Process

During the winterizing process, the discolored oil is stored for 24 h at a relatively low temperature, usually, 9 °C, to remove all possible solids that freeze the oil. These solids include high-melting glycerides and waxes. Thereafter, the high-pressure oil is pressed into the crystallized tanks with the help of air pressure to remove all solids from the oil, after which the pure oil is transferred into the deodorizing process of oil by filtrate operation Figure 5.

3.1.4. Deodorizing Process

Figure 6 shows a deodorizing process of oil in which the undesirable odor of oil is caused by ketones, lactones as well as free fatty acids. For removing these, first, the high-pressure oil is sprayed from the bottom into the odorless tower, which is used simultaneously to heat, steam, and vacuum to prevent oxidation and hydrolysis of the oil. The main purpose is to decrease the oil acid content to the standard level. After that, the oil enters to exchanger until it reaches a temperature of 30 to 40 °C. Then, turns into another exchanger until the oil temperature finally reaches 14 or 12 °C.

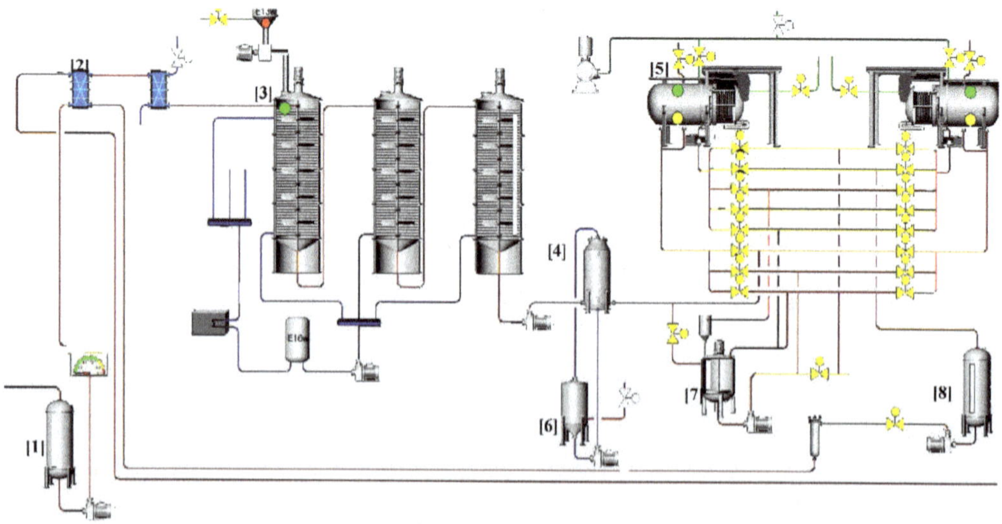

Figure 5. A schematic of the winterizing process: storage oil tank (1), exchanger (2), crystallization (3), exchanger tube shell (4), filtering tank (5), heater (6), keratinization (7), winterize oil (8).

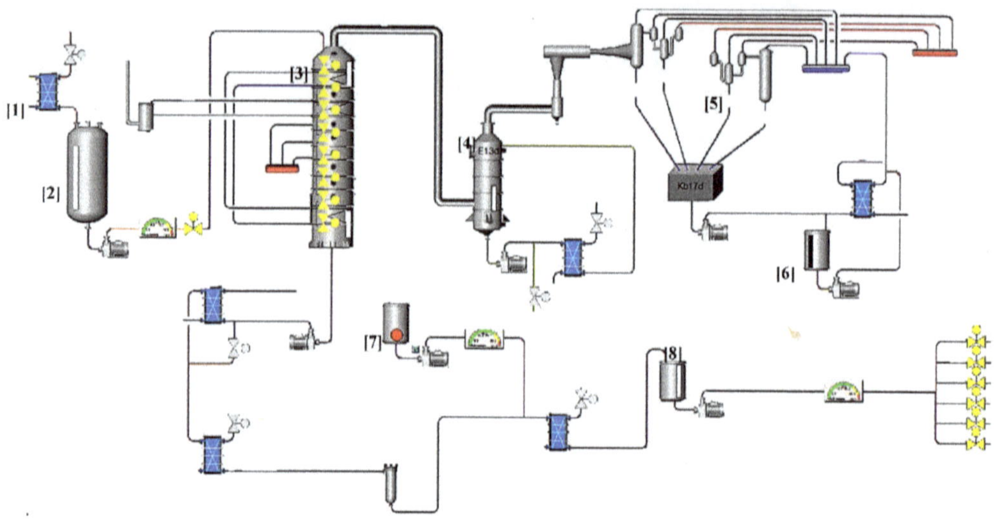

Figure 6. A schematic of the deodorizing process: exchanger (1), deaerator (2), deodorizing bridge (3), scrubber (4), a vacuum system (5), CIP tank (6), antioxidant tank (7), deodorized oil tank (8).

3.1.5. Potential Failures and Their Effects

In an edible oil purification plant in Iran, functional failures, causes, and their effects were discovered. To survey such items, a group of FMEA experts totally between 4 to 6 members is needed [21,56,57]. In this study, we have received the knowledge and experiences of four experts [two process engineers and two mechanical and electrical engineers], who were related and engaged in the whole process in edible oil-producing. So, based on the expert's knowledge and experiences, 67 failure modes of the process were derived. These failures are mainly caused by pumps, separators, chillers, boilers, dryers, compressors, valves, converters, mixers, electronic circuits, pipes, filters, tanks, and vacuum systems. Ultimately, this obtained information was used for estimating risk factors and RPN value.

3.2. *Second Step: Knowledge-Based Approach*

In this step, first, the risk parameters, e.g., S, O, and D are defined by the FMEA expert team and then the FIS structure, ANFIS, and SVM models based on FMEA models for risk analysis were programmed by MATLAB vR2020b (Math works Inc., Natick, MA, USA).

3.2.1. Risk Parameters Definition

The FMEA is a well-known risk analysis tool that is frequently used by RPN to assess the risk level of a process, rate failures, and prioritize maintenance actions [20]. To calculate the RPN value, a discrete ordinal scale of 1-10 is used to multiply three crisp values of the risk characteristics, namely occurrence (O), severity (S), and detection (D). Finally, the most critical failures can be found by sorting the RPNs in ascending order [21]. In the classical-FMEA, the risk parameters can be divided into five-linguistic terms including remote (R), low (L), moderate (M), high (H), and very high (VH). This attitude will help the FMEA team to prioritize the failure mods and their effects [58–60]. The linguistic scale of the risk characteristics and their fuzzy numbers in three class levels (3,5, and 10) for the present investigation was also provided by Soltanali et al. [17] in the FIS structure. The FMEA expert team also provided the necessary information on the severity of the failure and the inability to detect it. Finally, the failures were prioritized using the fuzzy risk numbers.

3.2.2. FIS Structure

FIS is a well-known intelligent risk analysis technique. Figure 7 depicts the FIS structure. The FIS environment is built in the first step using key elements including "and method," "or method," "implication method," and "aggregate method." The membership function of the input variables "risk parameters" was constructed in the second stage. The third step is to create the membership function for the output variable "FIS-RPN." Finally, the output control rules were defined. The Mamdani approach, which has been frequently utilized by others to build FIS boundaries which is used to evaluate the rules in the rule base [61]. The fuzzy logic system theory can be stated formally as Dağsuyu et al. [19] and Kumru and Kumru [37]. X be a nonempty set. A fuzzy set A in X is characterized by its membership function, i.e., $\mu_A : X \to [0, 1]$ and $\mu_A\ (x)$ is interpreted as the degree of membership of element x in the fuzzy set A for each $x \in X$. It is clear that A is completely determined by a set of tuples $A = ([u,\ \mu_A\ [u])/u \in X)$. Frequently, A (x) is used instead of $\mu_A\ (x)$. The family of all fuzzy sets in X is denoted by $F(x)$. If $X = (\ x_1,\ x_2,\ \ldots,\ x_n\ )$ is a finite set and A is a fuzzy set in X, the following notation can be used:

$$A = \frac{\mu_1}{x_1} + \frac{\mu_2}{x_2} + \ldots + \frac{\mu_n}{x_n} \tag{1}$$

where the term μ_i/x_i, i = 1,.., n signifies that μi is the grade of membership of x_i in A and the plus sign represents the union.

In this work, we looked at many types of membership functions such as Trim, Trapmf, Pimf, and Gaussmf, Gauss2mf, Gbellmf, Psigmf, and Dsigmf to produce fuzzy numbers using linguistic terms and fuzzy numbers for the risk parameters in the 10-class. Experts in the edible oil purification process determined the required rules such as 3, 5, and 10-class, appropriately, 27, 125, and 1000 rules. We used five defuzzification algorithms in the FIS environment's last stage to analyze the aggregating process and calculate the explicit RPN values.

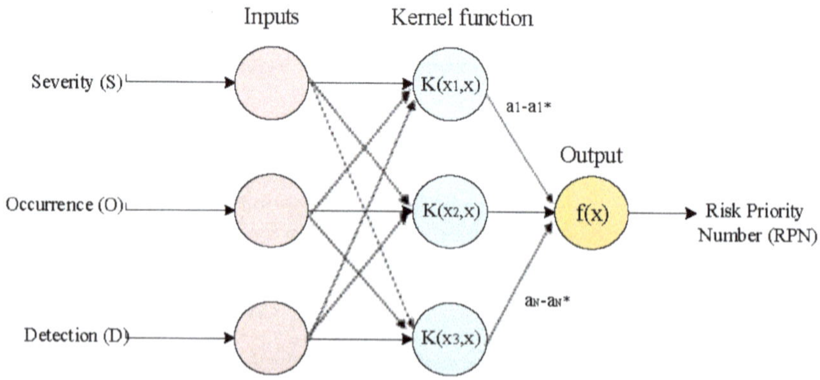

Figure 9. SVM algorithm structure.

3.2.5. Performance Criteria

Some metrics, such as mean absolute percentage error (MAPE), root mean square error (RMSE), efficiency (EF), and coefficient of variation (CV), are used in the literature to evaluate the performance of intelligent models for risk analysis [64,65]. These are their definitions:

$$\mathrm{MAPE} = \frac{1}{n}\sum_{j=1}^{n}\left|\frac{d_j - p_j}{d_j}\right| \times 100 \quad (12)$$

$$\mathrm{RMSE} = \sqrt{\frac{\sum_{j=1}^{n}(d_j - p_j)^2}{n}} \quad (13)$$

$$\mathrm{EF} = \frac{\sum_{j=1}^{n}(d_j - \bar{d})^2 - \sum_{j=1}^{n}(p_j - d_j)^2}{\sum_{j=1}^{n}(d_j - \bar{d})^2} \quad (14)$$

$$\mathrm{CV} = \frac{\sigma}{\mu} \times 100 \quad (15)$$

where d_j is the ith value of the desired (actual) output for the jth pattern; p_j is the predicted (fitted) output for the jth pattern, μ is the mean value and σ is the standard deviation.

4. Results

4.1. Classical FMEA Result

The results of the classical-FMEA model for three risk parameters and RPN values, based on experts' judgment, for an edible oil purification process, are addressed in Table 2. One of the model's drawbacks is its inability to rank the failure types in a unique and non-repetitive manner, as shown in Table 3. As a result, detecting high-risk failure modes and assigning appropriate maintenance duties is challenging. For example, (1st and 2nd), (5th and 10th) failure modes, and so on, all have the same RPNs and ranks, according to the first expert's assessment. According to the second expert, the failure modes are ranked in the same order for the (6th, 8th, and 9th), (4th and 25th), and so on. This issue can also be seen by third and fourth experts, resulting in a fundamental flaw in the risk analysis process.

Table 2. The classical FMEA result of S, O, D, RPN values.

System	FM	Expert 1				Expert 2				Expert 3				Expert 4			
		S	O	D	RPN	S	O	D	RPN	S	O	D	RPN	S	O	D	RPN
Tanks	1	4	2	4	32	3	1	5	15	3	2	4	24	4	3	4	48
	2	4	2	4	32	5	1	5	25	3	4	4	48	5	2	4	40
	3	8	6	7	336	10	7	8	560	6	5	7	210	7	5	6	210
	4	2	2	4	16	5	3	2	30	2	2	2	8	3	2	5	30
Pumps	5	10	6	8	480	10	4	8	320	8	7	8	448	9	6	8	432
	6	10	4	8	320	10	5	8	400	8	7	8	448	9	6	8	432
	7	10	10	5	500	10	8	2	160	8	8	2	128	8	9	3	216
	8	10	5	2	100	10	8	5	400	8	6	4	192	9	7	2	126
Separators	9	10	5	8	400	10	5	8	400	10	5	8	400	10	6	9	540
	10	10	6	8	480	10	6	6	360	10	4	6	240	8	5	4	160
	11	10	3	8	240	7	1	5	35	10	1	6	60	8	2	5	80
	12	10	5	5	250	10	8	3	240	10	4	5	200	9	6	5	270
	13	10	5	8	400	9	3	8	216	10	4	8	320	8	4	7	224
Filters	14	10	3	4	120	10	3	5	150	10	6	3	180	9	7	3	189
	15	10	7	5	350	10	5	5	250	10	5	3	150	9	6	2	108
	16	10	2	1	20	10	4	1	40	9	4	3	108	9	3	2	54
Chillers	17	10	3	2	60	8	4	5	160	8	3	8	192	9	3	7	189
	18	10	3	7	210	9	5	7	315	8	5	7	280	9	3	8	216
	19	10	6	9	540	9	5	10	450	9	3	8	216	8	4	8	256
	20	8	4	4	128	10	7	5	350	10	8	4	320	9	4	4	144
Mixers	21	1	9	5	45	10	4	7	280	10	5	7	350	10	6	6	360
	22	2	7	4	56	8	4	7	224	7	3	3	63	8	7	2	112
	23	3	6	5	90	5	1	7	35	5	1	7	35	8	7	6	336
	24	4	9	4	144	10	4	2	80	10	5	3	150	8	6	5	240
Dryers	25	7	3	1	21	10	3	1	30	8	2	3	48	8	2	2	32
	26	7	3	3	63	8	3	1	24	8	2	3	48	9	2	2	36
	27	7	3	5	105	9	5	5	225	6	4	3	72	7	3	2	42
	28	7	3	2	42	10	3	1	30	5	4	1	20	8	2	1	16
	29	5	3	1	15	5	3	1	15	6	2	1	12	5	3	1	15
	30	6	3	3	54	10	5	3	150	6	2	2	24	7	4	2	56
	31	6	2	1	12	10	1	1	10	6	2	1	12	8	2	1	16
	32	5	3	3	45	5	1	1	5	6	4	2	48	5	3	2	30
Boilers	33	4	1	1	4	10	3	1	30	10	3	2	60	8	2	1	16
	34	4	1	1	4	5	3	1	15	10	3	2	60	5	1	3	15
	35	10	3	1	30	10	3	1	30	10	3	2	60	10	2	4	80
	36	7	2	2	28	10	3	5	150	10	3	2	60	8	5	4	160
	37	7	2	4	56	5	1	2	10	7	3	3	63	8	3	3	72
	38	9	4	3	108	5	2	5	50	7	3	4	84	4	2	4	32
	39	8	3	4	96	10	4	5	200	8	3	4	96	10	4	5	200
	40	10	5	3	150	10	7	5	350	8	5	4	160	10	7	4	280
	41	2	2	1	4	2	1	1	2	5	3	2	30	2	1	3	6
	42	10	6	2	120	10	7	1	70	10	7	1	70	10	7	2	140
	43	10	4	2	80	10	3	1	30	10	3	1	30	10	5	1	50
Compressors	44	10	2	1	20	9	3	1	27	9	3	1	27	9	2	2	36
	45	10	2	8	160	10	3	8	240	10	3	6	180	10	3	5	150
	46	8	3	1	24	10	3	4	120	9	3	3	81	8	2	2	32
	47	6	2	3	36	8	5	3	120	8	6	1	48	7	6	2	84
	48	6	1	2	12	8	1	1	8	6	4	1	24	7	5	3	105
	49	8	3	4	96	8	5	4	160	8	4	2	64	9	3	4	108
	50	7	6	6	252	10	5	8	400	9	4	6	216	8	5	6	240

Table 2. *Cont.*

System	FM	Expert 1 S	Expert 1 O	Expert 1 D	Expert 1 RPN	Expert 2 S	Expert 2 O	Expert 2 D	Expert 2 RPN	Expert 3 S	Expert 3 O	Expert 3 D	Expert 3 RPN	Expert 4 S	Expert 4 O	Expert 4 D	Expert 4 RPN
Vacuum system	51	6	4	2	48	10	5	2	100	9	7	1	63	6	6	3	108
	52	7	5	3	105	10	7	4	280	9	8	2	144	7	7	3	147
	53	6	1	3	18	5	2	1	10	6	3	2	36	6	2	1	12
	54	10	3	3	90	10	5	4	200	8	8	2	128	8	6	3	144
	55	10	4	1	40	10	5	1	50	9	8	2	144	9	4	2	72
	56	8	6	4	192	10	5	3	150	9	8	2	144	8	6	2	96
	57	8	5	5	200	10	5	5	250	9	8	4	288	8	6	4	192
Exchangers	58	10	10	1	100	10	7	1	70	9	8	1	72	9	7	2	126
	59	10	6	5	300	10	5	7	350	7	4	3	84	8	6	5	240
Pipes	60	3	3	1	9	3	5	1	15	2	8	3	48	2	6	4	48
	61	3	3	1	9	5	7	4	140	2	8	4	64	2	6	4	48
PLCs	62	9	7	7	441	9	8	6	432	9	5	8	360	9	5	5	225
	63	8	5	7	280	9	6	6	324	10	6	8	480	9	7	6	378
Valves	64	8	5	2	80	10	5	1	50	8	4	3	96	5	3	2	24
	65	8	5	4	160	8	5	4	160	8	5	4	160	5	4	7	140
	66	5	8	2	80	5	8	2	80	5	8	3	120	8	3	3	72
	67	8	4	2	64	8	4	2	64	8	4	3	96	4	3	2	24

Table 3. The same RPN value issue of classical FMEA.

FM	RPN Expert 1	FM	RPN Expert 2	FM	RPN Expert 3	FM	RPN Expert 4
(6, 8, 9, 50)	400	(6, 8, 9, 50)	400	(5, 6)	448	(5, 6)	432
(20, 50, 40, 59)	350	(20, 40, 59)	350	(13, 20)	320	(24, 50, 59)	240
(21, 52, 15, 57)	280	(52, 21)	280	(19, 50)	216	(7, 18)	216
(15, 57)	250	(15, 57)	250	(8, 17)	192	(14, 17)	189
(12, 45)	240	(12, 45)	240	(14, 45)	180	(10, 36)	160
(39, 45)	200	(39, 54)	200	(40, 65)	160	(20, 54)	144
(7, 17, 49, 65)	160	(17, 49, 56)	160	(15, 24)	150	(42, 65)	140
(38, 55, 64)	50	(14, 30, 36, 56)	150	(52, 55, 56)	144	(15, 49, 51)	108
(11, 23)	35	(46, 47)	120	(7, 54)	128	(11, 35)	80
(4, 25, 28, 33, 35, 43)	30	(24, 66)	80	(39, 64)	96	(37, 55, 66)	72
		(42, 58)	70	(38, 59)	84	(1, 60, 61)	48
		(1, 29, 34, 60)	15	(27, 58)	72	(26, 44)	36
		(31, 37, 53)	10	(49, 61)	64	(25, 38, 46)	32
				(22, 37, 51)	63	(4, 32)	30
				(11, 33, 34, 35, 36)	60	(28, 31, 33)	16
				(2, 25, 26, 32, 47, 60)	48	(29, 34)	15
				(41, 43)	30		
				(1, 30, 48)	24		
				(29, 31)	12		

Using the geometric average method (GAM) to prioritize high-risk failures is one of the most common strategies to solve this issue in traditional FMEA. Figure 10 shows the results of the conventional RPN based on GAM-FMEA from the expert's assessment. Although this method has been able to address some of the shortcomings of the classical-FMEA (67 failure modes categorized into 59 classes), several of the failure modes still have the same RPN and rank values. The 4th and 33rd, 32nd and 60th, 1st and 44h failure modes, etc. are in the same classes. Therefore, to solve this outcome of classical and GAM- FMEAs, we have examined the potential of the intelligent models based on FMEA for classifying the failure modes during the risk analysis process.

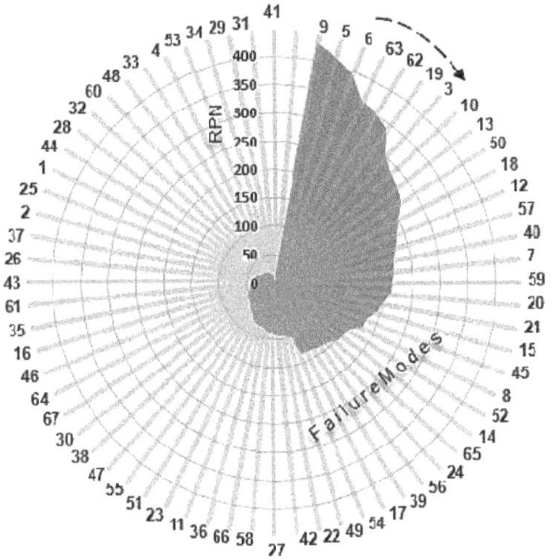

FM	GAM-RPN
(4, 33)	18
(32, 60)	24
(1, 44)	27
(37, 26)	40
(43, 61)	44
(35, 16)	46
(64, 67)	55
(11, 36)	80

Figure 10. The results of RPN value based on GAM- FMEA.

4.2. Intelligent- FMEA Results

In this subsection, the ability of three intelligent models such as FIS, ANFIS, and SVM-based FMEA to create a maximum class of risks for the edible oil purification process are examined. The results of the FIS-FMEA model based on three fuzzy scale classes are provided in Table 4. First, among the several defuzzification strategies, the centroid method was chosen as having the most potential for producing a maximum fuzzy number class. Various MFs combinations for the three-risk metrics and FIS-RPN (FRPN) of three classes were investigated as a result of this. The CV factor was used as the primary performance criterion to select the optimal combined MFs from 4096 combinations. The average CVs in all MFs for the three, five, and ten classes were 60.70, 53.64, and 50.76, respectively. It means that the 3-class with a high CV can provide maximum risk class numbers while avoiding repetitive clustering. As highlighted, two combinations of MFs in 3-class (27-rule) have greater potential to create 67 class numbers for 67 failure modes than other classes.

Table 4. The optimal MFs combination for risk clustering based on the FIS-FMEA model.

Number of Classes	Number of Rules	S	O	D	FRPN	CV (%)	Number of Cluster
3-class	(27)	Psigmf	Gauss2mf	Dsigmf	Trimf	58.30	67
		Dsigmf	Gauss2mf	Dsigmf	Trimf	58.30	67
		Trapmf	Dsigmf	Gaussmf	Dsigmf	56.94	66
5-class	(125)	Trimf	Gauss2mf	Psigmf	Trimf	56.77	66
		Trimf	Gauss2mf	Psigmf	Trapmf	56.77	66
		Trimf	Gauss2mf	Psigmf	Gaussmf	56.77	66
10-class	(1000)	Trimf	Gbellmf	Gaussmf	Gbellmf	56.42	64
		Trapmf	Gbellmf	Gaussmf	Gbellmf	56.42	64
		Gauss2mf	Gaussmf	Gaussmf	Trapmf	57.44	64

In the next step, the ability of the ANFIS-FMEA model for risk clustering of failure modes in the oil purification process was investigated in Table 5. For this purpose, the default values of the ANFIS network such as influential radius (IR), squash factor (SF), accept ratio (AR), and reject ratio (RR) in ANFIS are assumed as 0.5, 1.25, 0.5, and 0.15, respectively.

Additionally, in this study, two optimization methods such as hybrid and backpropagation were used for parameter training of membership functions. Following this, the performance of the ANFIS-FMEA model for risk clustering, under some well-known performance criteria, was evaluated. As seen, although most of the ANFIS optimization methods can create maximum risk clusters (67 failure modes in 67 clusters) in different fuzzy number classes, the hybrid model considering 5-class (125-rule) and 30 number epochs has been very successful in predicting the actual valves with the lowest errors (RMSE = 4.01 and MAPE = 4.25). To get better insight, Figure 11 shows that the total values of RMSE and MAPE (%) for the hybrid model with 5-class are lower than other fuzzy number classes and ANFIS models.

Table 5. The optimal performance criteria for risk clustering based on the ANFIS-FMEA model.

Optimization Method	Number of Class	Number of Rule	Number of Epoch	RMSE			MAPE (%)			EF (%)			Number of Cluster
				Train	Test	Total	Train	Test	Total	Train	Test	Total	
Hybrid model	3-class	(27)	10	5.84	5.75	7.08	11.85	12.15	13.09	99.00	99.00	99.00	67
		(27)	20	4.79	4.57	6.27	10.60	10.01	11.88	99.00	99.00	99.00	67
		(27)	30	3.84	3.23	5.19	8.43	6.92	9.30	99.00	99.00	99.00	67
	5-class	(125)	10	3.92	4.31	6.03	6.08	7.19	6.19	99.00	99.00	99.00	67
		(125)	20	3.14	3.69	5.08	4.60	8.27	5.30	99.00	99.00	99.00	67
		(125)	30	2.11	3.02	4.01	1.81	7.78	4.25	99.00	99.00	99.00	67
	10-class	(1000)	10	1.35	3.80	6.33	2.62	8.16	6.49	99.00	99.00	99.00	67
		(1000)	20	1.08	3.61	6.57	1.98	8.05	6.13	99.00	99.00	99.00	67
		(1000)	30	0.91	3.29	5.51	1.61	7.75	4.81	99.00	99.00	99.00	67
Back propagation	3-class	(27)	10	7.50	7.13	8.95	8.80	19.46	16.29	99.00	99.00	99.00	67
		(27)	20	6.45	8.61	9.03	7.71	20.25	15.19	99.00	99.00	99.00	67
		(27)	30	6.03	9.21	9.20	7.23	21.09	15.10	99.00	99.00	99.00	67
	5-class	(125)	10	7.39	7.40	7.56	9.53	16.23	9.71	99.00	99.00	99.00	67
		(125)	20	5.32	6.61	6.48	7.11	15.01	8.00	99.00	99.00	99.00	67
		(125)	30	4.72	6.80	6.37	6.20	14.92	7.42	99.00	99.00	99.00	67
	10-class	(1000)	10	2.66	2.55	4.91	6.76	2.99	6.15	99.00	99.00	99.00	67
		(1000)	20	2.57	2.69	4.88	6.49	3.13	5.97	99.00	99.00	99.00	67
		(1000)	30	2.49	2.82	4.86	6.22	3.24	5.80	99.00	99.00	99.00	67

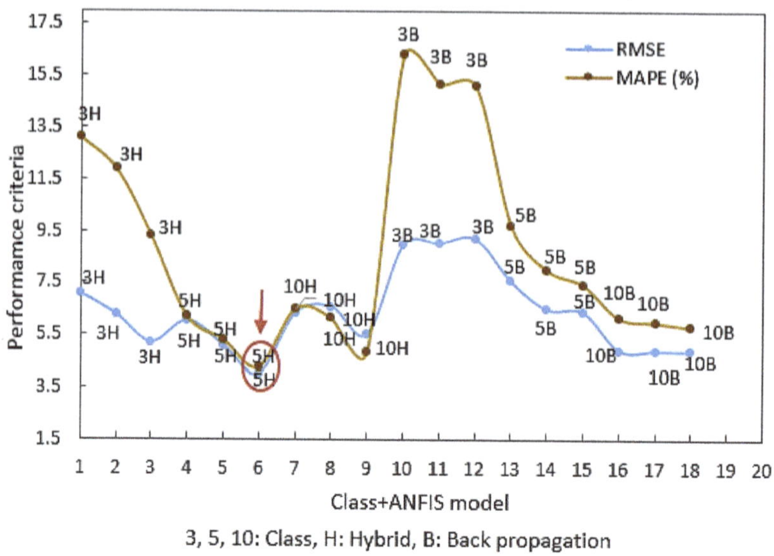

Figure 11. Total values of RMSE and MAPE (%) for two ANFIS models in three fuzzy classes.

Furthermore, the results of the SVM-FMEA algorithm as another intelligent model for risk analysis and to create the maximum risk clustering are presented in Table 6. As observed, the SVM algorithms such as sequential minimal optimization (SMO) and iterative single data algorithm (ISDA) can classify the 67 failure modes into 67 risk clusters. However, the ISDA algorithm using the polynomial-kernel function has higher accuracy to predict the actual values. In other words, this algorithm has been very effective in classifying the failure modes with the lowest errors (RMSE = 7.30 and MAPE = 13.19) and the highest performance (EF: 99%).

Table 6. The optimal performance criteria for risk clustering based on the SVM-FMEA model.

Solver Algorithm	Kernel Function	RMSE			MAPE (%)			EF (%)			Number of Cluster
		Train	Test	Total	Train	Test	Total	Train	Test	Total	
SMO algorithm	Gaussian	9.61	23.44	15.13	15.91	54.03	27.29	99.00	95.00	98.00	67
	Linear	28.25	41.37	32.72	41.91	124	66.42	0.94	0.85	0.92	67
	Rbf	9.65	23.70	15.26	15.94	55.52	27.67	99.00	95.00	98.00	67
	Polynomial	9.43	18.18	12.69	14.05	68.91	30.43	99.00	97.00	99.00	67
ISDA algorithm	Gaussian	8.36	21.00	13.44	17.08	30.73	21.16	99.00	96.00	99.00	67
	Linear	98.83	99.95	99.16	167.3	161.0	165.4	0.26	0.12	0.22	67
	Rbf	7.77	19.48	12.47	15.37	26.18	18.60	99.00	99.00	99.00	67
	Polynomi	7.30	7.31	7.30	14.04	11.17	13.19	99.00	99.00	99.00	67

4.3. Comparison Results

Figure 12 shows a comparison between the intelligent models such as FIS (Figure 12a), ANFIS (Figure 12b) and SVM (Figure 12c), and classical-FMEAs to identify the best model for raking the failures in an edible oil purification plant. As shown, the rank value of the SVM algorithm overlaps fairly well with the rank value of the classic model for most failure modes with that of other intelligent models. The error indices such as MAPE for FIS, ANFIS, and SVM were obtained as 21%, 4.64%, and 3.02%, respectively, and the values for RMSE were equal to 5.73, 2.85, and 1.12, respectively, to predict the classical rank value. Hence, it can be concluded that the SVM-FMEA model has a great potential for ranking all failure modes accurately with the lowest errors compared to other intelligent models. In the following, through the feedback of such model, a sensitivity analysis of risk parameters and alternatively the appropriate maintenance tasks were surveyed.

4.4. Sensitivity Analysis

To study the impact of risk parameters (S, O, D) on SVM-RPN in the edible oil purification process, a sensitivity analysis was performed. For example, in risk parameters, the S index represents the severity of the failure on the equipment or its impact on the entire process. The O index represents the chance of failure occurrence, and the D index represents how likely it is to identify the occurred failures. Figure 13 depicts the findings of the sensitivity analysis as surface plots. As shown in Figure 13a, the S index has a higher impact on the risk parameter than the O index because the slope change of SVM-RPN due to changes in the S (46°) index is greater than that due to changes in the O (18°) index. It means that the D index is the meaningful factor on the risk changes in the edible oil purification process. As a consequence, to improve the possibility of detecting the failures and to reduce the probability of failures, fault diagnosis tools and warning signs could be suggested for the edible oil purification process.

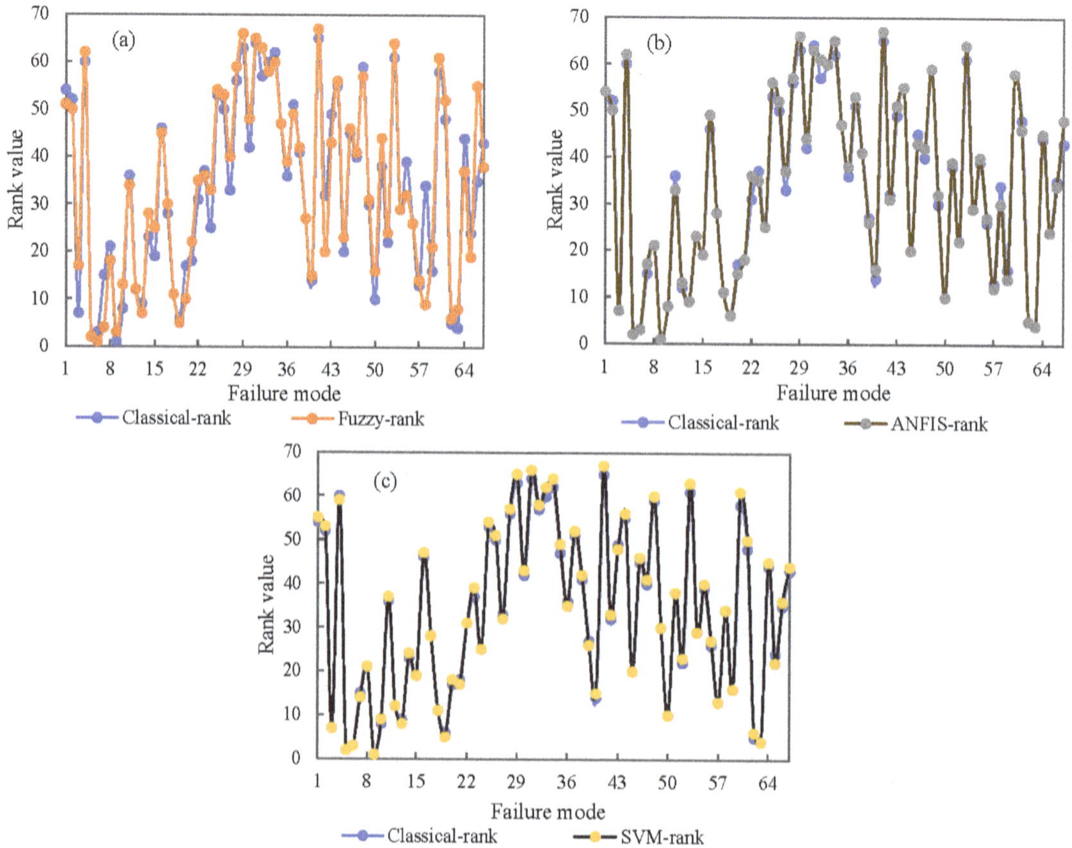

Figure 12. The comparison of failure mode ranking values between (**a**) classical-FMEA and Fuzzy model, (**b**) classical-FMEA and ANFIS model, and (**c**) classical-FMEA and SVM model.

4.5. Maintenance Activity

In this section, based on the best intelligent-FMEA model, the appropriate maintenance activities for the edible oil purification process were provided. Based on the results, the failure of bearings in separators and the failure of vanes and shafts in pumps were identified with the highest RPN values, e.g., 421, 409 and 391 as well as primary ranks, e.g., 1, 2 and 3, respectively. Because bearings are critical to achieving high operational dependability in separators, adopting robust inspection procedures and non-destructive tests weekly could be recommended. Furthermore, the majority of the operations in the purification process are associated with centrifugal pumps for moving fluids such as water and oils. As a result, appropriate maintenance chores such as monthly services such as checking lubricant levels and bearing operating temperature, vibration analysis of shafts, and changing the vanes and axis of shafts could be performed from quarterly to monthly. The failure of hydro-pneumatic valves and sensors in chillers, as well as the failure of programmable logic controller (PLC) circuits, were ranked as the second class and RPNs, respectively. The failure of O-rings and seals in hydro-pneumatic systems is the main cause of leakages due to the high pressure in the process. As a result, the key maintenance tasks may include increasing the frequency of O-ring and seal replacements from monthly to twice-weekly, utilizing higher-quality materials. The majority of sensors' failures in chillers are caused by excessive usage or function. As a result, a monthly replacement could be appointed. Meanwhile, different capability tests and well-timed inspections for PLCs, and timely replacement of the cables and wires could be taken before an irreparable fault occurs. As a result, the aforementioned maintenance operations would assist engineers in detecting and preventing unforeseen problems, resulting in increased safety and dependability in the edible oil purification process.

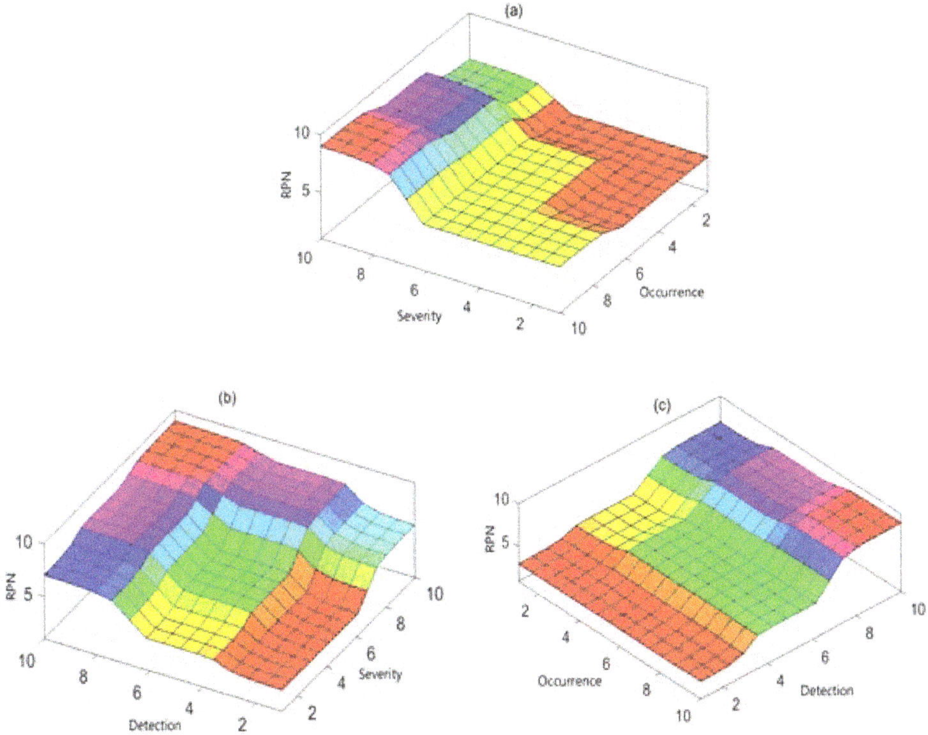

Figure 13. Risk parameters surface plots vs. SVM-RPN: (**a**) SVM-RPN slope change due to S parameter variations, (**b**) SVM-RPN slope change due to O parameter variations, and (**c**) SVM-RPN slope change due to D parameter variations.

5. Conclusions

This paper aimed to improve an intelligent-FMEA model for analyzing the risk and comparing the outcomes with the classical-FMEA, in the edible oil industries. To overcome the classical FMEA drawbacks, some well-known intelligent models such as FIS structure, ANFIS, and SVM models were carried out for risk analysis. To evaluate the accuracy prediction, the CV (%) factor for FIS structure, and some performance criteria such as RMSE, MAPE (%), and EF (%) for ANFIS and SVM models, were performed. Additionally, to determine the risk parameters and RPNs for the failure modes in the edible oil purification process, a knowledge-based approach was adapted. The results revealed that the 3-class (27-rule) in the FIS structure, and the 5-class (125-rule) in the hybrid-ANFIS network have high potential to create maximum risk number cluster of failure modes. Moreover, the results of the SVM algorithm indicated the ISDA algorithm using polynomial-kernel function has higher accuracy to predict the actual values and classify the failure modes. Based on the performance indicators, the SVM-FMEA algorithm has a great potential for ranking all failure modes accurately with the lowest errors compared to other intelligent models. According to the results of the 3-D sensitivity study, the detection index is more successful on SVM-RPN variation than on occurrence and severity. Finally, the authoritative control for the equipment with the highest risk within the edible oil purification was recommended through maintenance and inspection activities. In this study, knowledge-based methods for diagnosing failures and risk assessment were proposed due to a lack of sufficient and reliable operational data. As a result, future research can be expanded to evaluate and improve the accuracy of the proposed approach by establishing a trustworthy database in edible oil purification plants. Furthermore, the use of other hybrid models with data-driven based methods to automate risk monitoring within food processing systems can be recommended.

Author Contributions: Conceptualization, H.S., M.K.; methodology, H.S., M.K., J.T.F. and J.E.d.A.e.P.; software, H.S., M.K., J.T.F. and J.E.d.A.e.P.; validation, H.S., M.K., J.T.F. and J.E.d.A.e.P.; formal analysis, H.S., M.K., J.T.F. and J.E.d.A.e.P.; investigation, H.S., M.K., J.T.F. and J.E.d.A.e.P.; resources, H.S., M.K., J.T.F. and J.E.d.A.e.P.; data curation, H.S., M.K., J.T.F. and J.E.d.A.e.P.; writing—original draft preparation, H.S., M.K., J.T.F. and J.E.d.A.e.P.; writing—review and editing, H.S., M.K., J.T.F. and J.E.d.A.e.P.; visualization, H.S., M.K., J.T.F. and J.E.d.A.e.P.; supervision, H.S., M.K., J.T.F. and J.E.d.A.e.P.; project administration, H.S., M.K., J.T.F. and J.E.d.A.e.P.; funding acquisition, H.S., M.K., J.T.F. and J.E.d.A.e.P. All authors have read and agreed to the published version of the manuscript.

Funding: This research project was supported and funded by the Ferdowsi University of Mashhad, Iran (No.FUM-52316). Additionally, the research leading to these results has received funding from the European Union's Horizon 2020 research and innovation program, under the Marie Sklodowvska-Curie grant agreement 871284 project SSHARE, the European Regional Development Fund (ERDF) through the Operational Programme for Competitiveness and Internationalization (COMPETE 2020), under Project POCI-01-0145-FEDER-029494, and by National Funds through the FCT—Portuguese Foundation for Science and Technology, under Projects PTDC/EEI-EEE/29494/2017, UIDB/04131/2020, and UIDP/04131/2020.

Institutional Review Board Statement: Not applicable.

Informed Consent Statement: Not applicable.

Data Availability Statement: The data that support the findings of this study are available from the corresponding author, upon reasonable request.

Acknowledgments: The authors gratefully acknowledge the financial support from the Ferdowsi University of Mashhad, Iran (No.FUM-52316). Additionally, the research leading to these results has received funding from the European Union's Horizon 2020 research and innovation program, under the Marie Sklodowvska-Curie grant agreement 871284 project SSHARE, the European Regional Development Fund (ERDF) through the Operational Programme for Competitiveness and Internationalization (COMPETE 2020), under Project POCI-01-0145-FEDER-029494, and by National Funds through the FCT—Portuguese Foundation for Science and Technology, under Projects PTDC/EEI-EEE/29494/2017, UIDB/04131/2020, and UIDP/04131/2020.

Conflicts of Interest: The authors declare no conflict of interest for this article.

References

1. Chen, R.Y. Autonomous tracing system for backward design in the food supply chain. *Food Control* **2015**, *51*, 70–84. [CrossRef]
2. Papadopoulos, A.I.; Seferlis, P. Automation for the sustainable food industry: Computer-aided analysis and control engineering methods. In *Robotics and Automation in the Food Industry*; Woodhead Publishing: Cambridge, UK, 2013; pp. 441–485.
3. Rahimifard, S.; Woolley, E.; Webb, D.P.; Garcia-Garcia, G.; Stone, J.; Jellil, A.; Gimenez-Escalante, P.; Jagtap, S.; Trollman, H. Forging new frontiers in sustainable food manufacturing. In *International Conference on Sustainable Design and Manufacturing*; Springer: Berlin/Heidelberg, Germany, 2017; pp. 13–24.
4. Yi, Y.; Ke, Z.; Yi, C. An implementation of intelligent monitoring system for food processing. In Proceedings of the 2010 8th World Congress on Intelligent Control and Automation, Jinan, China, 7–9 July 2010; pp. 4225–4230.
5. Khan, Z.H.; Khalid, A.; Iqbal, J. Towards realizing robotic potential in future intelligent food manufacturing systems. *Innov. Food Sci. Emerg. Technol.* **2018**, *48*, 11–24. [CrossRef]
6. Abbas, H.; Maennel, O.; Assar, S. *Security and Privacy Issues in Cloud Computing*; Springer: Berlin/Heidelberg, Germany, 2017.
7. Iqbal, W.; Abbas, H.; Daneshmand, M.; Rauf, B.; Abbas, Y. An In-Depth Analysis of IoT Security Requirements, Challenges and their Countermeasures via Software Defined Security. *IEEE Internet Things J.* **2020**, *7*, 10250–10276. [CrossRef]
8. Carbone, T.A.; Tippett, D.D. Project risk management using the project risk FMEA. *Eng. Manag. J.* **2004**, *16*, 28–35. [CrossRef]
9. Bahill, A.T.; Smith, E.D. An industry standard risk analysis technique. *Eng. Manag. J.* **2009**, *21*, 16–29. [CrossRef]
10. Chemweno, P.; Pintelon, L.; Muchiri, P.N.; Van Horenbeek, A. Risk assessment methodologies in maintenance decision making: A review of dependability modelling approaches. *Reliab. Eng. Syst. Saf.* **2018**, *173*, 64–77. [CrossRef]

11. Alzghoul, A.; Backe, B.; Löfstrand, M.; Byström, A.; Liljedahl, B. Comparing a knowledge-based and a data-driven method in querying data streams for system fault detection: A hydraulic drive system application. *Comput. Ind.* **2014**, *65*, 1126–1135. [CrossRef]
12. Soltanali, H.; Garmabaki, A.H.; Thaduri, A.; Parida, A.; Kumar, U.; Rohani, A. Sustainable production process: An application of reliability, availability, and maintainability methodologies in automotive manufacturing. *Proc. Inst. Mech. Eng. Part O J. Risk Reliab.* **2019**, *233*, 682–697. [CrossRef]
13. Soltanali, H.; Rohani, A.; Abbaspour-Fard, M.H.; Farinha, J.T. A comparative study of statistical and soft computing techniques for reliability prediction of automotive manufacturing. *Appl. Soft Comput.* **2020**, *98*, 106738. [CrossRef]
14. Cho, W.I.; Lee, S.J. Fault tree analysis as a quantitative hazard analysis with a novel method for estimating the fault probability of microbial contamination: A model food case study. *Food Control* **2020**, *110*, 107019.
15. Song, Y.H.; Yu, H.Q.; Lv, W. Risk analysis of dairy safety incidents in China. *Food Control* **2018**, *92*, 63–71. [CrossRef]
16. Venkatasubramanian, V.; Rengaswamy, R.; Kavuri, S.N. A review of process fault detection and diagnosis: Part II: Qualitative models and search strategies. *Comput. Chem. Eng.* **2003**, *27*, 313–326. [CrossRef]
17. Soltanali, H.; Rohani, A.; Tabasizadeh, M.; Abbaspour-Fard, M.H.; Parida, A. An improved fuzzy inference system-based risk analysis approach with application to automotive production line. *Neural Comput. Appl.* **2019**, *32*, 10573–10591. [CrossRef]
18. Huang, J.; Xu, D.H.; Liu, H.C.; Song, M.S. A new model for failure mode and effect analysis integrating linguistic Z-numbers and projection method. *IEEE Trans. Fuzzy Syst.* **2019**, *29*, 530–538. [CrossRef]
19. Dağsuyu, C.; Göçmen, E.; Narlı, M.; Kokangül, A. Classical and fuzzy FMEA risk analysis in a sterilization unit. *Comput. Ind. Eng.* **2016**, *101*, 286–294. [CrossRef]
20. Silva, M.M.; de Gusmão, A.P.; Poleto, T.; e Silva, L.C.; Costa, A.P. A multidimensional approach to information security risk management using FMEA and fuzzy theory. *Int. J. Inf. Manag.* **2014**, *34*, 733–740. [CrossRef]
21. Soltanali, H.; Rohani, A.; Abbaspour-Fard, M.H.; Parida, A.; Farinha, J.T. Development of a risk-based maintenance decision making approach for automotive production line. *Int. J. Syst. Assur. Eng. Manag.* **2020**, *11*, 236–251. [CrossRef]
22. Ozilgen, S. Failure Mode and Effect Analysis [FMEA] for confectionery manufacturing in developing countries: Turkish delight production as a case study. *Food Sci. Technol.* **2012**, *32*, 505–514. [CrossRef]
23. Arvanitoyannis, I.S.; Varzakas, T.H. Application of failure mode and effect analysis [FMEA], cause and effect analysis and Pareto diagram in conjunction with HACCP to a potato chips manufacturing plant. *Int. J. Food Sci. Technol.* **2007**, *42*, 1424–1442. [CrossRef]
24. Varzakas, T.H.; Arvanitoyannis, I.S. Application of ISO22000 and Failure Mode and Effect Analysis [fmea] for Industrial Processing of Poultry Products. In *International Conference on Computer and Computing Technologies in Agriculture*; Springer: Boston, MA, USA, 2008; pp. 1783–1795.
25. Arvanitoyannis, I.S.; Varzakas, T.H. Application of ISO 22000 and failure mode and effect analysis [FMEA] for industrial processing of salmon: A case study. *Crit. Rev. Food Sci. Nutr.* **2008**, *48*, 411–429. [CrossRef] [PubMed]
26. Ebenezer, I.A.; Devadasan, S.R.; Sreenivasa, C.G.; Murugesh, R. Total failure mode and effects analysis in tea industry: A theoretical treatise. *Total Qual. Manag. Bus. Excell.* **2011**, *22*, 1353–1369. [CrossRef]
27. Kurt, L.; Ozilgen, S. Failure mode and effect analysis for dairy product manufacturing: Practical safety improvement action plan with cases from Turkey. *Saf. Sci.* **2013**, *55*, 195–206. [CrossRef]
28. Trafialek, J.; Kolanowski, W. Application of failure mode and effect analysis [FMEA] for audit of HACCP system. *Food Control* **2014**, *44*, 35–44. [CrossRef]
29. Özilgen, S.; Özilgen, M. General Template for the FMEA Applications in Primary Food Processing. In *Measurement, Modeling and Automation in Advanced Food Processing*; Springer: Berlin/Heidelberg, Germany, 2016; pp. 29–69.
30. Varzakas, T.H.; Arvanitoyannis, I.S. Application of Failure Mode and Effect Analysis [FMEA], cause and effect analysis, and Pareto diagram in conjunction with HACCP to a corn curl manufacturing plant. *Crit. Rev. Food Sci. Nutr.* **2007**, *47*, 363–387. [CrossRef] [PubMed]
31. Varzakas, T.; Manolopoulou, E. Comparison of HACCP and ISO 22000 in the ready-to-eat fruit and vegetable industry in conjunction with application of failure mode and effect analysis [FMEA] and Ishikawa diagrams. In *Minimally Processed Refrigerated Fruits and Vegetables*; Springer: Boston, MA, USA, 2017; pp. 685–721.
32. Rezaee, M.J.; Yousefi, S.; Valipour, M.; Dehdar, M.M. Risk analysis of sequential processes in food industry integrating multi-stage fuzzy cognitive map and process failure mode and effects analysis. *Comput. Ind. Eng.* **2018**, *123*, 325–337. [CrossRef]
33. Przystupa, K. The methods analysis of hazards and product defects in food processing. *Czech J. Food Sci.* **2019**, *37*, 44–50. [CrossRef]
34. Joshi, A.V. *Machine Learning and Artificial Intelligence*; Springer: Berlin/Heidelberg, Germany, 2020.
35. Mello, R.F.; Ponti, M.A. *Machine Learning: A Practical Approach on the Statistical Learning Theory*; Springer: Berlin/Heidelberg, Germany, 2018.
36. Kanimozhi, U.; Ganapathy, S.; Manjula, D.; Kannan, A. An intelligent risk prediction system for breast cancer using fuzzy temporal rules. *Natl. Acad. Sci. Lett.* **2019**, *42*, 227–232. [CrossRef]
37. Kumru, M.; Kumru, P.Y. Fuzzy FMEA application to improve purchasing process in a public hospital. *Appl. Soft Comput.* **2013**, *13*, 721–733. [CrossRef]

38. Şimşek, B.; Ic, Y.T. Fuzzy failure mode and effect analysis application to reduce risk level in a ready-mixed concrete plant: A fuzzy rule based system modelling approach. *Math. Comput. Simul.* **2020**, *178*, 549–587. [CrossRef]
39. Yucesan, M.; Gul, M.; Celik, E. A holistic FMEA approach by fuzzy-based Bayesian network and best–worst method. *Complex Intell. Syst.* **2021**, *7*, 1547–1564. [CrossRef]
40. Yousefi, S.; Valipour, M.; Gul, M. Systems failure analysis using Z-number theory-based combined compromise solution and full consistency method. *Appl. Soft Comput.* **2021**, *113*, 107902. [CrossRef]
41. Rezaee, M.J.; Yousefi, S.; Eshkevari, M.; Valipour, M.; Saberi, M. Risk analysis of health, safety and environment in chemical industry integrating linguistic FMEA, fuzzy inference system and fuzzy DEA. *Stoch. Environ. Res. Risk Assess.* **2020**, *34*, 201–218. [CrossRef]
42. Hassan, S.; Wang, J.; Kontovas, C.; Bashir, M. Modified FMEA hazard identification for cross-country petroleum pipeline using Fuzzy Rule Base and approximate reasoning. *J. Loss Prev. Process Ind.* **2022**, *74*, 104616. [CrossRef]
43. Araichi, S.; Almulhim, T. Vine copulas and fuzzy inference to evaluate the solvency capital requirement of multivariate dependent risks. *Appl. Econ.* **2021**, *53*, 6058–6074. [CrossRef]
44. Ivančan, J.; Lisjak, D. New FMEA Risks Ranking Approach Utilizing Four Fuzzy Logic Systems. *Machines* **2021**, *9*, 292. [CrossRef]
45. Sethukkarasi, R.; Ganapathy, S.; Yogesh, P.; Kannan, A. An intelligent neuro fuzzy temporal knowledge representation model for mining temporal patterns. *J. Intell. Fuzzy Syst.* **2014**, *26*, 1167–1178. [CrossRef]
46. Priya, P.I.; Ghosh, D.K.; Kannan, A.; Ganapathy, S. Behaviour analysis model for social networks using genetic weighted fuzzy c-means clustering and neuro-fuzzy classifier. *Int. J. Soft Comput.* **2014**, *9*, 138–142.
47. Boran, S.; Gökler, S.H. A Novel FMEA Model Using Hybrid ANFIS–Taguchi Method. *Arabian J. Sci. Eng.* **2020**, *45*, 2131–2144. [CrossRef]
48. Alawad, H.; An, M.; Kaewunruen, S. Utilizing an Adaptive Neuro-Fuzzy Inference System [ANFIS] for overcrowding level risk assessment in railway stations. *Appl. Sci.* **2020**, *10*, 5156. [CrossRef]
49. Ganapathy, S.; Yogesh, P.; Kannan, A. Intelligent agent-based intrusion detection system using enhanced multiclass SVM. *Comput. Intell. Neurosci.* **2012**, *2012*, 9. [CrossRef]
50. Vijilious, M.L.; Ganapathy, S.; Bharathi, V.S.; Kannan, A. A Novel Biometric Authentication using Contourlet Transform and Enhanced MSVM. *Eur. J. Sci. Res.* **2011**, *65*, 370–376.
51. Efe, M.O. A comparison of ANFIS, MLP and SVM in identification of chemical processes. In Proceedings of the 2009 IEEE Control Applications, (CCA) & Intelligent Control, (ISIC), St. Petersburg, Russia, 8–10 July 2009; pp. 689–694.
52. Okabe, T.; Otsuka, Y. Proposal of a Validation Method of Failure Mode Analyses based on the Stress-Strength Model with a Support Vector Machine. *Reliab. Eng. Syst. Saf.* **2021**, *205*, 107247. [CrossRef]
53. Yin, G.; Zhang, Y.T.; Li, Z.N.; Ren, G.Q.; Fan, H.B. Online fault diagnosis method based on incremental support vector data description and extreme learning machine with incremental output structure. *Neurocomputing* **2014**, *128*, 224–231. [CrossRef]
54. Mangeli, M.; Shahraki, A.; Saljooghi, F.H. Improvement of risk assessment in the FMEA using nonlinear model, revised fuzzy TOPSIS, and support vector machine. *Int. J. Ind. Ergon.* **2019**, *69*, 209–216. [CrossRef]
55. Ayodeji, A.; Liu, Y.K. Support vector ensemble for incipient fault diagnosis in nuclear plant components. *Nucl. Eng. Technol.* **2018**, *50*, 1306–1313. [CrossRef]
56. Soltanali, H.; Khojastehpour, M.; Farinha, J.T.; de Almeida e Pais, J.E. An Integrated Fuzzy Fault Tree Model with Bayesian Network-Based Maintenance Optimization of Complex Equipment in Automotive Manufacturing. *Energies* **2021**, *14*, 7758. [CrossRef]
57. Farinha, J.M.T. *Asset Maintenance Engineering Methodologies*; CRC Press: Boca Raton, FL, USA, 2018.
58. Guimarães, A.C.; Lapa, C.M. Fuzzy inference to risk assessment on nuclear engineering systems. *Appl. Soft Comput.* **2007**, *7*, 17–28. [CrossRef]
59. Preyssl, C. Safety risk assessment and management—The ESA approach. *Reliab. Eng. Syst. Saf.* **1995**, *49*, 303–309. [CrossRef]
60. Yazdi, M.; Soltanali, H. Knowledge acquisition development in failure diagnosis analysis as an interactive approach. *Int. J. Interact. Des. Manuf. (IJIDeM)* **2019**, *13*, 193–210. [CrossRef]
61. Chanamool, N.; Naenna, T. Fuzzy FMEA application to improve decision-making process in an emergency department. *Appl. Soft Comput.* **2016**, *43*, 441–453. [CrossRef]
62. Gerek, I.H. House selling price assessment using two different adaptive neuro-fuzzy techniques. *Autom. Constr.* **2014**, *41*, 33–39. [CrossRef]
63. Benkedjouh, T.; Medjaher, K.; Zerhouni, N.; Rechak, S. Remaining useful life estimation based on nonlinear feature reduction and support vector regression. *Eng. Appl. Artif. Intell.* **2013**, *26*, 1751–1760. [CrossRef]
64. Soltanali, H.; Nikkhah, A.; Rohani, A. Energy audit of Iranian kiwifruit production using intelligent systems. *Energy* **2017**, *139*, 646–654. [CrossRef]
65. Kutyłowska, M. Neural network approach for failure rate prediction. *Eng. Fail. Anal.* **2015**, *47*, 41–48. [CrossRef]

sustainability

Article

Occupational Risk Assessment in Native Rainforest Management (MIAR^{forest})—Parameters Definition and Validation

Killian Lima [1], Ana C. Meira Castro [2] and João Santos Baptista [1,*]

[1] Associated Laboratory for Energy, Transports and Aeronautics (LAETA-PROA), Faculty of Engineering, University of Porto (FEUP), Rua Dr. Roberto Frias, s/n, 4200-465 Porto, Portugal
[2] Centro de Recursos Naturais e Ambiente (CERENA-Porto), School of Engineering of the Polytechnic of Porto, Rua Dr. António Bernardino de Almeida, 431, 4200-072 Porto, Portugal
* Correspondence: jsbap@fe.up.pt

Abstract: Maintaining native rainforests as a sustainable ecosystem and their resilience to external pressures involves their economic profitability as a natural resource of unique and renewable products. For this purpose, new approaches have been developed and refined. This work seeks to contribute in this direction in the context of occupational safety and health (OSH) by presenting a new method for integrated assessment of risks for rainforests (MIAR^{forest}). The MIAR^{forest} is based on the MIAR, a method that has shown promising results in occupational risk assessment in different industrial sectors. Its parameters were discussed and assessed to improve their relevance, wording and risk assessment through the Delphi methodology by a panel of 62 experts in forestry and OSH who responded independently to questionnaires made available through Google Forms. A consensus of over 79% among the experts was reached in two rounds. This result highlights the high objectivity and the low percentage of dubious possible interpretations of the parameters and sub-parameters of this occupational risk assessment method.

Keywords: risk assessment; occupational risk; native forest; forest management; MIAR; Delphi

Citation: Lima, K.; Meira Castro, A.C.; Baptista, J.S. Occupational Risk Assessment in Native Rainforest Management (MIAR^{forest})—Parameters Definition and Validation. *Sustainability* **2023**, *15*, 6794. https://doi.org/10.3390/su15086794

Academic Editors: Esmaeil Zarei, Samuel Yousefi and Mohsen Omidvar

Received: 10 February 2023
Revised: 5 April 2023
Accepted: 14 April 2023
Published: 18 April 2023

1. Introduction

Tropical forests cover more than 9,300,000 km^2, of which the Amazonian forest occupies around 6,700,000 km^2, more than 70% of the total area. This vital forest occupies significant areas of Brazil, Bolivia, Colombia, Ecuador, French Guyana, Guyana, Peru, Venezuela and Suriname.

There are, however, other tropical forests that it is important to list. Of these, the largest is the Congo rainforest, which covers more than 1,800,000 km^2 between Cameroon, the Central African Republic, Congo, the Democratic Republic of Congo, Equatorial Guinea and Gabon. Other smaller but significant rainforests are the Papua New Guinea rainforest of approximately 545,000 km^2, the Borneo rainforest of approximately 290,000 km^2, the Xishuangbanna (China) rainforest of approximately 19,000 km^2 and the Daintree (Australia) rainforest of approximately 1200 km^2.

The sustainable exploitation of native rainforests is necessary for their resilience to external pressures and their maintenance as a unique and necessary ecosystem for the future of humanity [1]. Native rainforest exploitation is distinguished from industrial forest exploitation by its principles of biodiversity maintenance and respect for nature [2].

Sustainable forest management is a holistic approach that aims to ensure the use of planning practices and conservation principles so that a forest can continuously supply a given product or service [3]. It is believed to correspond to the management of the forest for obtaining economic, social and environmental benefits by following planning practices and nature conservation principles that guarantee that the forest is capable of supplying, on an ongoing basis, a specific product or service without jeopardising the sustainability of the ecosystem while being subject to management. The sustainable exploitation of the native rainforest is thus distinguished from industrial forest exploitation by following biodiversity and sustainability principles.

Professionals who carry out their work in forest exploitation are undeniably exposed to the risk of accidents and diseases inherent to the work [4–9]. Occupational risk assessment is a process that allows organisations to implement a proactive management policy in workplaces to prevent the occurrence of occupational accidents and diseases [10–12]. There are several occupational risk analysis techniques and methods for their assessment, from more generalist to more specific. The choice of the method to be used is typically based on its suitability for the activities under analysis and their correlated specificities. However, the use of matrix methods, generally not validated, has a weakness that emerges from its subjectivity. That is, the assessment depends on the experience and perception of the assessor [13,14]. In fact, to date, no well-established methodology in an aggregated form allows for the complete and simultaneous identification of all occupational aspects of a company's activity. Perhaps this is why it is common for organisations to develop their own safety management systems and, therefore, their own methods of occupational risk assessment [15–21].

Different approaches and methods are applied in the particular context of forestry activities. The most widely used are the AHP—analytic hierarchy process [10–21], the MMR—method of the magnitude of risk [22] and the PARCF—process of risk assessment in forest harvesting [23]. However, these methods are not specific to managing activities in native rainforests.

A method that has shown promising results in occupational risk assessment in different industrial sectors is the method for the integrated assessment of risk (MIAR) [24]. The MIAR follows control banding (CB) principles. CB is a risk-management strategy used to control occupational hazard exposure. It is a simplified approach which can be used to identify and implement appropriate control measures based on hazard levels and potential exposure.

Creating a new version of the method, specially developed for the evaluation of occupational risks in the native rainforest, will promote the sustainable use of these forests and thus contribute to their preservation by allowing workers to work with greater safety under the difficult conditions of this working environment.

Considering the above, it was defined as the objective of this work to identify the parameters and sub-parameters and perform their validation. In this way, we aim to contribute to the development of an occupational risk assessment methodology to support safety management systems in native rainforests by adopting the basic principles of the MIAR.

2. MIAR's Original Version—Short Presentation

The original version of the MIAR was developed to support the integration of management systems and allows the framing of the risk assessment of the environmental and occupational components. Its focus was on industrial activities in the chemical industry, creating synergies between the processes with the NP-EN-ISO 9001:2008—quality management system standard and the HAZOP—hazard and operability study method [24]. In this way, it allows organisations to improve their performance while at the same time being simple to apply and with reproducible and reliable results.

The first version of the MIAR has been investigated and applied in different industrial sectors such as metalworking [25], construction [26], industrial waste sorting [27], mining industry [28,29] and slaughterhouses [30]. These applications of the method always point towards high reproducibility, tending to be above 75%, and towards the reliability of the results, i.e., with the MIAR, the risks are assessed identically by different experts, and the results obtained are congruent with reality.

In the MIAR, the identification of hazards starts by identifying the sequence of industrial processes, sub-processes, activities and tasks, going down to the level considered adequate. It also includes the identification of materials and machinery used in activities, working conditions and constraints, the characteristics of the spaces where the activities occur and the surrounding spaces. Existing accident- and risk-protection equipment, minimisation procedures and potential failures are also checked [31].

In the MIAR, the risk is defined as a measure of the uncertainty of the occurrence of an event in a situation involving exposure to a hazard. The risk level (RL) is obtained as the product of two parameters, severity (S) and likelihood (Li), where Li is the product of the extent of impact (Ei) and the frequency of exposure (Fe) (Equations (1) and (2)). These parameters must be considered independently from each other.

$$\mathrm{Li} = \mathrm{Ei} \times \mathrm{Fe} \tag{1}$$

$$\mathrm{RL} = \mathrm{S} \times \mathrm{Li} \tag{2}$$

where accident severity corresponds to the likeliest consequence should the accident materialise, accident extent refers to the number of workers affected, and the frequency of exposure represents the time a worker is exposed to a given risk.

Within the scope of risk management, the prioritisation of interventions for risk mitigation considers another parameter, risk control (RC), calculated according to the ongoing organisational measures of accident prevention. Thus, after the valuation of the risk level, it is possible to estimate the weighted risk level (WRL) as a result of dividing the risk level by the risk control (Equation (3)):

$$\mathrm{WRL} = \mathrm{RL}/\mathrm{RC}, \tag{3}$$

In other words, WRL assesses the effectiveness of risk control processes (existing or soon to be implemented) according to a control hierarchy.

In all parameters, the valuation of the possible occurrence of an occupational accident is translated into the chromatic scale represented in Table 1.

Table 1. Association between colours and evaluation levels.

Absent/Very Low	Low	Moderate	High	Very High

3. Materials and Methods

The design of the new version of the MIAR for rainforests—the MIAR^{forest} (method for the integrated assessment of occupational risks in native rainforests)—was based on the MIAR's concepts defined in ISO 31000:2018 [32], namely those related to the three stages of the risk management program (identification, analysis and evaluation) and the concepts of hazard and risk, in accordance with ISO 45001:2018—the occupational health and safety (OHS) management system. Therefore, the MIAR^{forest} uses the equations of the MIAR (Equations (1)–(3)) and follows the principles of control banding (CB) methods, integrating information on potential hazards, levels of exposure and an assessment of occupational health and safety performance management systems. It seeks the latter to prioritise appropriate measures to minimise the impact of workers' exposure to hostile environments such as native rainforests.

To adapt the MIAR to the reality of native rainforest management activity, information was collected face-to-face in the Brazilian federal government forest holdings in the eastern Amazon. This data collection focused on the relevant hazards and risks at different stages of the management process [33]. Subsequently, a first draft of the MIAR^{forest} was developed based on this information and in the reference literature on this topic [7,13,14,34].

The validation process of each of the parameters and sub-parameters regarding their relevance and clarity of wording, as well as the risk-assessment scale considered, followed the premises of the Delphi methodology [35,36]. Therefore, a panel of experts, professionals in the forestry and OHS area, was invited to respond anonymously to the following questionnaire made available through Google Forms:

Do you agree with the proposed parameters and sub-parameters? Do you consider their description sufficiently clear?

- If you disagree with the presented proposal, what alternative wording do you propose?
- Do you agree with the assessment levels considered for the proposed parameters and sub-parameters?
- What valuation levels do you propose if you disagree with the proposal that has been made?

Do you consider that other sub-parameters should be added? Which ones and why?

The experts were informed that, when answering the survey, each parameter/sub-parameter should be assessed as if there were no influence on any other and that the chromatic scale used to assess the possibility of the occurrence of an occupational accident was chromatic (Table 1), with numerical assessment being performed at a later stage.

The answers obtained were analysed quantitatively and qualitatively, considering the experts' agreement with the proposals and their comments and suggestions. In accordance with the assumptions of the Delphi methodology, the parameters that did not obtain a consensus of opinions greater than 75% and/or received pertinent criticism regarding their definition were adjusted accordingly. A new proposal for the wording of the parameters and risk assessment was produced and submitted for consideration by the experts. In this new questionnaire, the questions to the experts were created in the same way as in the first round, with not only the new wording of the text to be evaluated being presented with all the changes made duly evidenced but also the version of the first round and the corresponding evaluation results.

This iterative process was stopped once the answers obtained reached a consensus of opinions higher than 75% [35,36]. Ten factors with direct influence on the severity (S) of accidents were identified:

- Two factors with global impact:
    i. Worker protection (WP): whether the worker is protected by personal protective equipment or by a collective protection system;
    ii. Forest typology (FT): tree density (in this research, only the ombrophilous forest is considered).
- Two controllable factors:
    iii. Machine- and tool-handling (MT): the level of protection that machines and tools have and the training that workers must have to use them;
    iv. Relationship between tasks (RBT): the number of and relationship between tasks being performed simultaneously on the same site.
- Six uncontrollable factors:
    v. Object fall (OF): the situation of a worker being hit not only by falling broken branches and/or trees but also by other objects such as logs and small tools/utensils that can fall during road and yard operations;
    vi. Terrain slope (TS): the ability/difficulty of maintaining the balance and the progress of workers both on foot and by vehicle within the forest;
    vii. Obstacles (Obst): the ability/difficulty of traversing vegetation, rivers and streams, fallen trees and rocks during road and yard operations;
    viii. Wild animals (WA): the presence of disease vectors, poisonous animals or predators that can cause severe injuries or death.
    ix. Precipitation intensity (PI): the feasibility of performing/or not performing work in rain;
    x. Wind intensity (WI): the feasibility of carrying out/or not carrying out work in windy conditions that contribute to the shaking of treetops as well as the falling of branches and objects onto the worker.

The parameters of exposure (E) and frequency (F) were renamed as, respectively, extent of impact (Ei) and frequency of exposure (Fe).

4. Delphi Rounds Results

The validation of the MIAR^{forest} was proposed to a group of 250 experts and was performed according to the Delphi approach. For the first round, 65 experts agreed to respond to the questionnaire; for the second round, there were 62 respondents. Of these professionals in the forestry and/or occupational safety area, 94% had university qualifications, and at least one had a postgraduate degree in occupational safety.

In the first round, the pertinence of the parameters, sub-parameters and respective levels was questioned. Experts' opinions and suggestions for changes were also collected.

The experts' suggestions considered relevant for the second round were introduced for the second version of the method. After this operation, the modified version was sent back to the experts to confirm their opinion on items that lacked consensus and validate the changes made.

The obtained results are shown in Table 2. All items with an approval rate greater than 75% could be considered validated at the end of the first round. However, only the sub-parameters of likelihood (Li)—(extent of impact (Ei) and frequency of exposure (Fe)), as well as the parameter of risk control (RC), were considered closed at the end of the first round and, due their pertinence, some changes suggested by the experts were introduced. The proposed modifications include the introduction of the type of protection for workers (individual PPE or collective cabin) and the subdivision of the sub-parameter "terrain characteristics" (TC) into two, the slope of the terrain (TS) and obstacles (Obst).

Table 2. Percentage of agreement obtained in the rounds.

Parameter/ Sub-Parameter		1st Round: Consideration of the Parameter	1st Round: Writing of Risk Levels	2nd Round: Parameter/ Sub-Parameter	2nd Round: Consideration of the Parameter	2nd Round: Writing of Risk Levels
S				WP	100.0	100.0
	FT	89.2	83.1			
	MT	89.2	86.2	MT	82.3	82.3
	RBT	89.2	86.2	RBT	82.3	83.9
				OF	88.7	88.7
	TC	93.8	93.8	TS	80.6	80.6
				Obst	83.9	82.3
	WA	95.4	90.8	WA	83.9	83.9
	PI	95.4	89.2	PI	80.6	80.6
	WI	96.9	89.2	WI	80.6	79.0
Li	Ei	89.2	84.6			
	Fe	95.4	89.2			
RC		100.0	95.4			

The introduced changes were presented to the experts in the second round, asking if they preferred the original or the new version with the changes. Most experts opted for the latest version (Table 2). The obtained results were considered sufficient to accept as finalising the process of defining the parameters through the Delphi methodology.

5. MIAR^{forest} in Detail

With a consensus among experts of over 79%, the MIAR^{forest} allows for identifying and assessing occupational risks arising from sustainable exploitation of native rainforests in different stages. In its design, simplicity was sought in the application of the method, associated with the quality of the results obtained from the risk assessment.

In the MIAR^{forest}, the parameters/sub-parameters assessment must be carried out considering the actual situation in which the work is performed. This means the tasks must be classified considering all safety measures already implemented during the evaluation.

The method generates risk assessment based on three groups of parameters corresponding to the severity, likelihood and capacity to control the risk of accident using the occupational health and safety management systems implemented in the organisation.

5.1. Severity

The severity of occupational accidents in rainforests depends on worker protection, which in turn depends on how the work is performed, that is, on whether the worker executes the activity while protected only by personal protective equipment (pedestrian work) or whether the worker executes the tasks while isolated from the environment (the forest) by a collective protection system (such as the cabin of a harvesting machine). These two scenarios will condition the accident's potential severity due to either controllable or uncontrollable factors. In addition, severity is also related to three classes of factors: meteorological, edapho-biological (ground, animals and vegetation) and operational (machinery and tools).

Thus, the MIAR^{forest} considers parameter severity (S) depending on ten sub-parameters and can be computed according to Equation (4). This equation was designed by assuming that all factors ("sub-parameters") significantly influence the severity of occupational accidents in native rainforests, each of them in their own particular way. Therefore, in order to avoid instability in results, the traditional computation using multiplication and sums of parameters was dropped, and a calculation that included a balance between median and maximum values was used instead. This option allows greater stability in the results and emphasises the sub-parameter with the highest potential for harm, whatever it may be.

The parameters used to assess the severity and the correspondent rating are summarised in Table 3.

$$S = WP \times FT \times \text{maximum}\ (MT, RBT, OF, TS, Obst, WA, PI, WI) \times \text{median}\ (MT, RBT, OF, TS, Obst, WA, PI, WI), \quad (4)$$

Table 3. MIAR^{forest} Severity parameters.

Subparameter	Level Description	Rating
Worker protection (WP)	Individual protection	1
	Collective protection	0.25
Forest typology (FT)	Submontane dense ombrophilous forest.	16
	Submontane open ombrophilous forest.	8
	Alluvial dense ombrophilous forest.	4
	Lowland dense ombrophilous forest.	2
	Lowland open ombrophilous forest.	1
Machinery and tools (MT)	Forest harvesting machine with a manual device, e.g., steel cable.	16
	Forest harvesting machine with a hydraulic device, e.g., grapple, blade.	8
	Portable forest harvesting machine, e.g., chainsaw.	4
	Hand tool, e.g., machete, wedge, sledgehammer.	2
	No use of tool or machine—situation without injury or damage.	1
Relationship between tasks (RBT)	>3 different tasks running simultaneously.	8
	Three different tasks running simultaneously.	4
	Two distinct and dependent tasks running simultaneously but lagged.	2
	Two separate and independent tasks running simultaneously.	1
	One task.	0.5
Precipitation intensity (PI)	Precipitation intensity > 0.5 mm/h.	256
	0 mm/h < precipitation intensity ≤ 0.5 mm/h.	2
	Without precipitation, precipitation probability—60% < pp ≤ 100%.	1
	Without precipitation, precipitation probability—0% < pp ≤ 60%.	0.5
	No precipitation, precipitation probability 0%.	0.25

Table 3. *Cont.*

Subparameter	Level Description	Rating
Wind intensity (WI)	Wind intensity > 40 km/h.	256
	20 km/h < wind intensity ≤ 40 km/h.	2
	10 km/h < wind intensity ≤ 20 km/h.	1
	0 km/h < wind intensity ≤ 10 km/h.	0.5
	No wind.	0.25
Object fall * (OF)	Fall of an object with sufficient energy to cause death or total permanent disability.	48
	Fall from an object with sufficient energy to cause severe injury with total temporary incapacity or partial but low-percentage permanent incapacity.	24
	Fall of an object with sufficient energy to cause minor injuries with partial temporary incapacity but low severity.	8
	Fall of an object with sufficient energy to cause minor injuries without any form of disability.	6
	Fall of an object without sufficient energy to cause injury to the worker.	3
Terrain slope (TS)	Strongly sloping surface (30–45%).	4
	Moderate sloping surface (8–30%).	2
	Smoothly sloping surface (3–8%).	1
	Flat surface (0–3%).	0.5
	Flat surface 0%.	0.25
Obstacles Obst	Surface with obstacles that are impossible to cross on foot.	4
	Surface with obstacles that are difficult to cross.	2
	Surface with obstacles that are easy to cross and/or remove.	1
	Surface with obstacles that are very easy to cross.	0.5
	Unobstructed surface.	0.25
Wild animals (WA)	Contact resulting in injury or damage by large mammals (e.g., *Panthera onca*), snakes with high venom inoculation (e.g., *Micrurus altirostris*), venomous spiders (e.g., *Loxosceles amazonica*) and swarms of bees.	4
	Contact resulting in injury or damage by mid-sized mammals in flocks (e.g., *Pecari tajacu*), snakes with moderate venom inoculation (e.g., *Bothrops jararaca* or *Lachesis muta*) and scorpions.	2
	Contact resulting in injury or damage by small mammals, snakes with low venom inoculation (e.g., *Helicops angulatus*).	1
	Contact resulting in injury or damage by isolated insects (e.g., *Paraponera clavata*).	0.5
	There is no contact with animals.	0.25

* not submitted for validation.

5.1.1. Forest Typology (FT)

The severity (S) of occupational accidents in a native rainforest depends on the characteristics of the forest where the worker carries out the activity. It was decided that S is higher if the activity is carried out in a forest with a high density of trees and other plants and at altitude. Since in the MIAR^{forest} only ombrophilous forest, characteristic of the Atlantic forest and Amazon biomes, is currently considered, the severity levels were defined based on the different characteristics that this type of forest may possess, depending on the altitude at which they are located [37,38].

5.1.2. Worker Protection (WP)

Only two basic situations in which work is executed in forestry operations will be considered for the work protection parameter. The first is pedestrian work, where the worker is equipped with personal protective equipment (PPE). The second refers to work inside a machine cabin. In addition to PPE, the worker can rely on the protection of the cabin itself, which functions as collective protection equipment against hazards existing in the surrounding environment.

5.1.3. Machines and Tools Handling (MT)

Regarding the use of machines and tools, the MIAR^{forest} states that the severity depends on the type of machine or tool the worker uses when carrying out the activity. The severity levels were defined according to the bibliography on this subject specifically applied to native forest exploitation [4,8,39–47]. In the valuation of this parameter, it was decided that the severity is higher if the activity is carried out with forestry machines such as, for example, a loader, a forest tractor, a tracked tractor or a logging truck. These machines may have steel cables, hydraulic clamps or other accessories attached. Only the chainsaw was considered a portable forest-harvesting machine, and manual tools without a motor, such as a machete, wedge or sledgehammer, were considered. The severity levels of occupational accident occurrence according to the type of machine or tool used and/or situations arising from their use were defined according to Roloff [48].

5.1.4. Relationship between Tasks of the Same Activity

In the MIAR^{forest}, the severity depends on the number of tasks that need to be performed as part of the same activity and their degree of dependence. An illustrative example of an activity with more than three distinct tasks running simultaneously is the activity of cubage, which the following professionals traditionally perform: note taker, chainsaw operator, loader operator, measurer and painter/planker. An illustrative example of an activity with only one task is the felling activity, traditionally performed by the following professionals: chainsaw operator and helper. The five levels of severity of occupational accident occurrence as a function of the number of tasks involved in the same activity were defined according to [33] and EMBRAPA guidelines [49].

5.1.5. Meteorological Conditions—Precipitation and Wind Intensity

In the MIAR^{forest}, two sub-parameters related to meteorology are considered—precipitation intensity (PI) and wind intensity (WI). For each of these sub-parameters, five severity levels were defined in accordance with the World Meteorological Organization (WMO) [50], the Beaufort wind scale [51], the specifications officially established by Brazilian Civil Defense [52] and the recommendations of the Tropical Forestry Institute on this subject [53].

In addition to the above references, in defining the severity levels, the opinion of experts in occupational safety in logging operations in native rainforests was also considered. According to the experience of these experts, activities should be wholly suspended during precipitation greater than 0.5 mm/h, as terrain and road conditions, if unpaved, become impractical for working safely. The same experts also stated the importance of considering wind when assessing safety conditions. However, the wind speed value at which safety conditions are compromised did not reach consensus (values ranged from 19 to 44 km/h). Thus, a 30 km/h speed was defined as the value at which the activities should be suspended. It was decided that above this value, the larger branches of trees become very agitated, and this agitation or resulting falls can cause severe accidents. The same can occur when the wind changes direction.

5.1.6. Object Fall (OF)

Occupational accident severity depends on the forest's characteristics where the worker carries out the activity. As the MIAR^{forest} considers ombrophilous forest with incidence of lianas or woody vines, this parameter is related to the movement of vegetation or parts of it (branch, trunk, etc.) and situations resulting from its movement in this forest typology, distinguishing situations that may eventually cause death, disability or superficial injury.

5.1.7. Site Characteristics—Terrain Slope (TS) and Obstacles (Obst)

The MIAR^{forest} states that severity is higher when work is carried out along a path with irregular terrain with a slope and obstacles to overcome and lower if the activity is carried out along a path on regular terrain with no slope and no obstacles to overcome. The severity levels of the occupational accident according to the characteristics of the path were defined according to specific bibliography on this topic, adapted to native forestry [8,39–46]. In the valuation of this parameter, small bridges, such as a bush or a *penguela*, were considered obstacles of difficult transposition. Trails constructed on waterlogged and unstable ground were also considered with the same severity level as trails with difficult obstacles. On the opposite side and with the lowest valuation are obstacles such as rocky outcrops of small size and/or fallen trunks/vegetation that can be easily transferred to another location, clearing the path.

The sub-parameter terrain slope (TS) definition met the experts' suggestions and the New Brazilian Forest Code, Law 12.651, of 25 May 2012, which defines 45° as the maximum slope limit allowed to operate in permanent preservation areas.

It was also decided that the severity is more significant in the case of the activity being developed with obstacles to overcome and lower otherwise, so large rocky outcrops were designated as obstacles of very complex transposition and occasionally requiring the use of external supports to overcome them, small bridges, such as a culvert or a footbridge, as obstacles of difficult transposition, small rocky outcrops and fallen trunks/vegetation which can easily be transferred to another location, clearing the path, as obstacles of easy transposition or removal and the paths performed on waterlogged and unstable ground as obstacles of very easy transposition.

5.1.8. Wild Animals

Regarding the presence (or not) of wild animals (WA) in the forest where the worker carries out the activity, it was decided that the severity is more significant if the activity is carried out within the presence of animals that are large and/or have a high potential for toxicity by bite or sting. The severity levels of the occurrence of occupational accidents due to the presence of wild animals was defined according to support from the SINAN—Sistema de Informação de Agravos de Notificação [54] and information from health professionals. The possibility of the occupational accident occurring not only by contact with venomous animals but also with other animals, such as mammals or birds, was considered.

5.1.9. Severity Bands

Severity is classified into five bands, as shown in Table 4. These bands comprise the values obtained according to Equation (3) and the values in the scores column in Table 4.

Table 4. MIAR^{forest} Severity bands.

Severity	Bands
Extreme	$S \geq 192$
High	$96 \leq S < 192$
Medium-high	$48 \leq S < 96$
Medium-low	$24 \leq S < 48$
Low	$S < 24$

To determine the limit of the bands, it was decided that all parameters would maintain the second lowest level of their scales, varying only the most harmful factor, falling objects.

1. Upper limit for "low severity" (<24)—all sub-parameters at the 2nd level of the respective scale;
2. Upper limit for "medium-low severity" (<48)—sub-parameter "object fall" at the 3rd level of the respective scale and the remaining parameters at the 2nd level of the respective scale;
3. Upper limit for "medium-high severity" (<96)—sub-parameter "object fall" at the 4th level of the respective scale and the remaining parameters at the 2nd level of the respective scale;
4. Upper limit for "high severity" (<192)—sub-parameter "object fall" at the 5th level of the respective scale and the remaining parameters at the 2nd level of the respective scale;
5. Lower limit for "extreme severity" (≥192)—sub-parameter "object fall" at the 5th level of respective scale and remaining parameters at the 2nd level of respective hierarchy.

5.2. *Likelihood*

The exposure value results from the association of two sub-parameters, the Extent of impact (Ei) and the Frequency of exposure (Fe), each with five bands (Table 5).

Table 5. MIAR^{forest} likelihood.

	Exposure—Ex—(Description)	Score
Extent of Impact Ei	>5 workers.	5
	4 workers.	4.9
	3 workers.	4.7
	2 workers.	4.4
	1 worker.	4
Frequency of Exposure Fe	Continuous (every day of the week).	5
	Usual (≥3 days/week).	4.5
	Partial (<3 days/week).	4
	Sporadic (≤1 day/week).	3.5
	Punctual (≤1 h/week).	3

The likelihood value is obtained by Equation (1) and is classified into five bands, as shown in Table 6.

Table 6. Likelihood bands.

Likelihood	Bands
Extreme	$E > 22$
High	$19.8 < E \leq 22$
Medium-high	$17.6 < E \leq 19.8$
Medium-low	$15.4 < E \leq 17.6$
Low	$E \leq 15.4$

5.2.1. Extent of Impact

The extent of impact is directly related to the number of workers that may be affected by the same occurrence during the activity(ies) they are performing. In this sub-parameter, five levels have been defined according to the number of workers potentially affected by an accident [33].

5.2.2. Frequency of Exposure

This sub-parameter is related to the length of time the worker is exposed to the risk of an occupational accident. Thus, allowing that the probability of an occupational accident increases with the time of exposure to the hazard [55], the five bands were defined considering different exposure periods in a continuous sequence. The attribution of the exposure time should be performed by attending to the following criteria:

- Continuous—the same activity is performed continuously and daily throughout the week;
- Usual—the same activity is performed for a period equal to or greater than half of the worker's weekly working hours;
- Partial—the same activity is performed during a period equal to or less than half of the worker's weekly working hours;
- Sporadic—the same activity is performed for a period equal to or less than one day during the working week;
- Occasional—the same activity is performed during a period equal to or less than one hour during the working week.

This classification is only valid for routine activities. Activities which are not included in the worker's weekly routine must be assessed separately.

5.2.3. Likelihood Bands Explanation

The likelihood bounds were calculated considering a task performed by a maximum of two workers for different exposure times, as explained below:

1. Upper limit for low likelihood—two workers exposed on an occasional basis (<1 h/week);
2. Upper limit for medium-low likelihood—two workers exposed on a sporadic basis (≤1 day/week);
3. Upper limit for medium-high likelihood—two workers exposed about half of the working week (<3 days/week);
4. Upper limit for high likelihood—two workers exposed about half of the working week (≥3 days/week);
5. Lower limit for extreme likelihood—two workers exposed continuously (every day of the week).

5.3. Risk Control

The risk to which workers are exposed when performing a task is not necessarily the same whether the task is performed in a company with an efficient risk management and control system or in a company without any risk control system. Therefore the MIAR and the $MIAR^{forest}$ take this into consideration when assessing risk. The $MIAR^{forest}$ considers various levels of implementation of risk management systems (Table 7) that should be included in the risk level assessment.

Table 7. Risk Control—Performance of prevention systems.

Performance of Prevention Systems (Description)	Score
There is no occupational health and safety management system or any control of occupational health and safety.	0.50
There is no occupational health and safety management system in place, and there is an occupational health and safety control system with visible flaws in its operation.	0.75
There is no occupational health and safety management system in place, but there is an occupational health and safety control system with evidence of operational practices.	1.00
There is an occupational health and safety management system, but there is no objective evidence of a continuous improvement culture.	1.50
There is a continuous improvement culture linked to an occupational health and safety management system with evidence of its functionality.	2.00

5.4. Risk Level

The $MIAR^{forest}$ states that the risk level (RL) is based on severity and likelihood. However, it also states that the resulting value must be weighted by the performance of the prevention systems existing in the organisation. Thus, the weighted risk level is determined according to Equation (5).

$$WRL = \frac{S \times Li}{RC}, \tag{5}$$

The rationale for determining each of the different risk bands is based on the fact that in native rainforest management operations, the vast majority of tasks are performed by three or fewer workers on a non-continuous basis. Starting from this reality, risk level band limits were calculated for these operational conditions of likelihood (maximum value—19.8) combined with the band limits already defined for severity. For all the levels of risk, the existence of an occupational health and safety control system, with evidence of operational practices (score 1—Table 7) is considered.

1. Risk Level 1 (RL1)—the band's upper limit represents the combination with a medium-low severity level (maximum value—24). To the result of this combination (475.2), the decimal digits have been truncated (475). Values lower than 475 obtained with other configurations are also included in this band.
2. Risk Level 2 (RL2)—the upper limit of the band is the combination of the defined likelihood ratio with a medium-low severity level (maximum value 48). To the result

of this combination (950.4), the decimal digits have been truncated (950). Scores of 950 or lower obtained with other configurations are also accepted.

3. Risk Level 3 (RL3)—the upper limit of the band is the combination of the defined likelihood ratio with a medium-high severity level (maximum value—96). To the result of this combination (1900.8), the decimal digits have been truncated (1900). Values lower than 1900 obtained with other configurations are also included in this band.
4. Risk Level 4 (RL4)—the band's upper limit is the combination of the defined likelihood ratio with a high level of severity (maximum value—192). The result of this combination (3801.6) has been rounded to 3800. Values lower than 3800 obtained with other configurations are also accepted in this band.
5. Risk Level 5 (RL5)—This band represents the maximum level of risk and includes all values above 3800.

The risk level bands and the respective assigned scores are presented in Table 8.

Table 8. MIAR^{forest} Risk level.

Risk Level 5 (RL5)	RL5 > 3800
Risk Level 4 (RL4)	1900 < RL4 ≤ 3800
Risk Level 3 (RL3)	950 < RL3 ≤ 1900
Risk Level 2 (RL2)	475 < RL2 ≤ 950
Risk Level 1 (RL1)	RL1 ≤ 475

5.5. Prioritisation of Control Measures

The prioritisation of control measures proceeds from risk level. That is, the highest risk level is the primary priority, and lower risk levels correspond to lower priority, as shown in Table 9.

Table 9. Prioritisation of control measures.

Priority	Risk Level	Description
V	RL5 > 3800	Unacceptable conditions. The activity/task should be suspended immediately. The activity/task should only be restarted after a detailed risk assessment and the definition and implementation of corrective actions and control measures.
IV	1900 < RL4 ≤ 3800	Critical conditions which require urgent correction. A detailed risk assessment and the short-term definition and implementation of corrective actions and control measures are required.
III	950 < RL3 ≤ 1900	Conditions to improve. Preventive and control measures should be taken/revised.
II	475 < RL2 ≤ 950	Conditions subject to surveillance. Possible improvements should be considered.
I	RL1 ≤ 475	Conditions in which no immediate intervention is required.

5.6. Control Measures

Mitigation and control of occupational risks resulting from native forest management activities shall incorporate good occupational health and safety practices. The implementation of control measures shall be in accordance with the following hierarchy:

1. Application of administrative controls, including worker procedures and training, emergency management and medical monitoring;
2. Implementation of missing personal protective equipment (PPE) as a complement to other control measures;
3. Modification of the process or process conditions, where possible, to enclose or isolate the worker from the environment to prevent exposure;
4. Implementation of engineering controls and external advice;
5. Elimination of the specific risks of a particular activity by avoiding carrying it out or replacing it with one of lower risk.

6. Conclusions

The process of timber forest management involves various risks related to the safety and health of the worker, which is known to result in a high number of accidents, generally serious ones. Workers' physical safety and health can be safeguarded by complying with legislation on health and safety at work and by implementing mechanisms for monitoring the forest management activity itself.

Assessing occupational risks allows employers to effectively protect workers from work accidents and occupational diseases. The MIAR^{forest} permits the assessment of occupational risk associated with the process of timber forest management in native rainforests in an appropriate way as it includes, among other aspects, the assurance that all relevant risks are taken into account and verifies the effectiveness of the safety measures adopted.

According to the results, the MIAR^{forest} seems to be a promising method for occupational risk assessment with the potential to be implemented strategically and systematically by the native rainforest industry.

The main added value of this method is that it is relatively simple to apply and allows reliable conclusions to be drawn. However, the MIAR^{forest} is not a closed method. Therefore, a website is being designed with two objectives, dissemination of the method and collection of suggestions for change in order to achieve a continuous improvement of the method.

Author Contributions: Conceptualisation, K.L., A.C.M.C. and J.S.B.; methodology A.C.M.C. and J.S.B.; software, K.L. and J.S.B.; validation, A.C.M.C. and J.S.B.; formal analysis, K.L., A.C.M.C. and J.S.B.; resources, J.S.B.; data curation, K.L.; writing—original draft, K.L.; writing—review and editing, A.C.M.C. and J.S.B. visualisation, A.C.M.C. and J.S.B.; supervision, A.C.M.C. and J.S.B.; project administration, A.C.M.C. and J.S.B.; funding acquisition, J.S.B. All authors have read and agreed to the published version of the manuscript.

Funding: This research was funded by the Doctoral Program in Occupational Safety and Health of the University of Porto, grant number demsso.ksf.PD9986 and The APC was funded by Biomechanics and Health Unit of the Associated Laboratory for Energy, Transports and Aeronautics (LAETA), FCT-UIDB/50022/2020.

Institutional Review Board Statement: Not applicable.

Informed Consent Statement: Not applicable.

Data Availability Statement: Non-confidential data is available on request.

Acknowledgments: The authors gratefully acknowledge the support of the CERENA Strategic project FCT-UIDB/04028/2020.

Conflicts of Interest: The authors declare no conflict of interest.

References

1. Gustafsson, L.; Baker, S.C.; Bauhus, J.; Beese, W.J.; Brodie, A.; Kouki, J.; Lindenmayer, D.B.; Lõhmus, A.; Pastur, G.M.; Messier, C.; et al. Retention Forestry to Maintain Multifunctional Forests: A World Perspective. *BioScience* **2012**, *62*, 633–645. [CrossRef]
2. *IN5—Instrução Normativa N° 5, de 11 de Dezembro de 2006*; DF, 13 dez; Diário Oficial da União: Brasília, Brazil, 2006; p. 155.
3. ISO 14001:2004 Environmental Management Systems—Requirements with Guidance for Use. Available online: https://www.iso.org/standard/31807.html (accessed on 1 January 2022).
4. Ackerknecht, C. Occupational Accidents Footprint: New Concept linked to Chain of Custody in Sustainable Forest Management. *Cienc. Trab.* **2014**, *16*, 131–137. [CrossRef]
5. Bermudes, W.; Barros, E. Work accident incidence rates from 2007 to 2013 in planted forests in the Brazilian states and the risks of this activity. *Vértices* **2016**, *18*, 53–64. [CrossRef]
6. EU-OSHA. E-Facts 29: Occupational Safety and Health in Europe's Forestry Industry. 2008. Available online: https://osha.europa.eu/en/tools-and-publications/publications/e-facts/efact29 (accessed on 1 January 2022).
7. ILO—International Labour Organization. Safety and Health in Forestry Work—ILO Codes of Practice. 1998. Available online: https://www.ilo.org/wcmsp5/groups/public/@ed_protect/@protrav/@safework/documents/normativeinstrument/wcms_107793.pdf (accessed on 1 January 2022).
8. Medeiros, J.; Jurado, S. Acidentes de trabalho em empresas florestais de plantio, cultivo e extração de madeira. *Rev. Agrogeoambiental* **2013**, *5*, 87–96. [CrossRef]

9. Viegas, J.; Freitas, L.; Santos, R.; Leite, A.; Fiedler, N. Work accidents in the Brazilian forest sector. *Floresta* **2017**, *47*, 561. [CrossRef]
10. Ackerknecht, C. Occupational Safety Indicators for Forest Operations, Sawmilling and Wood-Based Panels Manufacture; an International Benchmarking. *Cienc. Trab.* **2015**, *17*, 89–98. [CrossRef]
11. Albizu-Urionabarrenetxea, P.M.; Tolosana-Esteban, E.; Roman-Jordan, E. Safety and health in forest harvesting operations. Diagnosis and preventive actions. A review. *For. Syst.* **2013**, *22*, 392–400. [CrossRef]
12. Leka, S.; Wassenhove, W.V.; Jain, A. Is psychosocial risk prevention possible? Deconstructing common presumptions. *Saf. Sci.* **2015**, *71*, 61–67. [CrossRef]
13. EU-OSHA—Guidance on Risk Assessment at Work. 1996. Available online: https://op.europa.eu/en/publication-detail/-/publication/1a3462b0-728c-4a2b-88f0-6c641b91a86f (accessed on 24 October 2022).
14. ILO Guidelines for Labour Inspection in Forestry. 2005. Available online: https://www.ilo.org/safework/info/standards-and-instruments/WCMS_107610/lang--en/index.htm (accessed on 13 April 2023).
15. Unver, S.; Ergenc, I. Safety risk identification and prioritise of forest logging activities using analytic hierarchy process (AHP). *Alex. Eng. J.* **2021**, *60*, 1591–1599. [CrossRef]
16. Camara, G.; Assunção, A.; Lima, F. The limitations of the traditional approach to work accidents: The case of timber exploitation in Minas Gerais, Brazil. *Rev. Bras. de Saude Ocup.* **2007**, *32*, 41–51. [CrossRef]
17. Conway, S.H.; Pompeii, L.A.; Casanova, V.; Douphrate, D.I. A Qualitative Assessment of Safe Work Practices in Logging in the Southern United States. *Am. J. Ind. Med.* **2017**, *60*, 58–68. [CrossRef] [PubMed]
18. Enez, K.; Topbas, M.; Acar, H. An evaluation of the occupational accidents among logging workers within the boundaries of Trabzon Forestry Directorate, Turkey. *Int. J. Ind. Ergon.* **2014**, *44*, 621–628. [CrossRef]
19. Heck, S.; Oliveira, L. Evaluation of Health and Safety at Work: Chain Saw Operators in the Region of Campos Gerais in State of Paraná. *Espacios* **2015**, *36*, 11–20. Available online: http://www.revistaespacios.com/a15v36n08/15360811.html (accessed on 2 February 2021).
20. Lagerstrom, E.; Magzamen, S.; Rosecrance, J. A mixed-methods analysis of logging injuries in Montana and Idaho. *Am. J. Ind. Med.* **2017**, *60*, 1077–1087. [CrossRef] [PubMed]
21. Melemez, K. Risk factor analysis of fatal forest harvesting accidents: A case study in Turkey. *Saf. Sci.* **2015**, *79*, 369–378. [CrossRef]
22. Kuhn, G.C.; Kolodziej, S.F.; Cruz, E.R. Evaluation of risks in the forest terrestrial transportation. *Procedia Manuf.* **2015**, *3*, 4808–4815. [CrossRef]
23. Bermudes, W.L.; Minette, L.J.; Soranso, D.R.; Schettino, S. Aplicação do processo de avaliação de risco em atividades de colheita florestal semimecanizada e mecanizada. *Vértices* **2020**, *22*, 59–81. [CrossRef]
24. Antunes, F.A.; Baptista, J.S.; Diogo, M.T. Metodologia de avaliação integrada de riscos ambientais e ocupacionais. In *SHO 2010—International Symposium on Occupational Safety and Hygiene*; Carneiro, P., Perestelo, G., Baptista, J.S., Miguel, A.S., Arezes, P., Barroso, M.P., Costa, N., Eds.; Sociedade Portuguesa de Segurança e Higiene Ocupacionais: Guimarães, Portugal, 2010; pp. 75–79. Available online: https://repositorio-aberto.up.pt/bitstream/10216/85186/2/65660.pdf (accessed on 24 October 2022).
25. Silva, M. *Avaliação de Riscos no Trabalho como Instrumento de Gestão na Indústria Metalomecânica. Dissertação Apresentada para Obtenção do Grau de Mestre em Engenharia de Segurança e Higiene Ocupacionais*; Faculdade de Engenharia da Universidade do Porto–Universidade do Porto: Porto, Portugal, 2014. Available online: https://repositorio-aberto.up.pt/bitstream/10216/77228/2/33329.pdf (accessed on 24 October 2022).
26. Bessa, R.; Baptista, J.S.; Oliveira, M. *Comparing Three Risk Analysis Methods on the Evaluation of A Trench Opening in an Urban Site*; Occupational Safety and Hygiene, III; Arezes, P.M., Baptista, J.S., Barroso, M.P., Carneiro, P., Cordeiro, P., Costa, N., Melo, R., Miguel, A.S., Perestrelo, G., Eds.; Taylor & Francis Group: London, UK, 2015; ISBN 978-1-138-02765-7. Available online: https://www.taylorfrancis.com/chapters/edit/10.1201/b18042-87/comparing-three-risk-analysis-methods-evaluation-trench-opening-urban-site-bessa-santos-baptista-oliveira (accessed on 24 October 2022).
27. Botelho, R.M.R. *Avaliação de Riscos Pelos Métodos MIAR, NTP330 e WTF, Numa Empresa de Triagem de Resíduos Industriais. Dissertação, Engenharia de Segurança e Higiene Ocupacionais*; Faculdade de Engenharia da Universidade do Porto: Porto, Portugal, 2015. Available online: https://repositorio-aberto.up.pt/bitstream/10216/79942/2/36187.pdf (accessed on 24 October 2022).
28. Sousa, I. *Aplicação da Metodologia Integrada de Avaliação de Risco na Indústria Extrativa a céu Aberto. 2015. Dissertação Engenharia de Segurança e Higiene Ocupacionais*; Faculdade de Engenharia da Universidade do Porto: Porto, Portugal, 2015. Available online: https://repositorio-aberto.up.pt/bitstream/10216/80093/2/36352.pdf (accessed on 24 October 2022).
29. Branco, J.C. *Avaliação do Risco de Acidente Na Indústria Extrativa a céu Aberto. Tese, Engenharia de Segurança e Higiene Ocupacionais*; Faculdade de Engenharia da Universidade do Porto: Porto, Portugal, 2018. Available online: https://hdl.handle.net/10216/117873 (accessed on 24 October 2022).
30. Guimarães, H. *Avaliação do Risco de Exposição A Agentes Biológicos: Reprodutibilidade dos Métodos DGS, NTP 833 e MIAR (BIO) em Matadouros*; Departamento de Engenharia, Universidade do Porto: Porto, Portugal, 2016. Available online: https://repositorio-aberto.up.pt/bitstream/10216/85900/2/153123.pdf. (accessed on 1 June 2021).
31. Santos, M.; Almeida, A.; Lopes, C.; Oliveira, T. MIAR? (Método Integrado para a Avaliação de Riscos). *Revista Portuguesa de Saúde Ocup. Online* **2019**, *7*, 1–2. [CrossRef]
32. ISO 31000:2018—Risk Management—Guidelines. Available online: https://www.iso.org/obp/ui/#iso:std:iso:31000:ed-2:v1:en (accessed on 1 January 2022).

33. Lima, K.S.; Meira Castro, A.C.; Santos Baptista, J.; Silva, U. Wood-Logging Process Management in Eastern Amazonia (Brazil). *Sustainability* **2020**, *12*, 7571. [CrossRef]
34. Nogueira, M.; Lentini, M.; Pires, I.; Bittencourt, P.; Zweede, J. *Manual Técnico 1—Procedimentos Simplificados em Segurança e Saúde do Trabalho no Manejo Florestal*; Instituto Floresta Tropical: Belém, Brasil, 2010.
35. Linstone, H.A.; Turoff, M. (Eds.) The Delphi Method: Techniques and Applications. 2002. Available online: http://www.foresight.pl/assets/downloads/publications/Turoff_Linstone.pdf (accessed on 1 October 2022).
36. Marques, J.B.V.; Freitas, D. Método DELPHI: Caracterização e potencialidades na pesquisa em Educação. *Pro-Posições* **2018**, *29*, 389–415. [CrossRef]
37. Bispo, P.C.; Valeriano, M.M.; Kuplich, T.M. Variáveis geomorfométricas locais e sua relação com a vegetação da região do interflúvio Madeira-Purus (AM-RO). *Acta Amazônica* **2009**, *39*, 81–90. [CrossRef]
38. IBGE—Instituto Brasileiro de Geografia e Estatística. *Manual Técnico da Vegetação Brasileira*, 2nd ed.; IBGE: Rio de Janeiro, Brazil, 2012; Volume 271, ISSN 0103-9598. Available online: https://biblioteca.ibge.gov.br/index.php/biblioteca-catalogo?view=detalhes&id=263011 (accessed on 1 October 2022).
39. Canto, J.; Machado, C.; Souza, A.; Garlet, A.; Carvalho, R.; Noce, R. Evaluation of work safety conditions of timber harvesting and transport in fomented farms of the state of Espírito Santo, Brazil. *Rev. Árvore* **2007**, *31*, 513–520. [CrossRef]
40. David, H.; Fiedler, N.; Baum, L. Ergonomia e Segurança Na Colheita Florestal: Uma Revisão Ante a NR 17 e a NR 31. *Enciclopédia Biosf.* **2014**, *10*, 1537–1550. Available online: https://www.conhecer.org.br/enciclop/2014a/AGRARIAS/ergonomia.pdf (accessed on 24 October 2022).
41. Lopes, E.; Zanlorenzi, E.; Couto, L. Análise dos fatores humanos e condições de trabalho no processamento mecânico primário e secundário da madeira. *Ciência Florest.* **2003**, *13*, 177–183. [CrossRef]
42. Nascimento, K.; Catai, R. Risk sizing and classification of laboral risks of forest harvesting in slope relief. *BIOFIX Sci. J.* **2017**, *2*, 28–33. [CrossRef]
43. Pignati, W.; Machado, J. Riscos e agravos à saúde e à vida dos trabalhadores das indústrias madeireiras de Mato Grosso. *Ciência E Saúde Coletiva* **2005**, *10*, 961–973. [CrossRef]
44. Rodrigues, C. Colheita e Transporte Florestal, Curitiba, Brazil. 2018. ISBN 978-85-924196-2-2. Available online: https://www.passeidireto.com/arquivo/109434140/livro-transporteflorestal (accessed on 1 September 2022).
45. Souza, V.; Blank, V.L.G.; Calvo, M.C. Cenários típicos de trabalho na indústria madeireira. *Rev. Saúde Pública* **2002**, *36*, 702–708. [CrossRef]
46. Torres, A.; Pereira, T.; Almeida, R.; Cunha, F.; Nieri, E.; Melo, L. Análise de riscos associados a colheita florestal em áreas declivosas no Brasil. In *Conceitos e Conhecimentos de Métodos e Técnicas de Pesquisa Científica em Engenharia Florestal*; Atena Editora: Ponta-Grossa, Paraná, Brazil, 2021; Volume 2, pp. 91–107. [CrossRef]
47. Allman, M.; Allmanová, Z.; Jankovský, M. Is cable yarding a dangerous occupation? A Survey from the public and private sector. *Cent. Eur. For. J.* **2018**, *64*, 127–132. [CrossRef]
48. Roloff, C. Apostila de Mecanização Florestal. CEEPR, Visconde de São Leopoldo. 2019. Available online: http://www.ceepro.com.br/documentos/Apostila%20Mecanizacao%20Florestal%202019.pdf (accessed on 1 September 2022).
49. Fernandes, A.; Guimarães, P.; Braz, E.; Hoeflich, V.; Arce, J. Alternativas de Planejamento para A Exploração Florestal. *Embrapa Florestas* **2013**, *43*, 339–350. Available online: https://www.embrapa.br/busca-de-publicacoes/-/publicacao/975345/alternativas-de-planejamento-para-a-exploracao-florestal (accessed on 24 October 2022). [CrossRef]
50. WMO—World Meteorological Organization. *Guide to Instruments and Methods of Observation*; WMO: Geneva, Switzerland, 2018; Volume I, p. 506. Available online: https://library.wmo.int/index.php?id=12407&lvl=notice_display#.Y1ba7ezMKgQ (accessed on 24 October 2022).
51. Marinha do Brasil. Centro de Hidrografia da Marinha: Escala Beaufort. Available online: https://www.marinha.mil.br/chm/sites/www.marinha.mil.br.chm/files/u2035/escala_beaufort.pdf (accessed on 24 October 2022).
52. DCM-RJ Defesa Civil do Rio de Janeiro. Escala da Força dos Ventos de Beaufort. 2021. Available online: http://www0.rio.rj.gov.br/defesacivil/ventos.htm (accessed on 24 October 2022).
53. Pires, I.P.; Miranda, A.M.; Couto, C.S.; Lentini, M.W.; Zweede, J.C. Exploração de Impacto Reduzido em Período Chuvoso Em Florestas de Terra Firme da Amazônia Brasileira: Considerações Técnicas, Minimização de Impactos e Índices de Produtividade. Boletim Técnico do Instituto Floresta Tropical. 2015. Available online: http://ift.org.br/wp-content/uploads/2015/05/BOLETIM_TECNICO_07.pdf (accessed on 24 October 2022).
54. SINAN—Sistema de Informação de Agravos de Notificação. Acidente por Animais Peçonhentos. 2019. Available online: http://portalsinan.saude.gov.br/acidente-por-animais-peconhentos (accessed on 24 October 2022).
55. Cornelissen, I.R.; Jongeneelen, F.; Broekhuizen, V.; Broekhuizen, F.V. Guidance Working Safely with Nanomaterials and Nanoproducts, the Guide for Employers and Employees. 2011. Available online: http://www.etui.org/content/download/3553/39999/file/Guidance+on+safe+handling+nanomats&products.pdf (accessed on 24 October 2022).

sustainability MDPI

Article

Analysis of Factors Affecting Human Reliability in the Mining Process Design Using Fuzzy Delphi and DEMATEL Methods

Iraj Mohammadfam [1], Ali Asghar Khajevandi [1,*], Hesam Dehghani [2], Mohammad Babamiri [1] and Maryam Farhadian [1]

[1] Occupational Health and Safety Department, Hamadan University of Medical Sciences, Hamedan 65155, Iran; mohammadfam@umsha.ac.ir (I.M.); mohammad.babamiri@gmail.com (M.B.); m.farhadian@umsha.ac.ir (M.F.)

[2] Department of Mine Engineering, Hamedan University of Technology, Hamedan 65155, Iran; dehghani@hut.ac.ir

* Correspondence: khajevandi.asghar2008@gmail.com; Tel.: +98-910-180-1024

Abstract: Design errors have always been recognized as one of the main factors affecting safety and health management and sustainable development in surface mines. Unfortunately, scant attention is paid to design errors and the factors causing them. Therefore, based on expert opinions, this study aimed to identify, rank, and investigate cause-and-effect relationships among variables influencing human error in surface mine design in Iran. The study variables were identified by reviewing previous literature on "latent human errors" and "design errors." After specifying effective variables, two rounds of the Fuzzy Delphi study were carried out to reach a consensus among experts. Nineteen variables with an influencing score of 0.7 and higher were screened and given to the experts to be analyzed for cause-and-effect relationships by the fuzzy DEMATEL method. The results of the study revealed that the following variables were the major factors affecting human error as root causes: poor organizational management (0.62), resource allocation (0.30), training level (0.27), and experience (0.25). Moreover, self-confidence (−0.29), fatigue (−0.28), depression (−0.25), and motive (−0.23) were found to be effect (dependent) variables. Our findings can help organizations, particularly surface mines, to opt for effective strategies to control factors affecting design errors and consequently reduce workers' errors, providing a good basis for achieving sustainable development.

Keywords: design errors; sustainable development; accident; multi-criteria decision-making

Citation: Mohammadfam, I.; Khajevandi, A.A.; Dehghani, H.; Babamiri, M.; Farhadian, M. Analysis of Factors Affecting Human Reliability in the Mining Process Design Using Fuzzy Delphi and DEMATEL Methods. *Sustainability* **2022**, *14*, 8168. https://doi.org/10.3390/su14138168

Academic Editor: Edmundas Kazimieras Zavadskas

Received: 14 May 2022
Accepted: 1 July 2022
Published: 4 July 2022

1. Introduction

The mining industry is classified as one of the most dangerous and harsh work environments [1–3]. The consequences of mining accidents include occupational deaths and injuries, equipment damage [4], and environmental problems [5]. Besides, accidents and incidents in such a harsh work environment are very high (about 7–10 times) compared to other industries [1,6]. Identifying and eliminating the roots of mining accidents have always been one of the most important priorities of organizations and governments [7]. The analysis of mining accidents has shown that human error is the direct cause of 85% of these accidents. In recent years, many studies have been conducted to identify the factors affecting human error in mines. In most cases, the root cause of accidents resulting from human errors is a design error (DE), and thus the operator is just a victim of a poor design [8,9]. Liao asserts that despite efforts to reduce instances of human error by raising workers' awareness, not much progress has been achieved thus far. He regards DE as the main reason behind such a failure and states that it is one of the main causes of unsafe behaviors on workers' part in operational sectors [10]. DE is almost inevitable and can influence the safety of projects and their costs and timing [11]. More than 80% of the failures in buildings, bridges, hospitals, and civil engineering structures are caused by DE [12].

DE are important in various industries such as aviation [13], nuclear energy [14], process [15], and mining [16]. The diversity of mines, extensive operational space, and the extent of its consequences (occupational and public health, environmental, safety, social, and economic) have highlighted the role of design errors in this sector [17,18]. Unfortunately, the focus of human error studies for most of the 20th century has been on operational errors, which have been ignored [9]. The study by Thompson on road accidents in surface mines showed that design errors are the main causes of such accidents [19].

Flyrock is important in increasing the death rate and destroying mine equipment and structures. One of the main reasons for Flyrock production in blasting is DE in the blasting pattern [20]. Reason introduced this important construct as the latent human errors in 1998 because their consequences are not immediately known, and their identification takes longer. In other words, their identification needs a systematic approach [21]. Cho defines DE "as the result of a designer's actions and decisions in product development that lead to failure in the planned or intended outcome" [22].

Likewise, Mechlers believes that these latent human errors are cognitive processing errors, arguing that even the simplest forms of designs require cognitive functions [23]. From a cognitive psychological point of view, human error results from one or more deficits in human cognitive processes. Accidents happen due to perception, recognition, avoidance ability, and decision-making failures. Thus, failure in cognitive processes can lead to human errors and damage the system [24]. Studies show that design errors happen as a result of cognitive failure (CF) [25] influenced by individual, environmental, organizational, and task factors [26].

2. Review of Previous Research

In recent years, some studies have been conducted to find the effective factors behind human error in design. The results of the study by Kerli et al. [22] on DE showed that process (lack of design reviews), material (learning not shared amongst everyone), measurements (incomplete project tracking), tools (poor document traceability), people (loss of information and lack of making ability knowledge), and organization (scattered resources) are the main causes of such errors. Lopez et al. [12] reported that personal factors (loss of biorhythm and adverse behavior), organizational factors (training, experience, competitive professional fees, poor quality assurance), and project (time limitations and poor coordination) have a significant influence on DE in the construction sector. Some studies point out that errors result from individuals' tendency toward error or the conditions that induce error [27]. Also, some studies have classified the variables affecting DE into three groups: workplace, information flow, and organizational factors [28]. The study by Robert [29] revealed that designer knowledge, lack of standards, safety awareness, novel system, management of change, procedure, and lack of qualified staff were the most effective factors in design error. Zhaorong et al. [30] stated that defective workmanship, communication, lack of skill, contract issues, and external factors could lead to latent error and design error. Several studies have shown that these errors are influenced by individual, managerial, and social factors related to work, workplace, work methods and processes, task demands, workload, and physical work conditions [31]. However, DE has been considered as the major causes of accidents in many organizations [32]. There are many variables that directly or indirectly affect DE and are indeed the root causes of accidents. When a set of variables with complex relationships impact on a target variable, determining the most important variables requires extensive field studies, it is time-consuming and costly; and, moreover, the simultaneous controlling of all variables is not logical in system safety management and system safety engineering [33]. Therefore, using expert opinions to determine the most important variables based on scientific methods is a suitable strategy [34].

Multi-Criteria Decision-Making (MCDM) techniques are often adopted to solve complex problems based on experts' judgment. Previous studies have shown that MCDM methods, combined with a fuzzy set theory or other methods [35,36], can result in more reliable results. Several studies have used this approach in the areas of health [37], safety [38] and environment [39], and economy [40] for identifying and classifying relationships among variables. According to Fam et al. [41], the combination of fuzzy Delphi and DEMATEL is the best risk control strategy because DEMATEL can provide a cause-and-effect model. Similarly, in another study, Kumar et al. [42] reported that AHP and DEMATEL cannot determine the importance of the criteria. Therefore, the fuzzy Delphi method is very suitable to fill this gap. Renissa et al. [43] used the Delphi method and Fuzzy DEMATEL to identify the barriers to university technology transfer. Singh and Sardar [44] also used the Delphi method to determine the factors affecting sustainable product development and the Fuzzy DEMATEL method to illustrate the interrelationships among key factors by drawing a causal diagram in the automotive industry. The combination of these two methods can provide a deep understanding of a phenomenon.

Given the advantages of using fuzzy Delphi and DEMATEL methods and the lack of ample studies extensively surveying and prioritizing the factors affecting design error in Iranian surface mines, this study aimed to identify, rank, and investigate cause-and-effect relationships among variables influencing DE based on expert opinions.

Further, this study contributes to the literature in several ways:

(1) To our knowledge, this is one of the first studies investigating factors predicting DEs and their interactions. Thus, this study can contribute theoretically to the existing literature and fill the existing gaps in safety studies that addresses the role of latent errors in accidents;

(2) The proposed methodology of the present study provides a visual cause-and-effect model, which helps analyze DE. Mining managers and safety experts can update their goals and plan based on the results of the study;

(3) As a practical contribution, the study suggests strategic measures that may reduce DEs to avoid accidents; the study also presents evidence that helps improve health and safety at mines.

This study is organized as follows: Section 2 has the theoretical fundaments on DE, related literature gaps and the contribution of the study; in Section 3, the most important variables of DE in the mining design process are presented, followed by introducing Fuzzy Delphi and DEMATEL methods. The results and discussion are described in Sections 4 and 5; Section 6 specifies the conclusion and suggests future lines of research.

3. Materials and Methods

The methodology of this study comprised three phases: the identification of variables, the determination of effective variables via the Fuzzy Delphi method, and the analysis of cause-and-effect relationships among such variables via the Fuzzy DEMATEL method. The framework combining the two methods includes the following three phases, as shown in Figure 1.

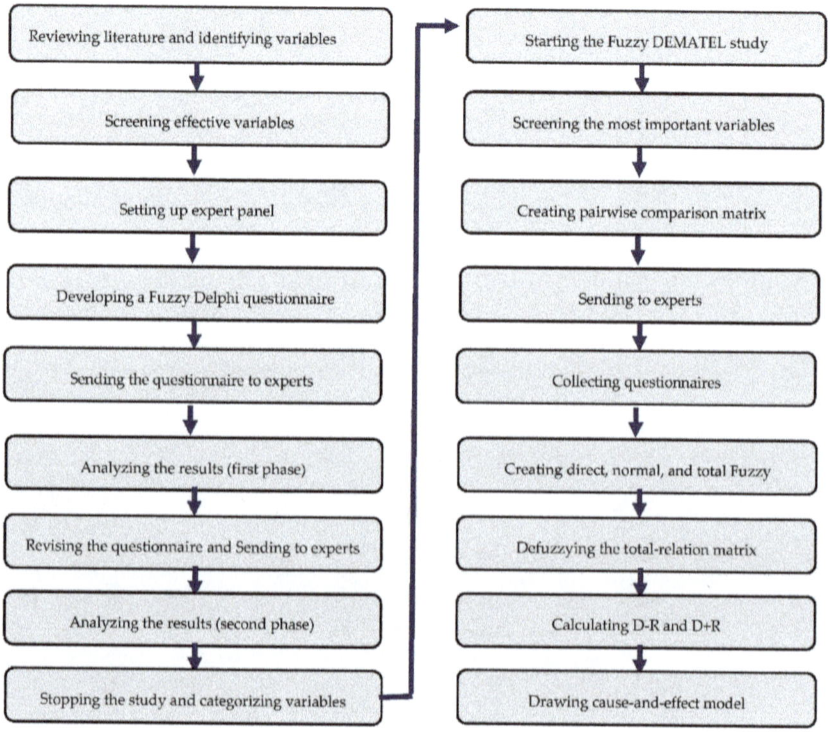

Figure 1. Research procedure.

3.1. *Searching and Classifying the Variables Affecting DE*

The important variables were first identified by a library research and literature review. Next, a panel of five experts in surface mine design was assigned to screen the most important variables and categorize them based on environmental, individual, external, organizational, and task factors for employing the Fuzzy Delphi method (Appendix A).

3.2. *Identifying the Variables Affecting DE Using the Fuzzy Delphi Method*

3.2.1. The Fuzzy Delphi Method

This method is a powerful tool used to reach a consensus based on expert opinions in a particular field of study [45]. In its classical form, the Delphi method makes use of expert opinions stated in the form of definite numbers. In this method, experts do not use their mental competence to state their opinions, showing a probability of uncertainty in the decisions made. Thus, to compensate for this drawback, a fuzzy set is used to collect the data in linguistic terms and interpret vague concepts stated by expert opinions [46,47]. Accordingly, the classical Delphi method was combined with fuzzy set theory to create the more effective Fuzzy Delphi method [48]. The Fuzzy Delphi method enjoys some advantages, including the unification of expert opinions to reach a consensus [49], the reduction of time and cost compared to the classical Delphi method [50], and the reduction of expert opinion collection rounds [41].

There are different types of fuzzy numbers, and this study used Triangular Fuzzy Numbers (TFN). In this study, TFN was shown using three real numbers M = (l, m, u), in which the upper bound is (u), lower bound is (l), infimum is (m), and 'M' is the most probable value of a fuzzy number [51]. TFN reflects the membership by the function, which can show the information of the experts more simply and accurately regarding a complex decision-making problem [52]. TFN has been applied in various domains, including risk, evaluation, anticipation, and expert systems [53].

3.2.2. Selection of Experts

In the MCDM method, the selection of experts is very important and vital. Powerful expert groups can ensure the accuracy of research results. Therefore, the expert panel in the study went through a rigorous selection process. In the first step, a database of experts active in surface mine design in Iran was collected. The inclusion criteria included being inclined to participate, having comprehensive knowledge, ample operational experience, and time adequacy. Due to the diversity of minerals, the difference in the size of the mines, the geography of the design environment, and the variety of techniques and tools used in the design, attempts were made to select decision-makers whose experience covered the listed items. Finally, out of 150 Iranian Open Mines Designers Association members, 25 were purposefully selected. The number of experts in the panel varies in various valid studies, and several studies have been conducted with fewer than 10 experts to higher numbers [54–56]. Among the experts, there were people with academic bachelor's degrees. These people are among the most famous mining designers in Iran who have a lot of experience in the field of exploration and extraction in surface mines. The demographic characteristics of the experts are shown in Table 1.

Table 1. Demographic characteristics of the experts.

	Delphi Study		DEMATEL Study	
Demographic Variables	**Total**	**Percentage**	**Total**	**Percentage**
Gender				
Male	16	84.21%	9	90.00%
Female	3	15.79%	1	10.00%
Educational				
Bachler	3	15.79%	-	-
Master	7	36.84%	2	20.00%
Doctoral	9	47.37%	8	80.00%
Experience in mine design				
<5 years	2	10.53%	-	-
5–15 years	6	31.58%	3	30.00%
>15 years	11	57.89%	7	70.00%

In line with previous literature using the Fuzzy Delphi method, a questionnaire with Likert-scale items was developed to be used in the study [50]. The expert panel was asked to review the developed semi-closed questionnaire and revise it by adding any important variables missing in the questionnaire.

3.2.3. First and Second Rounds Inquiry

Afterwards, the questionnaire was sent to three experts to be reviewed for face and content validities. Eventually, the finalized questionnaire was sent to 25 experts with a response rate of 76% (19 experts) in the first phase. In this phase, three new variables were suggested to be added to the questionnaire. After collecting expert opinions, the linguistic variables were changed into fuzzy numbers based on Table 2.

Table 2. Triangular fuzzy numbers corresponding to linguistic terms [54].

Linguistic Expressions	Triangular Fuzzy Numbers
No effect	(0, 0, 0.25)
Extremely weak effect	(0, 0.25, 0.5)
Weak effect	(0.25, 0.5, 0.75)
Strong effect	(0.5, 0.75, 1)
Extremely strong effect	(0.75, 1, 1)

The triangular fuzzy numbers set was measured for each expert's opinion based on Equation (1) [55]:

$$\widetilde{A}^{(i)} = \left(a_1^{(i)} \cdot a_2^{(i)} \cdot a_3^{(i)}\right) \; i = 1, 2, 3, \ldots, n. \tag{1}$$

Next, the mean of fuzzy numbers set ($\widetilde{A}_m^{(i)}$) out of all sets ($\widetilde{A}^{(i)}$) was measured based on Equation (2):

$$\widetilde{A}_m = (a_{m1} \cdot a_{m2} \cdot a_{m3}) = \left(\frac{1}{n}\sum_{i=1}^{n} a_1^i \cdot \frac{1}{n}\sum_{i=1}^{n} a_2^i \cdot \frac{1}{n}\sum_{i=1}^{n} a_3^i\right). \tag{2}$$

Then, the difference was calculated from the mean for each expert's opinion. After revisions and suggested variables were added, the questionnaire was re-sent to the experts to review and revise if needed. After collecting expert opinions in the second round based on Equations (1) and (2), expert opinions were aggregated, and their disagreements between the two rounds reached the minimum level of 0.2 [51]. At the end of the second round, the experts suggested that no new variable with disagreements reached the minimum level of 0.2. Accordingly, the Fuzzy Delphi study was stopped in this step [56].

3.2.4. Determination of the Most Important Variables

To defuzzy the numbers, the simple center of gravity method was used based on Equation (3):

$$S_j = \frac{l_j + m_j + u_j}{3}. \tag{3}$$

The ranking and determination of the most important variables were based on defuzzied scores: the higher the defuzzied score of a variable, the stronger the effect it exerted on human error, and hence more important. In this study, the screening process was conducted based on the 30–70 law, in which the threshold level for criterion acceptance was 7 [57]. Thus, if the amount of the defuzzied triangular number was found to be 0.7 or higher based on expert opinions, it was accepted as a criterion. Otherwise, it was removed from the study.

3.3. *Determining Cause-and-Effect Relationships between the Variables*

3.3.1. Fuzzy DEMATEL Method

Gabus et al. introduced a method called decision-making trial and evaluation laboratory (DEMATEL) in 1972 to analyze casual relationships and significant effects among variables with a strong validity [58]. This method works based on expert opinions expressed in linguistic terms; in order to avoid ambiguity and reach a unification of opinions, these linguistic terms need to be turned into fuzzy numbers. In 2008, Lin was the first person who used the DEMATEL method in a fuzzy environment [59]. The Fuzzy DEMATEL method investigates the relationships among criteria and sub-criteria and determines effective (cause) and affected (effect) criteria by the total-relation matrix [60,61]. This method is a multi-index decision-making technique [62]. One advantage of this method over other methods of investigation is that the process of decision-making is based on pairwise comparisons and the acceptance of relationships [63]. The Fuzzy DEMATEL method is frequently used in different fields of inquiry such as human resource management, risk assessment, and safety management system [24,64,65]. In this study, the following steps were taken to apply the Fuzzy DEMATEL method [66].

3.3.2. Setting up the Expert Panel

The first step aimed to identify experts qualified to participate in the inquiry process of the DEMATEL method. The respondent had to be a person who had adequate knowledge or experience related to the research problem. In this study, 15 experts with prominent experience and research history about mine design were selected, and the questionnaire was sent to them via email. Eventually, 10 experts collaborated in the study, performed the evaluation, and submitted the evaluation forms later.

3.3.3. Preparing Fuzzy DEMATEL Questionnaire

The Fuzzy DEMATEL questionnaire comprised a 20 × 20 matrix, which is not a symmetric matrix. The factors in these tables were assessed as a pairwise matrix. The experts used a 5-point Likert scale (Table 2) to express their opinions about the relationship among variables.

3.3.4. Analyzing the Data

(a) Based on experts' responses, the initial direct-relation fuzzy matrix was calculated

$$\widetilde{Z}_{ij}^{k} = \begin{pmatrix} 0 & \cdots & \widetilde{X}_{1n}^{k} \\ \vdots & \ddots & \vdots \\ \widetilde{X}_{n1}^{k} & \cdots & 0 \end{pmatrix}. K = 1,2,3,\ldots,P. \tag{4}$$

In this equation, P is the number of experts (10).
Then, using Equations (5)–(7) the aggregated mean of expert opinions was measured.

$$\widetilde{Z}_{ij} = \frac{\widetilde{X}^{1} + \widetilde{X}^{2} + \widetilde{X}^{3} + \widetilde{X}^{4} + \ldots + \widetilde{X}^{P}}{P}. \tag{5}$$

$\widetilde{X}^1$, $\widetilde{X}^2$, $\widetilde{X}^3$, and $\widetilde{X}^P$ are the pairwise comparison matrixes of the experts (expert 1, 2, 3, and P, respectively).

$$\widetilde{Z}_{ij} = \begin{pmatrix} 0 & \cdots & \widetilde{X}_{1n} \\ \vdots & \ddots & \vdots \\ \widetilde{X}_{n1} & \cdots & 0 \end{pmatrix}, \tag{6}$$

$$\widetilde{Z}_{ij} = (l_{ij} + m_{ij} + u_{ij}). \tag{7}$$

(b) Normalizing the direct-relation fuzzy matrix using Equations (8) and (9)

$$r = max \sum_{j=1}^{n} u'_{ij}, \tag{8}$$

$$\widetilde{H}_{ij} = \frac{\widetilde{z}_{ij}}{r} = \left(\frac{l'_{ij}}{r}.\frac{m'_{ij}}{r}.\frac{u'_{ij}}{r} \right) = \left(l''_{ij}.m''_{ij}.u''_{ij} \right). \tag{9}$$

(c) Determining the total-relation matrix.
The total-relation fuzzy matrix (T) was measured by the following Equations (12)–(14):

$$T = lim_{k \to \infty} \left(\widetilde{H}^{1} + \widetilde{H}^{2} + \widetilde{H}^{3} \right), \tag{10}$$

$$\widetilde{t}_{ij} = \left(l^{t}_{ij}.m^{t}_{ij}.u^{t}_{ij} \right), \tag{11}$$

$$\left[l^{t}_{ij} \right] = H_{l} \times (I - H_{l})^{-1}, \tag{12}$$

$$\left[m^{t}_{ij} \right] = H_{m} \times (I - H_{m})^{-1}, \tag{13}$$

$$\left[u^{t}_{ij} \right] = H_{u} \times (I - H_{u})^{-1}. \tag{14}$$

(d) Defuzzying the total-relation fuzzy matrix base on Equation (15)

$$t_{ij} = \frac{l_{ij}^t + 2m_{ij}^t + u_{ij}^t}{4}. \tag{15}$$

(e) Measuring the D-value and R-value based on extracted variables from the total-relation defuzzied matrix base on Equations (16) and (17):

$$D = \sum_{j=1}^{n} t_{ij}.(j = 1, 2, 3, \ldots, n), \tag{16}$$

$$R = \sum_{i=1}^{n} t_{ij}.(i = 1, 2, 3, \ldots, n). \tag{17}$$

To do so, the elements of each row (Di) and each column (Ri) were totaled out of the total-relation defuzzied matrix. The total number of elements in each row (D) for each factor shows the degree to which that factor affects other factors in the system. On the contrary, the total number of elements in each column (R) for each factor shows the degree to which that factor is affected by other factors in the system.

(f) In the end, D and R values were used to measure D + R and D − R values.

The D + R values show how much one factor affects and is affected by other factors. In other words, the higher the D + R value, the more interaction between the factor and other factors in a system. On the other hand, D − R values show how strongly one factor affects other factors in a system. In general, if D − R is positive, the variable is considered a cause variable, and if it is negative, it is considered an effect variable. After defuzzying numbers, a Cartesian coordinate system is drawn in which the x-axis shows D + R values, and the y-axis shows D − R values.

4. Results

First, the relevant literature on DE and human error variables was reviewed, and important variables were identified and extracted. These variables were then screened by experts and categorized into five factors: organizational, external, environmental, task, and individual.

4.1. Ranking Variables Affecting DE Based on the Fuzzy Delphi Method

After specifying effective variables, the two phases of the Fuzzy Delphi study were carried out to reach a consensus among experts. Accordingly, the semi-closed questionnaire with Likert-scale items was developed and given to the experts. After collecting the questionnaires, the mean triangular fuzzy value and defuzzied value were measured for each of the phases based on Equations (1)–(3). Table 3 shows the absolute mean of experts' agreement corresponding to the importance of each factor. The results revealed that the following variables strongly affected human error in mine design: technical knowledge (designing and safety), poor organizational management, resource allocation (hardware and software), and experience. Environmental factors, noise, indoor air quality in the workplace, and lighting exerted the strongest effects on DE.

As for task factors, mental workload, multitasking in designing projects, and an unclear work process strongly influenced DE. Finally, technical knowledge, experience, and depression were the most effective individual factors. Poor organizational management, resource allocation (hardware and software), and a safe design culture were the most effective organizational factors.

Table 3. Selected variables of the Fuzzy Delphi study for cause-and-effect analysis.

Subgroup	Identification Code	Variable	Defuzzied Number
Individual variable	Va1	Technical knowledge (safety and designing)	0.81
	Va2	Experience	0.78
	Va3	Depression	0.74
	Va4	Motive	0.72
	Va5	Self-confidence	0.72
	Va6	Financial satisfaction	0.72
	Va7	Stress	0.71
	Va8	Intelligence coefficient	0.71
	Va9	Fatigue	0.70
Task variables	Va10	Unclear work process	0.76
	Va11	Multitasking	0.70
	Va12	Workload	0.70
Environmental variables	Va13	Noise	0.73
	Va14	Poor indoor air quality	0.72
	Va15	Inappropriate lighting	0.71
Organizational variables	Va16	Poor management	0.81
	Va17	Resource allocation	0.79
	Va18	Safe designing culture	0.73
	Va19	Training	0.71

4.2. Determining Cause-and-Effect Relationships among Variables Affecting DE (CF)

In this phase, variables with an influencing score of 0.7 and higher were screened from each variable group (individual, organizational, external, task, and environmental) and given to the experts in the form of a pairwise-matrix questionnaire analyzed for cause-and-effect relationships. Table A1 demonstrates the list of variables selected for the Fuzzy DEMATEL study. After collecting expert opinions regarding the effects of variables on each other, the mean of opinions was acquired by forming the direct-relation fuzzy matrix. Next, the normalized direct-relation matrix was formed, followed by the total-relation matrix (Appendix B). The variables in each row were added to measure the D value (Figure 2), and the variables in each column were added to measure the R-value (Figure 3); eventually, using D and R values, the interaction of variables (D + R) or dominance matrix (Figure 4) and the relationship among variables or the influence of variables and their pure influenceability (D − R) or relationship matrix (Figure 5) were determined. Factors with a positive D − R relationship were considered effective (causes) and those with a negative D − R relationship were considered affected (effects).

Based on D + R values, unclear work process, CF, multitasking, and fatigue had the highest level of interaction with other variables; on the contrary, poor indoor air quality, inappropriate lighting, and noise had the lowest level of interaction with other variables. According to D − R values, poor organizational management, resource allocation (hardware and software), training level, and experience were the most effective variables respectively, less influenced by other variables. In other words, these variables had a strong guiding power with minor dependence on other variables. Thus, if these variables are fortified, failures in cognitive function are reduced, leading to a significant decrease in design errors. On the other hand, CF, self-confidence, depression, and motive were the most affected variables (effects) respectively, more affected by other cause variables.

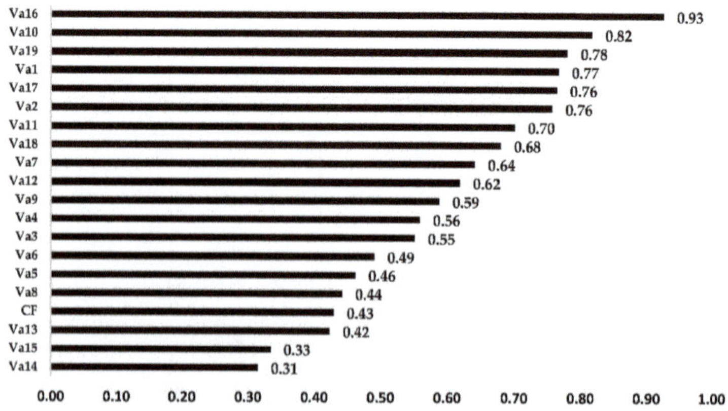

Figure 2. Influence of variable on other variables (D values).

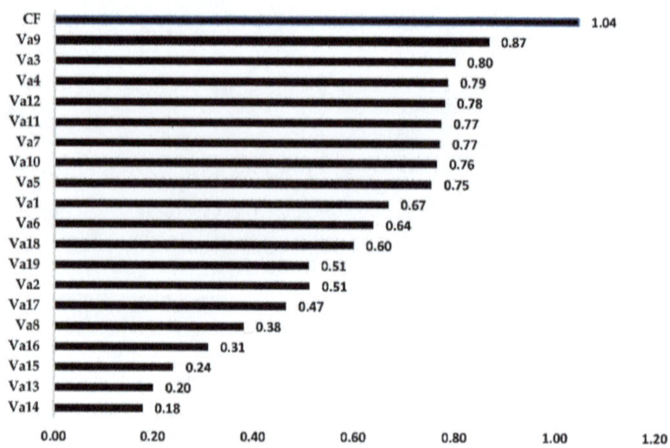

Figure 3. Influenced impact index variables (R values).

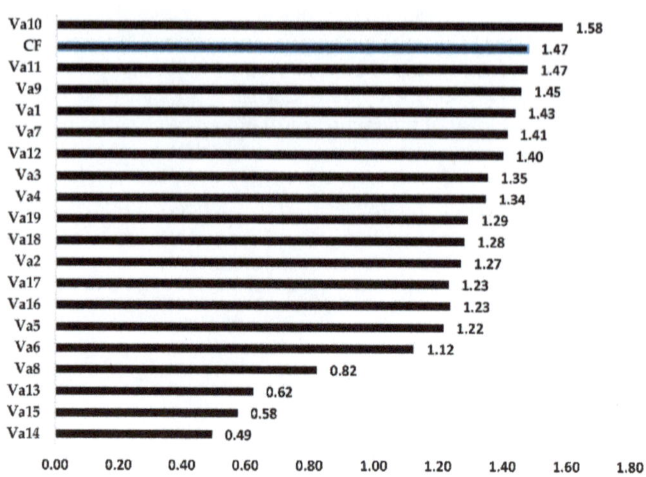

Figure 4. Interaction among variables (D + R values).

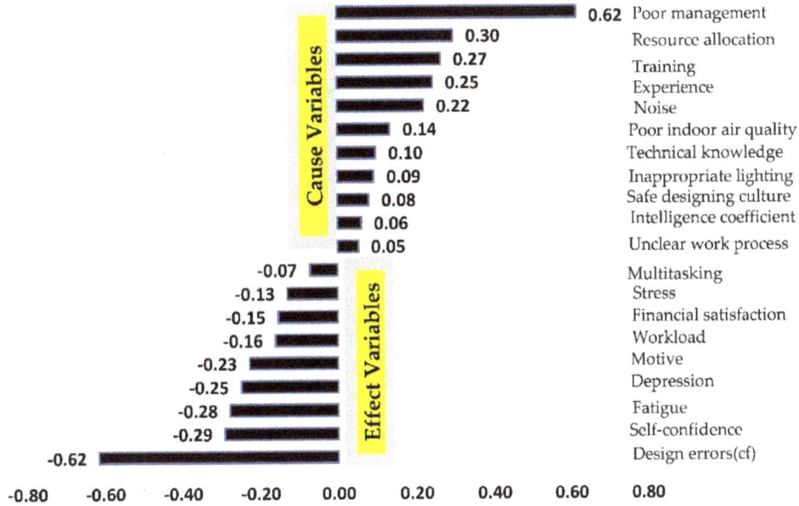

Figure 5. Cause-and-effect roles of the variables (D–R values).

According to the results of the cause-and-effect relationships presented in Figure 6, the variables of the study can be divided into four groups located in four different zones. The first group of cause (influencing) variables included poor organizational management, resource allocation (hardware and software), training, experience, technical knowledge (safety and designing), safe designing culture, and unclear work process. The second cause variables included noise, poor indoor air quality, and lighting. The third and fourth groups of variables were under the D + R axis including effect (influenced) variables. Financial satisfaction was the only variables present in this zone.

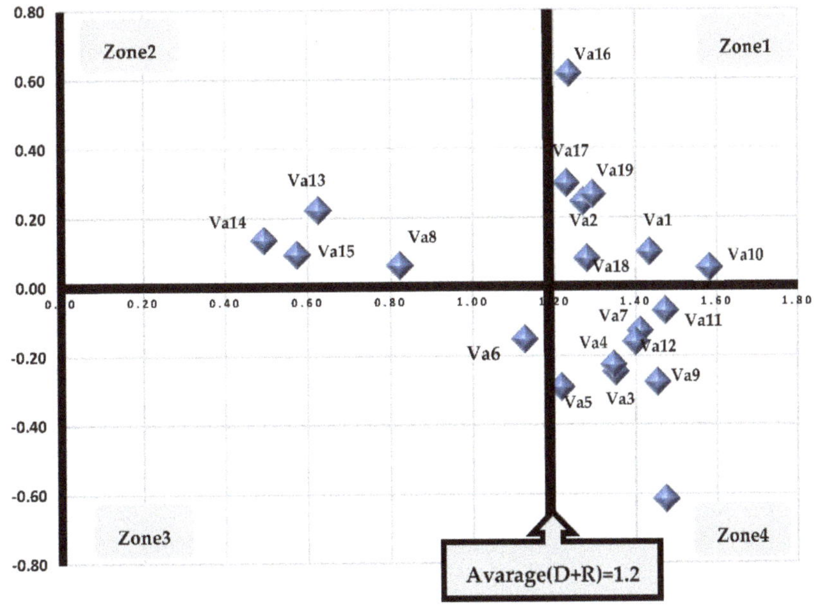

Figure 6. Cause-and-effect relationships among the variables.

This variable was affected by the variables of Zone 1 and Zone 2 but influenced variables of the fourth group. The fourth zone, however, as the most important group of effect variables, included CF, self-confidence, fatigue, depression, motive, workload, stress, and multitasking. Generally, to improve CF and reduce design errors, it is better to implement measures that consider Zone 1 variables followed by Zone 2 variables. If problems persist, the variable of Zone 3 needs to be considered. Zone 4 variables are the variables that are under the influence of variables present in previous zones, and thus no direct corrective action is performed for them. The study results demonstrate that environmental variables as one type of cause variables were the weakest variables in terms of affecting and being affected by other variables. In contrast, organizational variables were the strongest variables in terms of affecting other variables, showing that they are the most important variables affecting DE.

5. Discussion

In accordance with our findings, previous research also shows that organizational factors are one of the most important causes of DE [12,22]. Dedy et al. [67] named lack of training/education (training about design), poor resourcing, poor strategy and leadership, poor management, and lack of professionalism as the organizational factors affecting DE in construction projects. The results of Hafezi's study showed that organizational factors such as lack of training program for designers and poor use of technology had the highest priority in issues related to related to DE [68]. The results of Cho's study also revealed that poor management and lack of standard processes were the main organizational factors of DE [69]. Moreover, many studies focusing on human error have emphasized organizational factors, particularly management obligation [70–72], resource allocation [73], and safety culture [74].

The results of this study demonstrated that technical knowledge and experience were located in Zone 1; thus, these individual variables are the root cause of DE in surface mines. Similarly, the results of the study by Philemon et al. [75] showed that the lack of knowledge and experience of the design team was the most effective personal factor leading to DE and omission in construction projects in Tanzania. Lopez et al. also believe that employing inexperienced designers with low technical knowledge and engaging underqualified designers in important design projects are the main causes of DE in organizations [12]. Technical knowledge of designing, quality and quantity of training, and experience [76] are personal variables that can strongly affect cognitive function, especially in the early stages of detecting, noticing, understanding, and sense-making processes [77]. These two criteria are the most significant factors influencing cognitive function [78]. Continuous and adequate training and using experienced instructors are highly effective for preventing and controlling human errors on the one hand and reducing the risk of accidents on the other [79].

Environmental parameters such as noise, lighting, and indoor air quality were categorized into Zone 2 in this study, belonging to independent (cause) variables that could influence Zone 3 and Zone 4. To the researchers' knowledge, this important factor has been overlooked in DE studies. These factors can negatively influence the physiological balance of the human body; cognitive performance can cause stress, fatigue, depression, and workload, which in turn can result in the loss of focus and more human error [80,81]. Noise exposure can act as a stressor and increase mental workload, eventually impairing the mental performance required for one's responsibilities [82,83]. Noise can also lead to fatigue [84], significantly affecting one's performance while performing complex tasks requiring mental processing [85,86]. Appropriate lighting improves awareness and cognitive performance [87]. On the other hand, inappropriate lighting can result in depression, mental boredom, and sleep quality [88]. Research shows that indoor air quality in the workplace influences cognitive performance as chemical pollutants in the air, such as particles, and high levels of carbon dioxide in the air detrimentally affect cognitive performance [89].

Another dependent (effect) variable is financial satisfaction (Zone 3), which can affect a lot of variables in group 4, particularly cognitive function. For example, Tilley and McFallen conducted a study on Australian designers. They reported that most designers believed their payment was low despite their challenging job, which could eventually influence the quality of their designing performance [90]. Based on the study by Vaiana et al. [91], the contractor's lack of payment and inadequate cash flow was an important factor in increasing DE and accidents in Design and Build Projects in Malaysia. Financial dissatisfaction can demotivate designers, and low payments suggested by organizations can pave the way for inexperienced designers to take responsibility for important projects, increasing the risk of errors [12]. As for the fourth group of variables, the most important dependent (Zone 4) variables were located in this group, with the designer's cognitive function as the variable highly influenced by others. Other variables in this group, such as fatigue and depression, can also affect DE. From a cognitive point of view, chronic fatigue can lead to a decrease in the information processing capabilities of workers and designers and thus result in delayed reaction time, reduction in the field of vision, carelessness, unawareness, and lack of focus. Therefore, fatigue resulting from physical tiredness or insomnia negatively impacts cognitive resources and awareness [92]. Research shows that cognitive dissonance and well-being were the most important man factor of DE in the oil and gas industry [67]. Another variable belonging to Zone 4 was the workload. The increased workload can reduce mental health and stress, leading to cognitive overload, failures in cognitive performance, and increased human error [93,94]. Failures in cognitive function forge an important link between factors affecting performance and human error [26,95]. Thus, individual, environmental, task and organizational factors exert direct and indirect effects (fatigue, stress, demotivation, etc.) on the designer's cognitive function and lead to DEs eventually.

The comparison of the results of the abovementioned studies with the current study highlights some conflicting issues:

- Previous DE studies have focused on consequences such as rework, safety, and cost. Still, in mines, due to the diversity and wide operating spaces of the mines and the type and volume of equipment used, these consequences can be very significant. It can also have environmental, social, cultural, political, security and public health effects. Therefore, the role of design errors in this section is much more prominent than in other sections;
- Past studies focus only on identifying and categorizing the factors affecting design error. Still, in the present study, in addition to identifying and categorizing these factors, their relationships are also defined within a cause-and-effect model. This model aids decision-makers in focusing on the most important risks in mine design projects.

Based on the presented results, DE is one of the most important threats to sustainable development in mines. Therefore, identifying and prioritizing the factors affecting such errors is vital due to the financial and time constraints of organizations in eliminating and controlling them. This research proposes a comprehensive approach to managing design error in mines that, in addition to covering the existing theoretical gaps such as the lack of a comprehensive study in the field of design error and its factors affecting mines, provides important practical recommendations at all levels of the organization, especially for top management and mine safety experts. Concerning the findings of this study, inherently safe design culture, hardware and software resources, and individual factors such as insufficient experience and knowledge are the root causes of errors in mine design. Meanwhile, the role of top management is very important in developing, leading, and promoting an inherently safe design culture in the organization. The top management should be allocating the resources needed (hardware and software) to control errors in the design process, ensuring that engineers and designers are competent based on appropriate education, training, or experience, providing a safe and comfortable work environment based on ergonomic standards, and trying to improve the level of job satisfaction and motivation of the design team. In addition, based on the results, mining safety experts should pay special attention to design errors and predict the required resources in establishing objectives and planning to achieve them. They can reduce variables such as job stress, depression, and fatigue, and improve the designer's cognitive functions by conducting safe design training courses, implementing risk management and ergonomic programs, and monitoring physical factors in the workplace such as lighting and noise thermal comfort parameters. Moreover, the study can help legal organizations in mining safety to understand the nature of accidents and formulate strategic policies to implement safe design rules in the mining sector.

6. Conclusions and Future Research

Design error (DE), a latent human error, is a key factor behind many occupational accidents. Limited research, however, has been carried out investigating the relationship between the causes of human error and relevant negative consequences. This phenomenon is important, especially for the Iranian mining sector, which holds 7% of the mineral recourses in the world. Therefore, this study aimed to identify the most significant variables influencing surface mine designers' performance and investigate their cause-and-effect relationships. For this purpose, common effective factors were taken from the literature review and screened by the experts. One MCDM methodology, Fuzzy DEMATEL, was applied to investigate the relationships among variables and develop a cause-and-effect model. The results revealed that environmental variables (noise, lighting, and indoor air quality) had the weakest effects on other variables and were least affected by other ones; based on the cause-and-effect relationships model, it can be concluded that 'organizational factors' are vital for the DE control plan within the mining industry due to their effect on other factors.

Nevertheless, it should be noted that individual variables like training, experience, and technical knowledge were also found to influence DE. Similar to other studies, this study faced some limitations; therefore, this work can be extended in future studies. The most noticeable limitation is that the study is one of the first to study the most significant variables affecting DE in surface mines with the abovementioned methods. Hence it is not easy to generalize the findings to other industries. However, future studies may extend the research to different industries. In this study, the empirical analysis of the cause-and-effect relationship among variables was not conducted. For future research, empirical studies can be carried out to confirm the structural relationships found in the model. This study only investigated 19 variables, which are not exhaustive. More research should be conducted to determine the relationship between variables. Therefore, it is proposed that further studies should be done, focusing more on MCDM and new tools and approaches such as intuitionistic fuzzy set [96], type-2 fuzzy variable [97], and Rough interval [98], considering the challenges and control strategies for reaching a consensus via a group decision-making process [99,100]. In conclusion, the findings of this study can improve the status of health and environmental indicators and help achieve sustainable development goals in surface mines by identifying and prioritizing factors influencing DE and recommending practical solutions to eliminate and control such errors.

Author Contributions: Conceptualization, I.M. and M.B.; Data curation, A.A.K. and M.F.; Formal analysis, M.F.; Investigation, I.M., A.A.K., H.D. and M.B.; Methodology, A.A.K. and M.B.; Project administration, I.M. and A.A.K.; Resources, A.A.K. and H.D.; Supervision, I.M. and M.F.; Validation, H.D.; Writing—original draft, A.A.K.; Writing—review & editing, I.M. and M.B. All authors have read and agreed to the published version of the manuscript.

Funding: This research was funded by the Hamedan University of Medical Sciences and Health Services and grant number.140008257113.

Institutional Review Board Statement: The study was conducted in accordance with the Declaration of Helsinki, and approved by the Institutional Review Board (or Ethics Committee) of the national committee for ethics in medical research (protocol code IR.UMSHA.REC.1400.617 and date of approval 31 October 2021).

Informed Consent Statement: Not applicable.

Data Availability Statement: Not applicable.

Conflicts of Interest: The authors declare no conflict of interest.

Appendix A

Table A1. The Results of the Fuzzy Delphi Study Regarding the Rank of Variables Affecting DEs.

Subgroup	Variable
Individual factors	Technical knowledge (safety and designing)(0.81), Experience (0.78), Depression (0.73), Self-confidence (0.72), Financial satisfaction (0.72), Stress level (0.71), Intelligence coefficient (0.71), Work adaptation (0.69), Designing style (0.67), Fear of failure (0.65), Risk-taking (0.64), Understanding roles and responsibilities (0.62), Quality and quantity of sleep (0.62), Circadian rhythm (0.62), Risk Perception (0.61), Nutrition (0.52), Determinism (0.49), Disappointment (0.49), Personality type (0.35), Age (0.32), Lack of trust in performance (0.28), Gender (0.28).
Task factors	Workload (0.7), Multitasking (0.7), Time pressure (0.68), Instructions and procedure (0.67), Quality of human–system interaction (0.67), Lack of job security (0.65), Task complexity (0.62), Work posture (0.61), Work innovation (0.56), Freedom at work (0.55), Physical workplace (design) (0.51).
Organizational factors	Poor management (0.81), Resource allocation (0.79), Training (0.71), Employees' sense of belonging (0.69), Supervision level (0.63), Agreement between available and required information (0.62), Designers' sense of belonging (0.61).
Environmental factors	Noise (0.73), Poor indoor air quality (0. 0.72), Inappropriate lighting (0.71), Air circulation velocity (0.57), Hotness and coldness (0.56), Moisture (0.54), Radiation exposure (0.21).
External factors	Legal pressure (0.68), Conflict between work and family (0.51).

Appendix B

Table A2. Defuzzied Total-Relation Matrix.

CF	Va19	Va18	Va17	Va16	Va15	Va14	Va13	Va12	Va11	Va10	Va9	Va8	Va7	Va6	Va5	Va4	Va3	Va2	Va1	
0.07	0.05	0.06	0.04	0.02	0.02	0.01	0.01	0.05	0.06	0.03	0.05	0.02	0.05	0.04	0.05	0.05	0.04	0.04	0.02	Va1
0.06	0.05	0.05	0.03	0.01	0.01	0.01	0.01	0.05	0.06	0.03	0.05	0.03	0.05	0.03	0.07	0.04	0.04	0.01	0.05	Va2
0.05	0.02	0.02	0.02	0.02	0.01	0.01	0.01	0.04	0.03	0.02	0.04	0.02	0.03	0.02	0.05	0.05	0.03	0.02	0.03	Va3
0.05	0.03	0.03	0.02	0.02	0.01	0.01	0.01	0.05	0.03	0.02	0.05	0.02	0.03	0.03	0.04	0.02	0.02	0.03	0.04	Va4
0.05	0.02	0.02	0.01	0.01	0.02	0.01	0.01	0.03	0.04	0.02	0.03	0.03	0.03	0.02	0.01	0.03	0.03	0.02	0.03	Va5
0.06	0.02	0.06	0.04	0.05	0.01	0.01	0.01	0.05	0.06	0.05	0.04	0.03	0.04	0.02	0.05	0.05	0.04	0.05	0.06	Va6
0.06	0.04	0.02	0.01	0.01	0.02	0.01	0.01	0.05	0.06	0.03	0.06	0.03	0.02	0.03	0.04	0.05	0.05	0.02	0.03	Va7
0.04	0.02	0.01	0.01	0.01	0.01	0.01	0.01	0.02	0.03	0.02	0.03	0.01	0.02	0.03	0.04	0.03	0.03	0.03	0.03	Va8
0.06	0.03	0.02	0.02	0.02	0.01	0.01	0.01	0.04	0.05	0.03	0.02	0.02	0.05	0.03	0.03	0.04	0.05	0.03	0.05	Va9
0.05	0.03	0.05	0.03	0.01	0.01	0.01	0.01	0.04	0.03	0.26	0.06	0.02	0.04	0.02	0.03	0.03	0.03	0.02	0.03	Va10
0.06	0.05	0.03	0.03	0.01	0.01	0.01	0.01	0.06	0.02	0.03	0.06	0.03	0.06	0.04	0.05	0.05	0.05	0.03	0.04	Va11
0.05	0.03	0.03	0.03	0.01	0.01	0.01	0.01	0.02	0.06	0.02	0.04	0.02	0.05	0.02	0.04	0.05	0.05	0.03	0.04	Va12
0.04	0.02	0.01	0.01	0.01	0.01	0.01	0	0.04	0.02	0.01	0.04	0.01	0.04	0.02	0.03	0.04	0.04	0.02	0.02	Va13
0.03	0.01	0.01	0.01	0.01	0.01	0	0.01	0.02	0.01	0.01	0.04	0.01	0.02	0.01	0.01	0.02	0.03	0.01	0.01	Va14
0.04	0.02	0.01	0.01	0.01	0	0.01	0.01	0.03	0.01	0.01	0.04	0.01	0.03	0.02	0.02	0.02	0.03	0.01	0.01	Va15
0.07	0.07	0.06	0.06	0.01	0.02	0.02	0.03	0.05	0.05	0.05	0.06	0.02	0.06	0.04	0.05	0.06	0.06	0.04	0.06	Va16
0.07	0.04	0.06	0.01	0.01	0.02	0.02	0.02	0.05	0.03	0.04	0.05	0.02	0.05	0.03	0.05	0.06	0.06	0.03	0.05	Va17
0.06	0.04	0.02	0.05	0.05	0.02	0.01	0.02	0.03	0.03	0.04	0.04	0.01	0.03	0.03	0.03	0.04	0.04	0.03	0.05	Va18
0.04	0.01	0.02	0.02	0.01	0.01	0.01	0.01	0.04	0.03	0.02	0.02	0.02	0.04	0.02	0.04	0.05	0.04	0.02	0.02	V19
0.02	0.03	0.02	0.02	0.01	0.01	0.01	0.01	0.02	0.05	0.02	0.04	0.01	0.04	0.03	0.02	0.02	0.02	0.02	0.02	CF

References

1. Kumar, P.; Gupta, S.; Agarwal, M.; Singh, U. Categorization and standardization of accidental risk-criticality levels of human error to develop risk and safety management policy. *Saf. Sci.* **2016**, *85*, 88–98. [CrossRef]
2. Aliabadi, M.M.; Mohammadfam, I.; Soltanian, A.R.; Najafi, K. Human Error Probability Determination in Blasting Process of Ore Mine Using a Hybrid of HEART and Best-Worst Methods. *Saf. Health Work* 2022, *in press*.
3. Permana, H. Risk Assessment as a Strategy to Prevent of Mine Accidents in Indonesian Mining. *Rev. Min. Min. Rev.* **2012**, *18*, 43–49.
4. Onder, S.; Mutlu, M. Analyses of non-fatal accidents in an opencast mine by logistic regression model–a case study. *Int. J. Inj. Control Saf. Promot.* **2017**, *24*, 328–337. [CrossRef]
5. Mojarradi, G.; Rezaei, R.; Ketabi, A. Negative impacts of mine exploitations on rural regions of Tekab Township. *J. Min. Environ.* **2016**, *7*, 57–66.
6. Aliabadi, M.M.; Aghaei, H.; Kalatpour, O.; Soltanian, A.R.; Nikravesh, A. Analysis of human and organizational factors that influence mining accidents based on Bayesian network. *Int. J. Occup. Saf. Ergon.* **2020**, *26*, 670–677. [CrossRef]
7. Floris, L.M.; Carvalho, E.G.; Faustino, B.C.R.; Calegario, C.L.L. Workplace Accidents as a Consequence of Human Error: An Empirical Study in a Gold Mine/Os Acidentes de Trabalho como Consequência dos Erros Humanos: Um Estudo de Caso em uma Mina de Ouro no Brasil. *Rev. FSA Cent. Univ. St. Agostinho* **2021**, *18*, 74–99.
8. Kim, S.; Wallace, K.M. Identifying Design Errors and Human Errors Using Automatic Classifications. In Proceedings of the ICED 2007, the 16th International Conference on Engineering Design, Paris, France, 28–31 July 2007; pp. 439–440.
9. Salvendy, G.; Karwowski, W. *Handbook of Human Factors and Ergonomics*; John Wiley & Sons: Hoboken, NJ, USA, 2021.
10. Liao, P.-C.; Luo, X.; Wang, T.; Su, Y. The mechanism of how design failures cause unsafe behavior: The cognitive reliability and error analysis method (CREAM). *Procedia Eng.* **2016**, *145*, 715–722. [CrossRef]
11. Peansupap, V.; Ly, R. Evaluating the impact level of design errors in structural and other building components in building construction projects in Cambodia. *Procedia Eng.* **2015**, *123*, 370–378. [CrossRef]
12. Lopez, R.; Love, P.E.; Edwards, D.J.; Davis, P.R. Design error classification, causation, and prevention in construction engineering. *J. Perform. Constr. Facil.* **2010**, *24*, 399–408. [CrossRef]
13. Zhefeng, J.; Dayong, D.; Yinbo, Z.; Hua, M. Design and Evaluation Methodology for Cockpit Human Factor of Civil Transport Aircraft. In Proceedings of the International Conference on Applied Human Factors and Ergonomics, Washton, DC, USA, 24–28 July 2019; pp. 269–275.
14. Stevenson, J.D. An estimate and evaluation of design error effects on nuclear power plant design adequacy. In *Structural Engineering in Nuclear Facilities*; Springer: Berlin/Heidelberg, Germany, 1984; Volume 2.
15. Taylor, J.R. *Human Error in Process Plant Design and Operations: A Practitioner's Guide*; CRC Press: Boca Raton, FL, USA, 2016.
16. Danko, G. Quality evaluation and error mitigation for mine climate models. *Calitatea* **2019**, *20*, 123.
17. Lööw, J.; Johansson, B.; Andersson, E. *Designing the Safe and Attractive Mine*; Luleå Tekniska Universitet: Luleå, Sweden, 2016.
18. Stacey, T. The importance of engineering design with regard to safety in mining. In Proceedings of the Hard Rock Safe Safety Conference, Sun City, South Africa, 28 September 2009.
19. Thompson, R. Mine haul road design and management best practices for safe and cost-efficient truck haulage. In Proceedings of the Society for Mining, Metallurgy and Exploration 2010 Conference Proceedings, Lexington, KY, USA, 26–29 April 2010; pp. 1–10.
20. Norouzi Masir, R.; Ataei, M.; Mottahedi, A. Risk assessment of Flyrock in Surface Mines using a FFTA-MCDM Combination. *J. Min. Environ.* **2021**, *12*, 191–203.
21. Reason, J. *Managing the Risks of Organizational Accidents*; Ashgate Publishing: Farnham, UK, 1997.
22. Kustola, K. Mitigating Design Error Root-Causes in Product Development. 2018. Available online: https://publications.lib.chalmers.se/records/fulltext/255252/255252.pdf (accessed on 29 June 2022).
23. Melchers, R. Human error in structural reliability assessments. *Reliab. Eng.* **1984**, *7*, 61–75. [CrossRef]
24. Fang, D.; Jiang, Z.; Zhang, M.; Wang, H. An experimental method to study the effect of fatigue on construction workers' safety performance. *Saf. Sci.* **2015**, *73*, 80–91. [CrossRef]
25. Smidts, C.; Shen, S.; Mosleh, A. The IDA cognitive model for the analysis of nuclear power plant operator response under accident conditions. Part I: Problem solving and decision making model. *Reliab. Eng. Syst. Saf.* **1997**, *55*, 51–71. [CrossRef]
26. Ekanem, N.J.; Mosleh, A.; Shen, S.-H. Phoenix–a model-based human reliability analysis methodology: Qualitative analysis procedure. *Reliab. Eng. Syst. Saf.* **2016**, *145*, 301–315. [CrossRef]
27. Mohammadfam, I.; Aliabadi, M.M.; Soltanian, A.R.; Mahdinia, M. Modeling the causes-effect relationships among major accident predictors based on a fuzzy multi-criteria decision-making method. *Work* **2020**, *67*, 313–321. [CrossRef]
28. Al Hattab, M.; Hamzeh, F. Using social network theory and simulation to compare traditional versus BIM–lean practice for design error management. *Autom. Constr.* **2015**, *52*, 59–69. [CrossRef]
29. Taylor, J.R. Statistics of design error in the process industries. *Saf. Sci.* **2007**, *45*, 61–73. [CrossRef]
30. Zhaorong, W.; Zekavat, P.; Moon, S.; Tang, L.; Sherif, M. Investigating The Root Cause Of Legal Disputes Over Defective Construction In Proceedings of the 43rd AUBEA: Australasian Universities Building Education Association Conference Proceedings, Noosa, QLD, Australia, 6–8 November 2019; RACV Noosa Resort.

31. Islam, R.; Khan, F.; Abbassi, R.; Garaniya, V. Human error assessment during maintenance operations of marine systems–What are the effective environmental factors? *Saf. Sci.* **2018**, *107*, 85–98. [CrossRef]
32. Athar, M.; Shariff, A.A.; Buang, A.; Shaikh, M.S.; Khan, M.I. Review of process industry accidents analysis towards safety system improvement and sustainable process design. *Chem. Eng. Technol.* **2019**, *42*, 524–538. [CrossRef]
33. Wudhikarn, R. Improving the intellectual capital management approach using the hybrid decision method. *J. Intellect. Cap.* **2018**, *19*, 670–691. [CrossRef]
34. Ebrahimnezhad, S.; Gitinavard, H.; Sohrabvandi, S. A new extended analytical hierarchy process technique with incomplete interval-valued information for risk assessment in IT outsourcing. *Int. J. Eng.* **2017**, *30*, 739–748.
35. Wudhikarn, R.; Chakpitak, N.; Neubert, G. Improving the strategic benchmarking of intellectual capital management in logistics service providers. *Sustainability* **2020**, *12*, 10174. [CrossRef]
36. Wudhikarn, R.; Chakpitak, N.; Neubert, G. Use of an analytic network process and Monte Carlo analysis in new product formula selection decisions. *Asia-Pac. J. Oper. Res.* **2015**, *32*, 1550007. [CrossRef]
37. Kumar, A.; Kaviani, M.A.; Bottani, E.; Dash, M.K.; Zavadskas, E.K. Investigating the role of social media in polio prevention in India: A Delphi-DEMATEL approach. *Kybernetes* **2018**, *47*, 1053–1072. [CrossRef]
38. Mohandes, S.R.; Sadeghi, H.; Fazeli, A.; Mahdiyar, A.; Hosseini, M.R.; Arashpour, M.; Zayed, T. Causal analysis of accidents on construction sites: A hybrid fuzzy Delphi and DEMATEL approach. *Saf. Sci.* **2022**, *151*, 105730. [CrossRef]
39. Thakur, V.; Mangla, S.K.; Tiwari, B. Managing healthcare waste for sustainable environmental development: A hybrid decision approach. *Bus. Strategy Environ.* **2021**, *30*, 357–373. [CrossRef]
40. Lu, M.; Wudhikarn, R. Using the best-worst method to develop intellectual capital indicators in financial service company. In Proceedings of the 2022 Joint International Conference on Digital Arts, Media and Technology with ECTI Northern Section Conference on Electrical, Electronics, Computer and Telecommunications Engineering (ECTI DAMT & NCON), Chiang Rai, Thailand, 26–28 January 2022; pp. 81–86.
41. Mohammadfam, I.; Aliabadi, M.M.; Soltanian, A.R.; Tabibzadeh, M.; Mahdinia, M. Investigating interactions among vital variables affecting situation awareness based on Fuzzy DEMATEL method. *Int. J. Ind. Ergon.* **2019**, *74*, 102842. [CrossRef]
42. Kumar, A.; Pal, A.; Vohra, A.; Gupta, S.; Manchanda, S.; Dash, M.K. Construction of capital procurement decision making model to optimize supplier selection using Fuzzy Delphi and AHP-DEMATEL. *Benchmarking Int. J.* **2018**, *25*, 1528–1547. [CrossRef]
43. Quiñones, R.S.; Caladcad, J.A.A.; Himang, C.M.; Quiñones, H.G.; Castro, C.J.; Caballes, S.A.A.; Abellana, D.P.M.; Jabilles, E.M.Y.; Ocampo, L.A. Using Delphi and fuzzy DEMATEL for analyzing the intertwined relationships of the barriers of university technology transfer: Evidence from a developing economy. *Int. J. Innov. Stud.* **2020**, *4*, 85–104. [CrossRef]
44. Singh, P.K.; Sarkar, P. A framework based on fuzzy Delphi and DEMATEL for sustainable product development: A case of Indian automotive industry. *J. Clean. Prod.* **2020**, *246*, 118991. [CrossRef]
45. Häder, M.; Häder, S. Delphi und Kognitionspsychologie: Ein Zugang zur theoretischen Fundierung der Delphi-Methode. *ZUMA Nachr.* **1995**, *19*, 8–34.
46. Azar, A.; Faraji, H. *Fuzzy Management Science*; Institute Mehraban book publisher: Tehran, Iran, 2010.
47. Zhou, Q.; Huang, W.; Zhang, Y. Identifying critical success factors in emergency management using a fuzzy DEMATEL method. *Saf. Sci.* **2011**, *49*, 243–252. [CrossRef]
48. Murray, T.J.; Pipino, L.L.; Van Gigch, J.P. A pilot study of fuzzy set modification of Delphi. *Hum. Syst. Manag.* **1985**, *5*, 76–80. [CrossRef]
49. Kuo, Y.-F.; Chen, P.-C. Constructing performance appraisal indicators for mobility of the service industries using Fuzzy Delphi Method. *Expert Syst. Appl.* **2008**, *35*, 1930–1939. [CrossRef]
50. Bavafa, A.; Mahdiyar, A.; Marsono, A.K. Identifying and assessing the critical factors for effective implementation of safety programs in construction projects. *Saf. Sci.* **2018**, *106*, 47–56. [CrossRef]
51. Cheng, C.-H.; Lin, Y. Evaluating the best main battle tank using fuzzy decision theory with linguistic criteria evaluation. *Eur. J. Oper. Res.* **2002**, *142*, 174–186. [CrossRef]
52. Wang, F. Preference degree of triangular fuzzy numbers and its application to multi-attribute group decision making. *Expert Syst. Appl.* **2021**, *178*, 114982. [CrossRef]
53. Zhang, X.; Ma, W.; Chen, L. New similarity of triangular fuzzy number and its application. *Sci. World J.* **2014**, *2014*, 1–7.
54. Li, R.-J. Fuzzy method in group decision making. *Comput. Math. Appl.* **1999**, *38*, 91–101. [CrossRef]
55. Hsu, Y.-L.; Lee, C.-H.; Kreng, V.B. The application of Fuzzy Delphi Method and Fuzzy AHP in lubricant regenerative technology selection. *Expert Syst. Appl.* **2010**, *37*, 419–425. [CrossRef]
56. Cafiso, S.; Di Graziano, A.; Pappalardo, G. Using the Delphi method to evaluate opinions of public transport managers on bus safety. *Saf. Sci.* **2013**, *57*, 254–263. [CrossRef]
57. Kosmidou, K. The determinants of banks' profits in Greece during the period of EU financial integration. *Manag. Financ.* **2008**, *34*, 146–159.
58. Saraswathi, A. A fuzzy-trapezoidal DEMATEL approach method for solving decision making problems under uncertainty. In Proceedings of the AIP Conference Proceedings, Chennai, India, 8–10 April 2019; p. 020076.
59. Wu, W.-W.; Lee, Y.-T. Developing global managers' competencies using the fuzzy DEMATEL method. *Expert Syst. Appl.* **2007**, *32*, 499–507. [CrossRef]

60. Li, Y.; Hu, Y.; Zhang, X.; Deng, Y.; Mahadevan, S. An evidential DEMATEL method to identify critical success factors in emergency management. *Appl. Soft Comput.* **2014**, *22*, 504–510. [CrossRef]
61. Yazdi, M.; Nedjati, A.; Zarei, E.; Abbassi, R. A novel extension of DEMATEL approach for probabilistic safety analysis in process systems. *Saf. Sci.* **2020**, *121*, 119–136. [CrossRef]
62. Aghaee, M.; Aghaee, R. Selection of logistics personnel by using and hybrid Fuzzy DEMATEL and Fuzzy ANP. *Int. Res. J. Manag. Sci.* **2016**, *4*, 14–22.
63. Akyuz, E.; Celik, E. A fuzzy DEMATEL method to evaluate critical operational hazards during gas freeing process in crude oil tankers. *J. Loss Prev. Process Ind.* **2015**, *38*, 243–253. [CrossRef]
64. Gholamizadeh, K.; Zarei, E.; Omidvar, M.; Yazdi, M. Fuzzy Sets Theory and Human Reliability: Review, Applications, and Contributions. *Linguist. Methods under Fuzzy Inf. Syst. Saf. Reliab. Anal.* **2022**, *414*, 91–137.
65. Omidvar, M.; Zarei, E.; Ramavandi, B.; Yazdi, M. Fuzzy Bow-Tie Analysis: Concepts, Review, and Application. In *Linguistic Methods under Fuzzy Information in System Safety and Reliability Analysis*; Springer: Berlin/Heidelberg, Germany, 2022; pp. 13–51.
66. Lin, R.-J. Using fuzzy DEMATEL to evaluate the green supply chain management practices. *J. Clean. Prod.* **2013**, *40*, 32–39. [CrossRef]
67. Fuadie, D.F.; Rahmawati, Y.; Utomo, C. Factors of Design Errors in Construction Project (A Review). *IPTEK J. Proc. Ser.* **2017**, *3*, 3263. [CrossRef]
68. Hafezi, M.; Hsseinalipour, M.; Ghafoori, S.P. Design Errors in Construction Projects: The Meaning, Reasons, Consequences and Solutions. *Soffeh* **2019**, *29*, 21–34. [CrossRef]
69. Chao, L.P.; Ishii, K. Design process error-proofing: Benchmarking gate and phased review life-cycle models. In Proceedings of the International Design Engineering Technical Conferences and Computers and Information in Engineering Conference, Long Beach, CA, USA, 24–28 September 2005; pp. 301–310.
70. Franciosi, C.; Di Pasquale, V.; Iannone, R.; Miranda, S. A taxonomy of performance shaping factors for human reliability analysis in industrial maintenance. *J. Ind. Eng. Manag.* **2019**, *12*, 115–132. [CrossRef]
71. Mohammadfam, I.; Ghasemi, F.; Kalatpour, O.; Moghimbeigi, A. Constructing a Bayesian network model for improving safety behavior of employees at workplaces. *Appl. Ergon.* **2017**, *58*, 35–47. [CrossRef] [PubMed]
72. Abbassi, R.; Khan, F.; Garaniya, V.; Chai, S.; Chin, C.; Hossain, K.A. An integrated method for human error probability assessment during the maintenance of offshore facilities. *Process Saf. Environ. Prot.* **2015**, *94*, 172–179. [CrossRef]
73. Nkosi, M.; Gupta, K.; Mashinini, M. Causes and impact of human error in maintenance of mechanical systems. In Proceedings of the MATEC Web of Conferences, Palu, Indonesia, 25–27 April 2020; p. 05001.
74. Rollenhagen, C. Can focus on safety culture become an excuse for not rethinking design of technology? *Saf. Sci.* **2010**, *48*, 268–278. [CrossRef]
75. Philemon, E.; Msomba, Z.; Samson, M.; Ramadhan, S.M. Identification of Crucial Risks Categories in Construction Projects in Tanzania. *Int. J. Eng. Res.* **2018**, *7*, 62.
76. Delikhoon, M.; Zarei, E.; Banda, O.V.; Faridan, M.; Habibi, E. Systems Thinking Accident Analysis Models: A Systematic Review for Sustainable Safety Management. *Sustainability* **2022**, *14*, 5869. [CrossRef]
77. Dhillon, B. Human error in maintenance: An investigative study for the factories of the future. In Proceedings of the IOP Conference Series: Materials Science and Engineering, London, UK, 22–24 July 2014; p. 012031.
78. Franciosi, C.; Iung, B.; Miranda, S.; Riemma, S. Maintenance for sustainability in the industry 4.0 context: A scoping literature review. *IFAC-Pap.* **2018**, *51*, 903–908. [CrossRef]
79. Andi, A.; Minato, T. Representing causal mechanism of defective designs: Exploration through case studies. *Constr. Manag. Econ.* **2004**, *22*, 183–192. [CrossRef]
80. Amiri, F.; Zamanian, Z.; Mani, A.; Hasanzadeh, J. Effects of combined exposure to noise, heat and lighting on cognitive performance. *Iran Occup. Health* **2015**, *12*, 10–20.
81. Vatanpour, S.; Hrudey, S.E.; Dinu, I. Can public health risk assessment using risk matrices be misleading? *Int. J. Environ. Res. Public Health* **2015**, *12*, 9575–9588. [CrossRef] [PubMed]
82. Szalma, J.L.; Hancock, P.A. Noise effects on human performance: A meta-analytic synthesis. *Psychol. Bull.* **2011**, *137*, 682. [CrossRef] [PubMed]
83. DiGiovanni, J.; Riffle, T.L.; Lynch, E.E.; Nagaraj, N.K. Noise characteristics and their impact on working memory and listening comprehension performance. *Acoust. Soc. Am.* **2017**, *141*, 050007. [CrossRef]
84. Beyan, A.C.; Demiral, Y.; Cimrin, A.H.; Ergor, A. Call centers and noise-induced hearing loss. *Noise Health* **2016**, *18*, 113. [CrossRef]
85. Hebb, D.O. Drives and the CNS (conceptual nervous system). *Psychol. Rev.* **1955**, *62*, 243. [CrossRef]
86. Jafari, M.J.; Sadeghian, M.; Khavanin, A.; Khodakarim, S.; Jafarpisheh, A.S. Effects of noise on mental performance and annoyance considering task difficulty level and tone components of noise. *J. Environ. Health Sci. Eng.* **2019**, *17*, 353–365. [CrossRef]
87. Asadzadeh, H.; Farokhzadeh, N.; Moghadam, Y.A. The Effect of training materials to help color and light, on the amount of attention and learning Persian language lessons in elementary school girls in the first grade in Sharoud city. *Life Sci. J.* **2013**, *10*, 246–251.
88. Pirmoradi, Z.; Golmohammadi, R.; Motamedzade, M.; Faradmal, J. Assessing lighting and color Temperature in the Office Workplaces and Relationship to Visual Comfort. *Iran Occup. Health J.* **2020**, *17*, 1–10.

89. Künn, S.; Palacios, J.; Pestel, N. Indoor Air Quality and Cognitive Performance. 2019. Available online: https://www.iza.org/publications/dp/12632/indoor-air-quality-and-cognitive-performance (accessed on 29 June 2022).
90. Efficiency, C.P. Design and Documentation Quality Survey Comparison of Designers' and Contractors' Perspectives. 2000. Available online: https://www.researchgate.net/publication/301655224_Design_and_Documentation_Quality_Survey_-_Designers_Perspective_BCE_Doc_00-113 (accessed on 29 June 2022).
91. Saaidin, S.; Endut, I.R.; Samah, A.; Akmar, S.; Ridzuan, A.R.M. Risk Factors for Design and Build Projects in Malaysia-Project Manager's Perception. *Pertanika J. Sci. Technol.* **2017**, *25*, 185–190.
92. Sneddon, A.; Mearns, K.; Flin, R. Stress, fatigue, situation awareness and safety in offshore drilling crews. *Saf. Sci.* **2013**, *56*, 80–88. [CrossRef]
93. Young, M.S.; Brookhuis, K.A.; Wickens, C.D.; Hancock, P.A. State of science: Mental workload in ergonomics. *Ergonomics* **2015**, *58*, 1–17. [CrossRef] [PubMed]
94. Mohamed, S.; Ali, T.H.; Tam, W. National culture and safe work behaviour of construction workers in Pakistan. *Saf. Sci.* **2009**, *47*, 29–35. [CrossRef]
95. Kim, Y.; Park, J. Suggestions of HRA method improvement for the practical assessment of human reliability. *J. Ergon. Soc. Korea* **2018**, *37*, 229–241.
96. Vafadarnikjoo, A.; Mobin, M.; Firouzabadi, S. An intuitionistic fuzzy-based DEMATEL to rank risks of construction projects. In Proceedings of the 2016 International Conference on Industrial Engineering and Operations Management, Detroit, MI, USA, 23–25 September 2016.
97. Seikh, M.; Karmakar, S.; Castillo, O. A novel defuzzification approach of type-2 fuzzy variable to solving matrix games: An application to plastic ban problem. *Iran. J. Fuzzy Syst.* **2021**, *18*, 155–172.
98. Seikh, M.R.; Dutta, S.; Li, D.F. Solution of matrix games with rough interval pay-offs and its application in the telecom market share problem. *Int. J. Intell. Syst.* **2021**, *36*, 6066–6100. [CrossRef]
99. Gao, Y.; Zhang, Z. Consensus reaching with non-cooperative behavior management for personalized individual semantics-based social network group decision making. *J. Oper. Res. Soc.* **2021**, *72*, 1–18. [CrossRef]
100. Zhang, Z.; Li, Z.; Gao, Y. Consensus reaching for group decision making with multi-granular unbalanced linguistic information: A bounded confidence and minimum adjustment-based approach. *Inf. Fusion* **2021**, *74*, 96–110. [CrossRef]

sustainability

Article

A Statistical Framework for Evaluating the Effectiveness of Vegetation Management in Reducing Power Outages Caused during Storms in Distribution Networks

William O. Taylor [1,2,*], Peter L. Watson [1,2], Diego Cerrai [1,2] and Emmanouil Anagnostou [1,2,*]

1 Department of Civil & Environmental Engineering, University of Connecticut, Storrs, CT 06269, USA; peter.watson@uconn.edu (P.L.W.); diego.cerrai@uconn.edu (D.C.)
2 Eversource Energy Center, University of Connecticut, Storrs, CT 06269, USA
* Correspondence: william.o.taylor@uconn.edu (W.O.T.); manos@uconn.edu (E.A.)

Abstract: This paper develops a statistical framework to analyze the effectiveness of vegetation management at reducing power outages during storms of varying severity levels. The framework was applied on the Eversource Energy distribution grid in Connecticut, USA based on 173 rain and wind events from 2005–2020, including Hurricane Irene, Hurricane Sandy, and Tropical Storm Isaias. The data were binned by storm severity (high/low) and vegetation management levels, where a maximum applicable length of vegetation management for each circuit was determined, and the data were divided into four bins based on the actual length of vegetation management performed divided by the maximum applicable value (0–25%, 25–50%, 50–75%, and 75–100%). Then, weather and overhead line length normalized outage statistics were taken for each group. The statistics were used to determine the effectiveness of vegetation management and its dependence on storm severity. The results demonstrate a higher reduction in damages for lower-severity storms, with a reduction in normalized outages between 45.8% and 63.8%. For high-severity events, there is a large increase in effectiveness between the highest level of vegetation management and the two lower levels, with 75–100% vegetation management leading to a 37.3% reduction in trouble spots. Yet, when evaluating system reliability, it is important to look at all storms combined, and the results of this study provide useful information on total annual trouble spots and allow for analysis of how various vegetation management scenarios would impact trouble spots in the electric grid. This framework can also be used to better understand how more rigorous vegetation management standards (applying ETT) help reduce outages at an individual event level. In future work, a similar framework may be used to evaluate other resilience improvements.

Keywords: electric power distribution; power outages; reliability; resilience; severe weather; tree trimming; vegetation management

Citation: Taylor, W.O.; Watson, P.L.; Cerrai, D.; Anagnostou, E. A Statistical Framework for Evaluating the Effectiveness of Vegetation Management in Reducing Power Outages Caused during Storms in Distribution Networks. *Sustainability* **2022**, *14*, 904. https://doi.org/10.3390/su14020904

Academic Editors: Esmaeil Zarei, Samuel Yousefi and Mohsen Omidvar

Received: 20 December 2021
Accepted: 12 January 2022
Published: 13 January 2022

1. Introduction

Trees are one of the leading causes of outages in electric distribution systems [1], and the leading cause for some utilities [2,3]. One study over the Eastern United States and Canada found tree growth to be the most frequent cause of preventable outages [3], and these outages prove costly. A study from 2001 shows even a one-second outage can cost individual businesses an average of $1477 [4], highlighting the benefit of improving system resilience. Another report from 2015 reinforced the idea of high costs for short duration outages, as it evaluated 34 datasets from 10 utilities across 1989 to 2012 and estimated the cost of momentary outages to small commercial and industrial (C&I) customers (defined as under 50,000 kilowatt-hours annually) as $412 per event, and medium and large C&I customers as $12,952 per momentary outage event [5]. During storms in the Northeastern United States (US), trees are responsible for a particularly high percentage of outages, with Eversource (a major power utility that services more than 3 million electric customers

This study focuses on grid resilience during storms as it is of critical importance to understand any differences in the efficacy of vegetation management for different storm severity (lower impact, more frequent storms versus stronger, less frequent events) in order to conduct comprehensive resilience and economic analyses. This is the first of three main contributions of this study to the existing literature, as to the authors' knowledge, it is the first study that looks at the effectiveness of vegetation management in reducing trouble spots for varying storm severity. This study addresses the gap in the existing literature by performing a statistical analysis that looks at the effectiveness of implementing more rigorous vegetation management standards (by applying ETT) in reducing trouble spots in the electric distribution grid while binning and normalizing storm events by their severity. The second contribution this study provides is a quantitative tool that provides the ability to retrospectively quantify return on investment by evaluating various scenarios of ETT for historical events and also the approximate return on investment for future climate storm scenarios. The tool can be used to analyze individual events, such as hurricanes, as well as perform more comprehensive analyses such as evaluating the total trouble spots reduced over the domain in a given year. This will help stakeholders (municipalities, utilities, regulators) to more fully understand the impacts of vegetation management, allow for more accurate economic analysis, and provide useful information to help guide resilience planning and policy decisions such as optimizing resilience budgets between activities, including reducing the time between regular maintenance tree trimming (by applying SMT more frequently), enacting more rigorous vegetation management standards (by extending ETT), performing pole upgrades, performing wire upgrades, and undergrounding lines. The third contribution of this study is the framework developed, as it is extendable to evaluate the effectiveness of other resilience and grid hardening efforts such as pole replacements/upgrades and reconductoring wires.

2. Data and Methods

2.1. Study Area

The study area for this analysis is the Eversource Energy service area in Connecticut, which serves 1.2 million electric customers [6] and covers over 4400 square miles across 149 of 169 towns in the state [31]. The topography of the state can be generally regarded as hilly with elevation ranging from sea level to approximately 750 m [12].

2.2. Data

2.2.1. Outage Data and Storm Events

The outage data used in this study were provided by Eversource Energy from their outage management system and includes starting time and grid circuit (operational units of the distribution network) on which each outage occurred. Historical storms were identified by using METeorological Aerodrome Reports (METARs) data collected at Connecticut airport stations. The identified storms were dynamically simulated at 4-km grid spacing using the 3.8.1 version of the Advanced Research (ARW) core of the Weather Research and Forecasting (WRF) model, initialized with the North American Mesoscale Forecasting System's (NAM) initial and boundary conditions. The WRF model configuration used in this study is described in detail in [32]. Outages were analyzed for 173 storm events with varying type and size, and for each event, all outages that occurred during the storm window were analyzed. The events occurred between the years of 2005 and 2020, and include Hurricane Irene, Hurricane Sandy, and Tropical Storm Isaias. The additional events display rain and wind conditions consistent with extratropical frontal systems. The outage data from the utility have an assigned field for the circuit to which the piece of infrastructure belongs, the start date , start time, and the duration, among other information. The start date and time were used to assign outages to storm events. For each event, a window for associated outages is created based on the timing of weather conditions and the magnitude of the storms. The window length is 48 h for small events, which is increased in length for more extreme events. The outage duration is increased as for larger events, with many outages, there are often nested outages that take some time to be discovered and reported. Nested outages are those where an outage downstream (the nested outage) in the electric grid is discovered when a trouble spot upstream is repaired, and not all of the expected customers have power returned. For large storms, the window to count outages is expanded until the active outages are reduced to a background noise level. The longest window for recording outages is 240 h for Tropical Storm Isaias. To obtain the total outages for each circuit for each event, the damage locations (trouble spots) associated with each circuit during each storm window were summed.

As the focus of this study is on outage reductions due to vegetation management, it is important to understand how many of the outages analyzed are in fact caused by vegetation. Of the 118,227 trouble spots in the electric grid across the 173 storm events used in this study, 100,956 or 85.4% had vegetation-related cause labels, which is comparable to the 90% tree-caused outage value for storms with heavy rain or snowfall reported by Eversource [7]. Of the remaining outages, only 1058 or 0.89% were assigned an outage caused by lightning. The remaining outages had various causes including "unknown", "patrolled [and] nothing found", and "miscellaneous". As a high percentage of outages were caused by vegetation, and knowing that some of the other outages in other categories such as "miscellaneous" and "unknown" may have also been caused by vegetation, all outages were used for the analysis conducted in this study. There were no outage data with a missing value for outage cause.

2.2.2. Infrastructure, Vegetation, and Vegetation Management Data

Infrastructure data were provided by the utility including the precise geographic position of the overhead power lines for the 957 circuits in the Eversource Connecticut domain. Additionally, the utility (Eversource) provided the vegetation management data for each circuit extending backward to the beginning of the ETT program (2009), meaning no ETT had been performed on any circuits (all values of 0). The tree height data used in the study are available at a 30 m-resolution and were developed through integration between the Global Ecosystem Dynamics Investigation (GEDI) light detection and ranging (lidar) data and Landsat analysis-ready data [33].

2.3. Methodology

For each circuit, the lengths of primary overhead lines, lateral lines, and backbone lines were measured. The overhead line locations were used in combination with the raster of vegetation height [33] to determine the maximum potential ETT length (maxETTlength) for each circuit. A mask of the raster of vegetation height was created, only keeping those cells with vegetation height above 6 m (19.69 feet), and buffering those cells out 20 m (65.62 feet) to account for any potential imprecision in the data. The overhead lines were overlaid with this raster and trimmed to keep only the overhead lines where they intersect the tree height mask. We then measured the remaining length of overhead lines for each circuit. This maximum trimmable length (maxETTlength) for each circuit is combined with the vegetation management data to create the vegetation management variable used for analysis. The vegetation management data were obtained from Eversource Energy in two formats, as the collection methodology was updated during the life of the ETT program. The length of ETT performed on each circuit was available at a monthly resolution for the most recent years of data, 2016–2019. For the earlier years of ETT data (2009–2015), the ETT data were available for each circuit at an annual resolution. These data were temporally downscaled to monthly values using weights derived from the relative percentage of tree trimming completed monthly in 2016. To perform the downscaling of the data, the miles trimmed in each month during 2016 were divided by the total miles trimmed in 2016. The monthly percentages were multiplied by the total annual trimming for each circuit in the years 2009 to 2015, to approximate the portion of annual trimming completed in each month.

The full monthly ETT dataset (2009–2019) was then used to create a variable called `instETT`. The `instETT` variable is the cumulative value of ETT that had been performed on a given circuit between the start of the ETT program in 2009 and the starting time of each storm, divided by the calculated maximum potential ETT length (maxETTlength) for that circuit. The equation for `instETT` for a circuit for one event can be seen in Equation (1) below, where $ETTcumulative$ is the cumulative ETT that had been performed on the given circuit at the time of the storm since the beginning of the program (at a monthly temporal resolution), and $maxETTlength$ is the maximum applicable length of the circuit over which ETT can be applied, as previously described in this section of the paper.

$$instETT = \frac{ETTcumulaive}{maxETTlength} \tag{1}$$

In order to account for any data irregularities where the cumulative ETT performed on a circuit was greater than the calculated maximum potential ETT length for that circuit, the `instETT` variable was capped at a value of 1.

Over the Eversource Connecticut domain, there exist 957 circuits. However, for the purposes of this analysis, circuits with less than 1/4th of a mile of overhead line length where ETT is applicable were removed as very little trimming on these circuits leads to large swings in the `instETT` value, which may influence the statistical analysis. Due to having a short span of wire where ETT is applicable, when only short stretches of wire are trimmed, the `instETT` values for these circuits rapidly approach 100%. This results in a distribution of `instETT` values that is more discrete in appearance—with large gaps between values—than for the circuits with trimmable lengths above 1/4th of a mile. Moreover, below 1/4th of a mile of trimmable length there is little signal, as despite making up 6.17% of the circuits in the service territory, these low trimmable length circuits only account for 345 outages, or 0.29% of all outages in the dataset. There is also a high density of instances of circuits with less than 1/4th of a mile of trimmable length with `instETT` values close to 100%. Thus, in order to create a more representative statistical analysis for those circuits in the highest `instETT` bin (75–100%) and mitigate the analysis being skewed by an unrepresentative subset of the data in terms of trimmable length, circuits with less than 1/4th of a mile of overhead line length where ETT is applicable were removed. This results in 898 circuits for use in the analysis after removing the aforementioned circuits with short overhead line lengths where ETT is applicable.

In order to compare the effectiveness of various tree trimming levels for differing storm severity, we divided the data in several ways. First, the data for the 173 storms were split at the 90th percentile of exceedance probability in terms of trouble spots caused in the distribution network. The top ten percent, or seventeen events, are considered the high-severity events, with the other 156 classified as low-severity. The split point and trouble spots for each event are displayed in Figure 2.

For each storm class (high- and low-severity), the data were subset three times based on their `instETT` values to evaluate various percentages of applicable ETT performed. Each of the three subsets compare the outage rates of circuits from the lowest `instETT` bin (0–25%) to one of the higher ETT bins (25–50%, 50–75%, or 75–100%). In order to try and make the comparison more fair in terms of data samples, only circuits were kept that had `instETT` values in both bins at the time of one or more storms from the storm class of interest (high- and low-severity). Only keeping circuits that have data samples in both bins being compared helps to control for other factors such as location, tree cover, circuit size, and infrastructure among other variables.

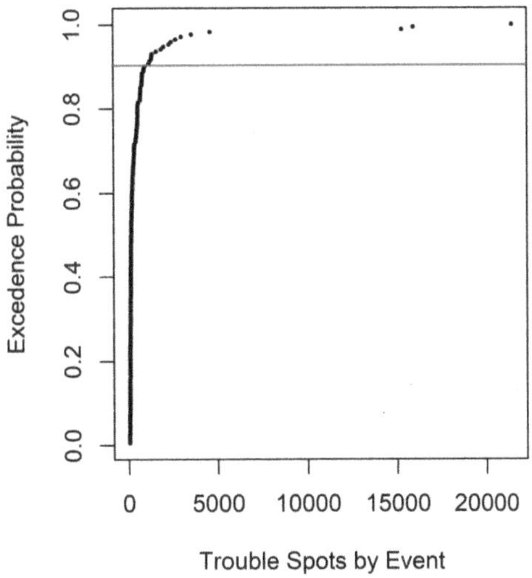

Figure 2. Exceedance Probability by Trouble Spots per Event. Exceedance probability based on the 173 storms included in the dataset. The break point in classification between low- and high-severity lines (the 90th percentile) is represented by the horizontal line.

This control is demonstrated by examining a comparison between two sample groups in further detail, such as the comparison of circuits with 0% to 25% ETT versus those with 25% to 50% ETT for high-severity storms, which we will call Experiment A. First, the data are subset to only include the data for each circuit from the seventeen high-severity events. Next, the data are further subset to only keep circuits that have `instETT` values in both of the ranges being compared (0–25% ETT and 25–50% ETT in this case). For example, if a circuit has `instETT` values between 0% and 25% for seven of the seventeen high-severity storms, and `instETT` values between 25% and 50% for another seven events, those fourteen rows of data are included in the analysis, whereas the remaining three rows of data for the given circuit where a high-severity event occurred but the `instETT` value was above 50% are excluded.

For high-severity events, the same binning process was used to compare circuits with `instETT` values between 0% and 25% to those with `instETT` values between 50% and 75% (Experiment B), as well as to compare circuits with `instETT` values between 0% and 25% to those with `instETT` values between 75% and 100% (Experiment C). Experiments A through C were repeated for the low-severity events (Experiments D–F) where D, E, and F correspond to the same `instETT` bin comparisons as Experiments A, B, and C, respectively. The resulting bin sizes for each data subset for high- and low-severity events can be seen in Table 1. For high-severity events, the number of rows of data in Experiment C, where we compare circuits that have the lowest percentages of applicable ETT performed (0% to 25%) against circuits with the highest levels of ETT (75% to 100%), is notably smaller than the other data subsets in the comparison groups, but this imbalance is a result of the available data. There are not many circuits that have had high-severity events occur at both times when a low amount of ETT had been performed and again after significant trimming. This is likely due to the fact that there are only seventeen events classified as high-severity, and a minority of circuits have reached `instETT` values between 75% and 100%. The average amount of applicable ETT performed on circuits has risen significantly in recent years, but was still only 32% at the time of Tropical Storm Isaias, which is the most recent storm in the data.

Table 1. Size of comparison bins for each Experiment, A-F, and average Kinetic Energy Proxies for the `instETT` bins in each experiment.

Experiment	Severity	instETT Values	Rows of Data	# of Unique Circuits	% of Circuits (of 898)	Kinetic Energy Proxy [(m/s)^2]
A	High	0–25%	4453	465	51.8%	130.1
A	High	25–50%	2896	465	51.8%	145.6
B	High	0–25%	1188	140	15.6%	117.5
B	High	50–75%	854	140	15.6%	140.0
C	High	0–25%	428	48	5.3%	123.4
C	High	75–100%	347	48	5.3%	141.5
D	Low	0–25%	31,408	486	54.1%	76.4
D	Low	25–50%	36,196	486	54.1%	62.5
E	Low	0–25%	6390	144	16.0%	77.0
E	Low	50–75%	9299	144	16.0%	58.6
F	Low	0–25%	2014	47	5.2%	78.6
F	Low	75–100%	4153	47	5.2%	58.5

To mitigate any impacts that may come from the varying average overhead line lengths in the ETT bins being compared, we normalize the outages in each circuit by miles of overhead line length to generate an outage rate per unit of infrastructure. We do this because if all else was equal, circuits with longer overhead line length would be at higher risk of outages as there is more infrastructure with the potential to be damaged.

In addition to normalizing outages by overhead line length for each circuit, for some results we also normalized by a proxy for the average kinetic energy of each storm using the wind speed 10 m above the ground. This step was taken to help control for the differences between the risk presented by various intensities of storms. This is particularly important for high-severity events, as there are only seventeen total and the events inside the high-severity category do not have homogeneous weather. Some high-severity events such as the hurricanes have much stronger winds and weather, and an order of magnitude greater impact than the other events in the same category.

The low versus high `instETT` cumulative density functions for each of the three high-severity Experiments (A–C) can be seen in Figure 3. It can be seen in the figure that for approximately the lower 80% of quantiles for each of the three comparisons, the bins representing the higher `instETT` values have higher average kinetic energy proxy. However, the data groups with lower `instETT` values have the higher kinetic energy proxy for approximately the top 20% of quantiles.

Due to the difference in kinetic energy proxy distributions for the lower and higher ETT bins in each comparison, we normalize outages by this proxy for each bin. To obtain a value to normalize by, the mean of the squared maximum 10-m wind speeds for each circuit is taken, where Equation (2) below represents the kinetic energy proxy for one circuit. The average of the kinetic energy proxies for each circuit is then taken in a given comparison bin:

$$KE_{proxy} = (V_{max})^2 \tag{2}$$

In the above equation, the 10-m wind speed (V_{max}) is squared as velocity is in the equation for kinetic energy. Values of the average kinetic energy proxy for each `instETT` comparison bin for high- and low-severity events can be seen in Table 1.

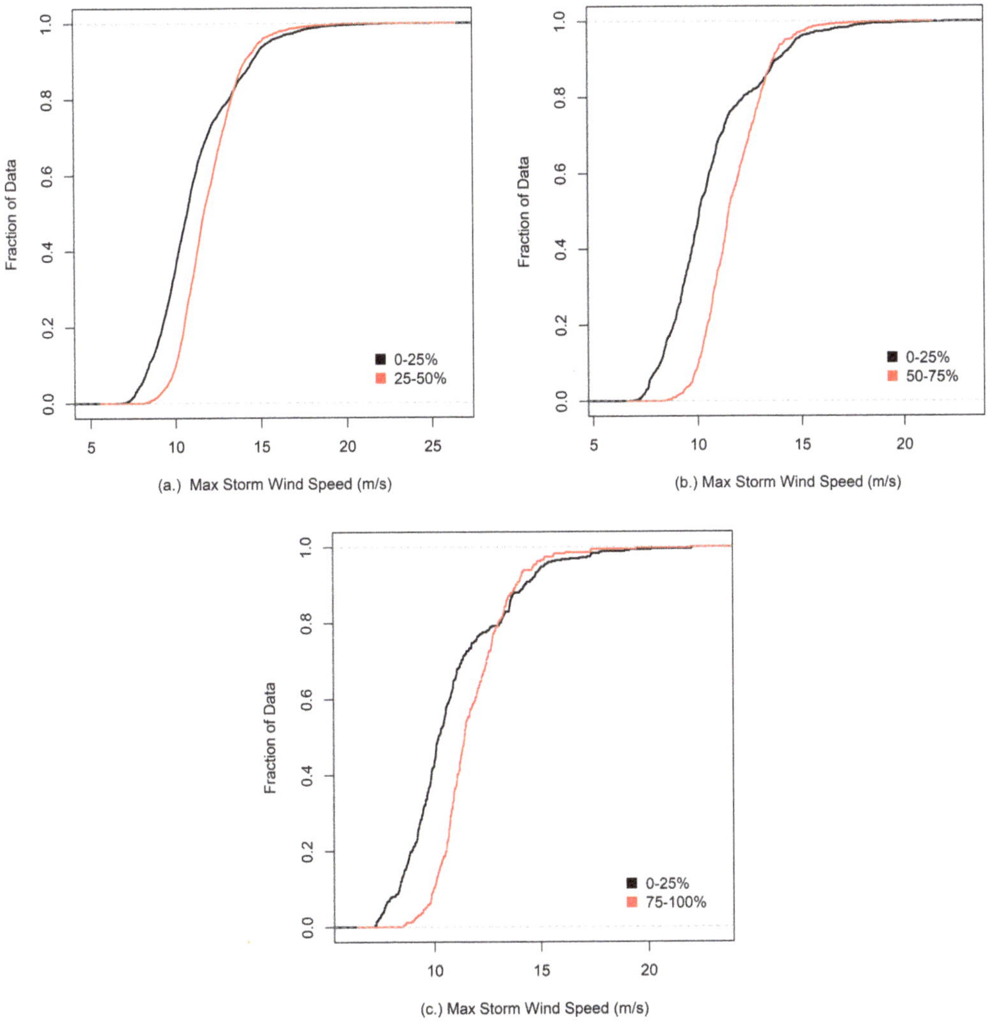

Figure 3. Cumulative distribution functions (CDFs) of Kinetic Energy Proxy. Comparisons between low and higher ETT data subsets for each comparison group for Severe Storms. (**a**) 0–25% vs. 25–50% ETT groups—Experiment A. (**b**) 0–25% vs. 50–75% ETT groups—Experiment B. (**c**) 0–25% vs. 75–100% ETT groups—Experiment C.

Each of the data bins were also tested for the normality of the distribution, to confirm which statistical test to use for determining if the difference between trouble spots in each ETT group comparison is significant. The Shapiro–Wilk normality test was used [34–38]. For bins with under 5000 data samples, all data were used in the test. For the bins with over 5000 samples, 5000 data points were randomly selected and used for the analysis. All of the bins were found to not have normally distributed data. As the data subsets are not normal in their distributions, a nonparametric Wilcoxon–Mann–Whitney test (alpha of 0.05) was used to compare the means for significant differences [35,39–41].

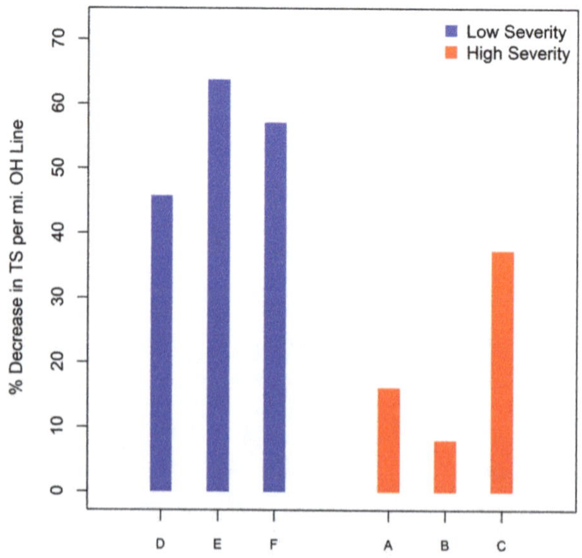

Figure 4. Percent decrease in normalized trouble spots from low `instETT` to higher `instETT` bins for each experiment. The left (blue) three bars represent the percent change in average kinetic energy proxy and overhead line normalized trouble spots from the 0–25% `instETT` bin to each higher `instETT` bin for low-severity events (Experiments D–F). The right (red) three bars represent the percent change in average kinetic energy proxy and overhead-line-normalized trouble spots from the 0–25% `instETT` bin to each higher `instETT` bin for high-severity events (Experiments A–C).

To focus on the cases of the most extreme impacts, the percent changes in kinetic energy normalized trouble spots were used to estimate trouble spot values for each circuit for the three tropical storms (Irene, Sandy, Isaias) under two hypothetical conditions: 1, if no ETT had been performed prior to the storms; 2, if every circuit had an `instETT` value between 75% and 100% prior to the storms. For each circuit and for each of the three storms, the actual `instETT` bin was determined. If the `instETT` value was between 0% and 25%, no adjustment was made. Otherwise, the expected outages if `instETT` was between 0% and 25% were calculated as follows. The trouble spot value for the given circuit and event was divided by 1 minus the expected percentage decrease for the circuit's `instETT` value and the storm severity, where the expected decreases are the percentages shown in Figure 4. Equation (3) outlines the trouble spot adjustment, where O is the outage value for a single circuit for a single event, and P_{adj} is the adjustment percentage derived from previous analysis dependent on the circuit `instETT` value and the event severity. In this case, the three storms being analyzed are all high-severity. The calculated values for each circuit with 0% to 25% ETT were then used to obtain outage values if the `instETT` values were 75–100% for each circuit by multiplying by 1 less 0.373, where 0.373 is the expected percentage decrease in trouble spots for high-severity storms if `instETT` is increased from 0–25% to 75–100% (Experiment C), in decimal form.

$$O_{adj} = \frac{O}{(1 - \frac{P_{adj}}{100})} \tag{3}$$

This same calculation methodology is used for each of the 173 storms to determine the trouble spots if no ETT had been performed prior to each storm, and determine the trouble spots if every circuit had an `instETT` value between 75% and 100% before the storm.

A representation of the domain and methodology can be seen in Figure 5, where the data are split by event severity and then again by ETT levels to run six experiments. The expected statistical reduction in outages is calculated for each experiment, and then those values are applied on a circuit by circuit basis to evaluate the effectiveness of different vegetation management scenarios in terms of reducing trouble spots in the electric grid.

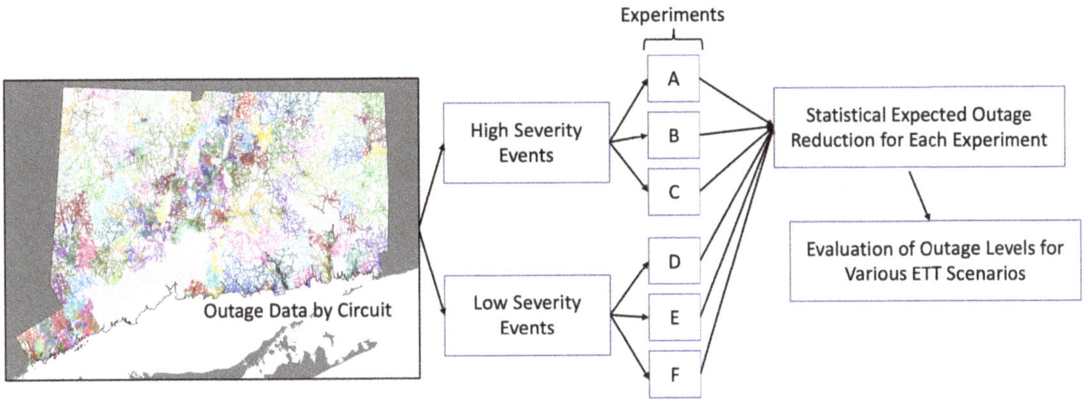

Figure 5. Study framework. The left-hand figure depicts the domain, with the overhead lines colored by circuit ID. The experiments correspond to the groups in Table 1. The statistical outage reductions include controls for storm intensity (kinetic energy proxy) and overhead line length.

3. Results

From Experiments A–C in Table 1, it is evident that for high-severity events, the kinetic energy proxies are higher for the bins of data with higher `instETT` values. This is important to note, as stronger storms tend to produce more outages, so without normalizing for kinetic energy, ETT impacts may appear less beneficial than they would if the storm strength was equivalent across the comparison bins. The percentage changes in average kinetic energy from the low (0–25%) to high `instETT` bins from Experiments A–C are displayed in Figure 6. Conversely, for the lower-severity storms, all of the average kinetic energy proxies are higher in the lower ETT bins versus their higher ETT comparison bins.

The differences in trouble spots between the lower and higher ETT bins that can be seen in Figure 7 are significant when tested with a Wilcoxon–Mann–Whitney test, except for Experiment C, which compares the lowest and highest levels of ETT. This is likely due to the small size of the groupings compared in that experiment.

When looking at the results in Figure 4, there are several noteworthy findings. The first is that for low-severity events (Experiments D–F), ETT is effective for each of the vegetation management levels (`instETT` bins) analyzed, with normalized trouble spot reductions ranging between 45.8% and 63.8%. This may be due to the utility having targeted particularly vulnerable areas, or areas with particularly heavy tree cover first. Further, in smaller storms, lower winds and kinetic energy are generally experienced, making it more likely that branches fall, as opposed to entire trees uprooting and falling on power lines. As ETT is more rigorous than other vegetation management strategies at clearing branches above power lines, but not necessarily at removing entire trees that could fall on power lines during more intense storms, this may partially explain the greater effectiveness of ETT for lower-severity storms. While we do see a reduction in outages across all ETT levels for high-severity storms as well, there is a large increase in effectiveness between the two lower ETT Experiments (A,B) versus the highest ETT Experiment (C), with the results suggesting that the highest ETT level reduces trouble spots by 37.3% while the lower bins reduce outages by between 8–16.1%.

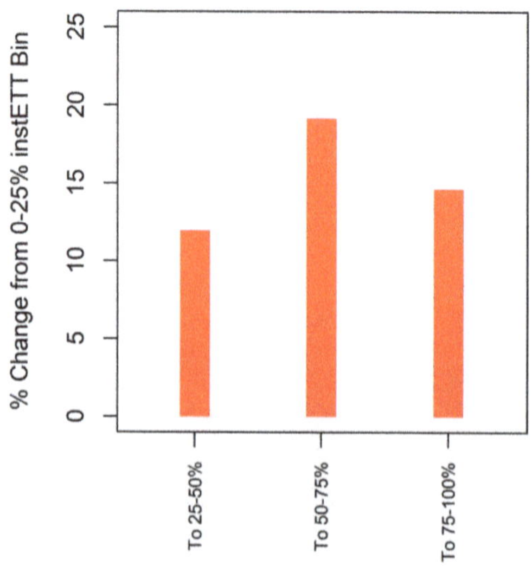

Figure 6. Percent Change in Average Kinetic Energy Proxies. Percentage change from Low ETT (0–25%) to High ETT Bins for Severe Storms (Experiments A–C).

Looking at the results of the percent decreases in trouble spots for each experiment in Figure 4, the percent decreases are not monotonically increasing with higher `instETT` for either high- or low-severity, despite normalizing outages by miles of overhead line and kinetic energy proxy. However, this is likely due to the fact that a simple linear normalization scheme was applied to account for kinetic energy while the relationship between kinetic energy and outages is not linear [11,26], and that the interaction between storm conditions and power outages is affected by many more variables than just the kinetic energy of the storm. Some other variables that affect the number of outages in a storm include precipitation, drought, and leaf area among others [11,13,26,32,42–46]. In the case of severe storms, the results demonstrate a smaller reduction in trouble spots when completing 50–75% of applicable ETT versus 25–50% of applicable ETT; however, this is partly explained by Figure 6, which demonstrates that of the high-severity experiments, the largest percent difference in average kinetic energy between the low and high `instETT` bins of each experiment is for Experiment B. As the data in Experiment B have the highest average increase in kinetic energy between its low and high ETT comparison groups, a nonlinear relationship between kinetic energy and trouble spots may be driving the results, which show ETT is less effective for Experiment B compared with Experiment A, when only 25–50% of the applicable length of the circuit has ETT performed on it. This is a logical result if the kinetic energy disparity is not fully compensated for by the linear normalization.

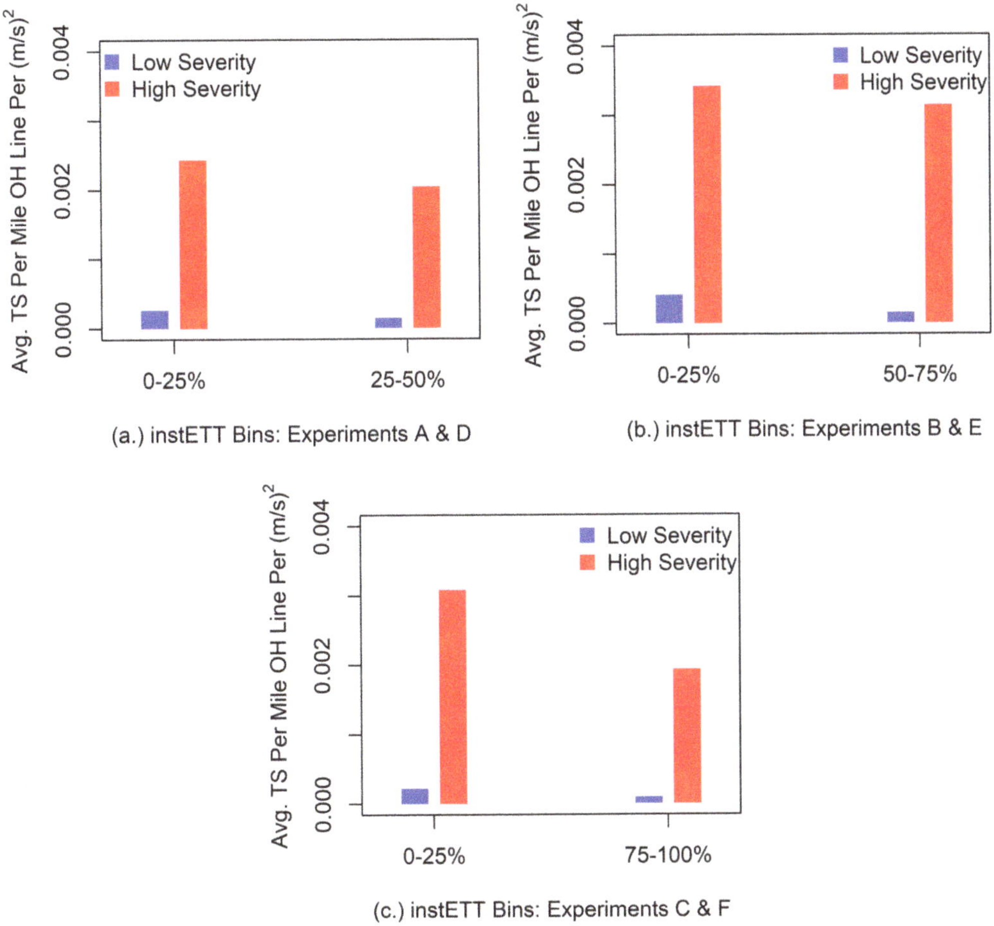

Figure 7. Average outages per mile of overhead line, normalized for kinetic energy proxy. (**a**) Comparing 0–25% `instETT` to 25–50% `instETT`—Experiments A and D. (**b**) Comparing 0–25% `instETT` to 50–75% `instETT`—Experiments B and E. (**c**) Comparing 0–25% `instETT` to 75–100% `instETT`—Experiments C and F.

4. Discussion

While the results do suggest that ETT reduces trouble spots by a greater percentage for low-severity events when compared with high-severity events, particularly at lower applicable ETT completion percentages, it is imperative to look at large and small storms in combination to understand the comprehensive benefits of ETT, as demonstrated in Figure 8. While the largest storms such as Hurricane Sandy, Hurricane Irene, and Tropical Storm Isaias produce a large number of outages that can take over a week to fully restore, these types of storms do not occur nearly as frequently as smaller storms. By improving the reliability and resilience of the electric grid to smaller storms through ETT, outages are reduced, which in turn reduces the necessary spending on crews to restore outages. As utilities may have set annual budgets, by reducing spend on restoration for many smaller storms, more resources are left available for preparing and responding to the most extreme events when they do occur. Further, when outages occur, there is an economic cost to society in addition to the utility costs, and by reducing outages for the more frequent smaller storms, this cost is reduced. Figure 8 demonstrates the annual sums of trouble spots if no ETT had been performed before each storm, and the trouble spots if the `instETT` value for each circuit was between 75–100% before each storm, where the values for each scenario are calculated using the expected percent differences in trouble spots dependent on the `instETT` value from Figure 4. Figure 8 demonstrates that when considering the low and high severity events together, there is a sizable reduction in outages for each year of the study, with annual reductions ranging between 37.3% and 57.1%. It is noted that the reductions shown in Figure 8 represent an underestimation of the actual reduction of outages, as the list of events used in the study is not comprehensive in that some storm events that are small in magnitude are not included in the dataset. Further, the system exclusively focuses on rain and wind storms, excluding thunderstorms and winter storms, which also introduce significant outage events.

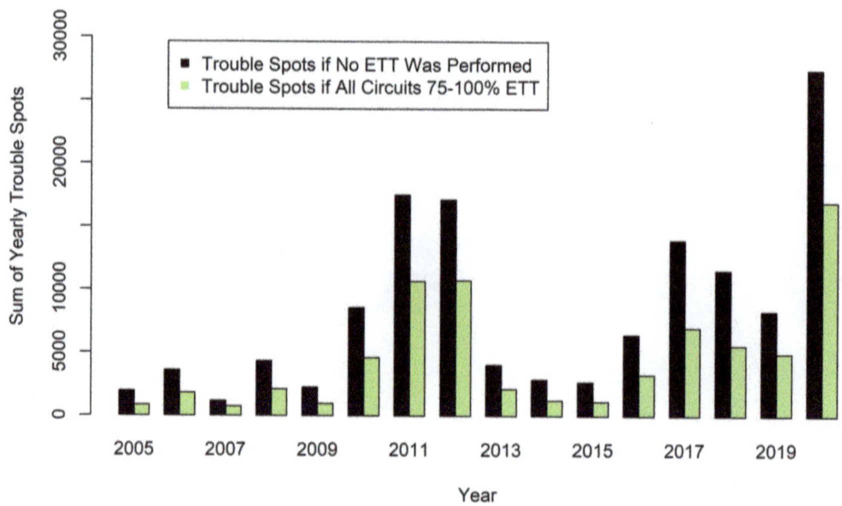

Figure 8. Sum of annual trouble spots. The left (black) bars represent the trouble spots if no ETT was performed prior to each storm. The right (light green) bars represent if the `instETT` value was between 75–100% before each storm occurred.

To examine the impact of high amounts of applicable ETT (`instETT`) on some of the most extreme storms, we similarly use the expected change in trouble spots from Figure 4 to approximate how many outages would have occurred had no applicable ETT been performed on each circuit before each storm, as well as if 75–100% of applicable ETT had been performed on every circuit. These values are compared with the actual trouble spots observed for the storms in Table 2. As seen in the table, there is not much difference between the actual `instETT` trouble spot value and the trouble spot value if no applicable ETTs were performed for Irene and Sandy since they took place in the early years of the study, when not much ETT had been implemented. However, as Isaias took place much more recently, we see a reduction of over 2500 trouble spots in the electric grid from the expected value had no ETT been performed. There is also a large estimated reduction in trouble spots for each of the three major storms had extensive ETT been performed for each circuit before the storm. Information on reduced trouble spots for major events for various ETT scenarios may be useful to utilities and regulators to optimize a resilience strategy between grid hardening efforts, vegetation management, and increased crew response, helping to reduce restoration times and save money. Additionally, the results give an ability to retrospectively quantify return on investment. By looking at historical storms with various ETT scenarios, it is possible to perform economic analyses on projected future return on investment under different vegetation scenarios, which can aid in development of resilience plans. This study acts as the first step in quantifying the return on investment of vegetation management while considering storm severity and provides a tool that can also be applied to future climate scenario storm events.

While the results do show reductions in trouble spots for high and low storm severity, as well as for individual extreme events, there are some limitations to the study that should be noted. First, areas with different vegetation cover or predominant storm types may see different results, as the reduction in outages is dependent upon the physical environment and storm characteristics. Secondly, normalization for other hardening techniques that may have been previously applied is not performed. Nevertheless, these techniques (pole upgrades, undergrounding in previously nonundergrounded locations) have been applied in much more isolated fashion and on a much smaller scale across the domain. Further, the stress applied to electrical infrastructures is not only dependent on wind speed, and may be dependent on other factors such as line length. To help account for this issue, the study normalized outages by overhead line length. To help control for other factors that may influence outages, such as infrastructure age or tree height in the surrounding area, each binned comparison only includes circuits that had data in each bin. This means that for the inclusion of a circuit into each binned analysis, at least one storm of the severity being analyzed must have occurred over the domain when the `instETT` value for the given circuit was inside each of the two `instETT` ranges being compared (0–25%, 25–50%, 50–75%, or 75–100%).

Table 2. Actual and Estimated Trouble Spots for Major Storms for Various ETT Scenarios.

Storm	No ETT	Actual ETT	75–100% ETT
Irene	15,980	15,932	10,012
Sandy	15,530	15,282	9703
Isaias	24,217	21,473	15,173

5. Conclusions

Through various statistical analyses, we have been able to demonstrate and statistically model the relationship between vegetation management and outages in the electric grid for storms of different severity. The results demonstrate that enhanced vegetation management is particularly helpful in reducing trouble spots for lower severity storms, with reductions between 45.8% and 63.8%, and substantially reduces trouble spots during the most severe events when vegetation management is particularly comprehensive, demonstrating a 37.3% reduction when compared with circuits with little to no enhanced tree trimming. These reductions in power outages can be seen for individual storms as well as in annual totals.

When compared with previous studies in the effectiveness of vegetation management activities, this analysis provides a better understanding of how more rigorous vegetation management standards (applying ETT) help reduce outages at an individual event level, for both the more frequent events and those less frequent, stronger storms, which may also occur more frequently in the future due to climate change. This analysis may also provide insight to be used when training machine learning outage prediction models, as future models may see benefits from including vegetation management data and focusing explicitly on low- or high-severity storms, or feature engineering new input variables that combine storm intensity and vegetation management information. The results of this study also provide useful information on annual trouble spots in the electric grid, taking into consideration vegetation management data and storm intensity, to provide a retrospective look at how different vegetation management levels and schemes would have impacted trouble spots. This information is useful to various stakeholders in performing cost-benefit analysis when developing vegetation management, or more broadly, resilience plans or budgets. Specifically, outputs from this or a similar analysis can be used in economic analysis to optimize vegetation management efforts and compare and contrast short- and long-term costs versus other resilience efforts such as wire and pole upgrades, or undergrounding wires, and is a recommended inclusion into such analyses. In future works, this model can be adapted for other types of storms such as thunderstorms and winter storms.

Another potential research focus to expand on this work is controlling for possible overlap in historical resilience upgrades to the distribution grid, including pole and wire upgrades, which may have been performed in tandem with ETT in some locations. However, these other hardening techniques are more isolated and are typically applied much less broadly than vegetation management by US power utilities. Additionally, a similar statistical framework can be used to analyze the effectiveness of other resilience efforts for varying storm intensities, where the data are similarly available at a circuit level, including efforts such as reconductoring wires and pole upgrades or replacements. These results may also be useful to stakeholders including utilities, regulators, and municipalities in understanding if it is worthwhile to partake in expensive grid hardening measures such as undergrounding wire, and where such activities may be the most impactful. In this way, the results of this study and future works utilizing the same framework can be used to optimize grid resilience to storms and climate changes ensuring the reliable delivery of power long into the future.

Author Contributions: Conceptualization, E.A. and W.O.T.; methodology, W.O.T., P.L.W., D.C., and E.A.; software, W.O.T.; validation, W.O.T., P.L.W., D.C., and E.A.; formal analysis, W.O.T.; investigation, W.O.T.; resources, W.O.T. and P.L.W.; data curation, W.O.T. and P.L.W.; writing—original draft preparation, W.O.T.; writing—review and editing, W.O.T., P.L.W., D.C., and E.A.; visualization, W.O.T.; supervision, E.A.; project administration, E.A.; funding acquisition, E.A. All authors have read and agreed to the published version of the manuscript.

Funding: This research was funded by Eversource Energy Connecticut through the Eversource Energy Center at the University of Connecticut, USA.

Institutional Review Board Statement: Not applicable.

Informed Consent Statement: Not applicable.

Data Availability Statement: Not applicable.

Acknowledgments: This publication uses classified electric utility data. The authors have full access to all of the data in this study and we take complete responsibility for the integrity of the data and the accuracy of the data analysis.

Conflicts of Interest: The funders had no role in the design of the study; in the analyses, or interpretation of data; in the writing of the manuscript, or in the decision to publish the results. P.L.W., D.C., and E.A. hold stock in Whether Inc.

References

1. Appelt, P.; Goodfellow, J. Research on How Trees Cause Interruptions—Applications to Vegetation Management. In Proceedings of the Rural Electric Power Conference, Scottsdale, AZ, USA, 25–25 May 2004; pp. 6–10. [CrossRef]
2. Novembri, R. *Utility Vegetation Management Final Report*; Technical Report FERC-03AL-30574; Federal Energy Regulatory Commission: Washington, DC, USA, 2004.
3. Simpson, P.; Van Bossuyt, R. Tree-Caused Electric Outages. *J. Arboric.* **1996**, *22*, 117–121.
4. Lineweber, D.; McNulty, S. The Cost of Power Disturbances to Industrial & Digital Economy Companies. 2001. Aviliable online: https://www.epri.com/research/products/3002000476 (accessed on 13 December 2021).
5. Sullivan, M.; Schellenberg, J.; Blundell, M. *Updated Value of Service Reliability Estimates for Electric Utility Customers in the United States*; Technical Report; Lawrence Berkeley National Lab. (LBNL): Berkeley, CA, USA, 2015.
6. Eversource. Customer Profile. Available online: https://www.eversource.com/content/nh/about/about-us/about-us/customer-profile (accessed on 1 July 2021).
7. Eversource. Understanding Vegetation Managment. Available online: https://www.eversource.com/content/docs/default-source/my-account/veg-mgmt-guide.pdf (accessed on 5 December 2021).
8. Peracchio, D. Connecticut's 2020 Forest Action Plan. 2020. Available online: https://portal.ct.gov/DEEP/Forestry/CT-Forest-Action-Plan (accessed on 13 December 2021).
9. Forest Area (% of Land Area)—United States. Available online: https://data.worldbank.org/indicator/AG.LND.FRST.ZS?name_desc=true&locations=US (accessed on 10 June 2021).

10. *New England Forests: The Path to Sustainability*; Technical Report Chapter 1: Keep New England Forested; New England Forestry Foundation: Littleton, MA, USA, 2014.
11. Wanik, D.; Parent, J.; Anagnostou, E.; Hartman, B. Using vegetation management and LiDAR-derived tree height data to improve outage predictions for electric utilities. *Electr. Power Syst. Res.* **2017**, *146*, 236–245. [CrossRef]
12. Parent, J.R.; Meyer, T.H.; Volin, J.C.; Fahey, R.T.; Witharana, C. An analysis of enhanced tree trimming effectiveness on reducing power outages. *J. Environ. Manag.* **2019**, *241*, 397–406. [CrossRef] [PubMed]
13. Cerrai, D.; Watson, P.; Anagnostou, E.N. Assessing the effects of a vegetation management standard on distribution grid outage rates. *Electr. Power Syst. Res.* **2019**, *175*, 105909. [CrossRef]
14. Campbell, R.J. Weather-Related Power Outages and Electric System Resiliency. 2012. Available online: https://digital.library.unt.edu/ark:/67531/metadc122249/ (accessed on 13 December 2021).
15. *Economic Benefits of Increasing Electric Grid Resilience to Weather Outages*; Technical Report; Executive Office of the President: Washington, DC, USA, 2013.
16. Simpson, P.O. Tree Damage to Electric Utility Infrastructure Assessing and Managing the Risk from Storms. In Proceeding of the 10th Annual International Cold Regions Engineering Conference, Lincoln, NH, USA, 16–19 August 1999; pp. 768–778. Available online: https://cedb.asce.org/CEDBsearch/record.jsp?dockey=0118612 (accessed on 5 December 2021).
17. Poulos, H.M.; Camp, A.E. Decision Support for Mitigating the Risk of Tree Induced Transmission Line Failure in Utility Rights-of-Way. *Environ. Manag.* **2010**, *45*, 217–226. [CrossRef] [PubMed]
18. Poulos, H.M.; Camp, A.E. Mapping Threats to Power Line Corridors for Connecticut Rights-of-Way Management. *Environ. Manag.* **2011**, *47*, 230–238. [CrossRef] [PubMed]
19. Dokic, T.; Kezunovic, M. Predictive Risk Management for Dynamic Tree Trimming Scheduling for Distribution Networks. *IEEE Trans. Smart Grid* **2019**, *10*, 4776–4785. [CrossRef]
20. Kuntz, P.; Christie, R.; Venkata, S. Optimal vegetation maintenance scheduling of overhead electric power distribution systems. *IEEE Trans. Power Deliv.* **2002**, *17*, 1164–1169. [CrossRef]
21. Louit, D.; Pascual, R.; Banjevic, D. Optimal interval for major maintenance actions in electricity distribution networks. *Int. J. Electr. Power Energy Syst.* **2009**, *31*, 396–401. [CrossRef]
22. Guikema, S.; Davidson, R.; Liu, H. Statistical Models of the Effects of Tree Trimming on Power System Outages. *IEEE Trans. Power Deliv.* **2006**, *21*, 1549–1557. [CrossRef]
23. Najafi Tari, A.; Sepasian, M.S.; Tourandaz Kenari, M. Resilience assessment and improvement of distribution networks against extreme weather events. *Int. J. Electr. Power Energy Syst.* **2021**, *125*, 106414. [CrossRef]
24. Ma, S.; Chen, B.; Wang, Z. Resilience Enhancement Strategy for Distribution Systems Under Extreme Weather Events. *IEEE Trans. Smart Grid* **2018**, *9*, 1442–1451. [CrossRef]
25. Bhattacharyya, A.; Yoon, S.; Hastak, M. Optimal strategy selection framework for minimizing the economic impacts of severe weather induced power outages. *Int. J. Disaster Risk Reduct.* **2021**, *60*, 102265. [CrossRef]
26. Nateghi, R.; Guikema, S.; Quiring, S.M. Power Outage Estimation for Tropical Cyclones: Improved Accuracy with Simpler Models: Power Outage Estimation for Tropical Cyclones. *Risk Anal.* **2014**, *34*, 1069–1078. [CrossRef] [PubMed]
27. Hughes, W.; Zhang, W.; Bagtzoglou, A.C.; Wanik, D.; Pensado, O.; Yuan, H.; Zhang, J. Damage modeling framework for resilience hardening strategy for overhead power distribution systems. *Reliab. Eng. Syst. Saf.* **2021**, *207*, 107367. [CrossRef]
28. DiFalco, S.; Morzillo, A.T. Comparison of Attitudes towards Roadside Vegetation Management across an Exurban Landscape. *Land* **2021**, *10*, 308. [CrossRef]
29. Gazzea, M.; Pacevicius, M.; Dammann, D.O.; Sapronova, A.; Lunde, T.M.; Arghandeh, R. Automated Power Lines Vegetation Monitoring using High-Resolution Satellite Imagery. *IEEE Trans. Power Deliv.* **2021**. [CrossRef]
30. Graziano, M.; Gunther, P.; Gallaher, A.; Carstensen, F.V.; Becker, B. The wider regional benefits of power grids improved resilience through tree-trimming operations evidences from Connecticut, USA. *Energy Policy* **2020**, *138*, 111293. [CrossRef]
31. Service Territory. Available online: https://www.eversource.com/content/ct-c/about/about-us/about-us/service-territory (accessed on 5 December 2021).
32. Watson, P.L.; Cerrai, D.; Koukoula, M.; Wanik, D.W.; Anagnostou, E. Weather-related power outage model with a growing domain: Structure, performance, and generalisability. *J. Eng.* **2020**, *2020*, 817–826. [CrossRef]
33. Potapov, P.; Li, X.; Hernandez-Serna, A.; Tyukavina, A.; Hansen, M.C.; Kommareddy, A.; Pickens, A.; Turubanova, S.; Tang, H.; Silva, C.E.; et al. Mapping global forest canopy height through integration of GEDI and Landsat data. *Remote Sens. Environ.* **2021**, *253*, 112165. [CrossRef]
34. Shapiro, S.S.; Wilk, M.B. An analysis of variance test for normality (complete samples). *Biometrika* **1965**, *52*, 591–611. [CrossRef]
35. R Core Team. *R: A Language and Environment for Statistical Computing*; R Core Team: Vienna, Austria, 2021. Available online: https://www.gbif.org/tool/81287/r-a-language-and-environment-for-statistical-computing (accessed on 13 December 2021).
36. Royston, J.P. An Extension of Shapiro and Wilk's W Test for Normality to Large Samples. *Appl. Stat.* **1982**, *31*, 115. [CrossRef]
37. Royston, J.P. Algorithm AS 181: The W Test for Normality. *Appl. Stat.* **1982**, *31*, 176. [CrossRef]
38. Royston, P. Remark AS R94: A Remark on Algorithm AS 181: The W-test for Normality. *Appl. Stat.* **1995**, *44*, 547. [CrossRef]
39. Mann, H.B.; Whitney, D.R. On a Test of Whether one of Two Random Variables is Stochastically Larger than the Other. *Ann. Math. Stat.* **1947**, *18*, 50–60. [CrossRef]
40. Bauer, D.F. Constructing Confidence Sets Using Rank Statistics. *J. Am. Stat. Assoc.* **1972**, *67*, 687–690. [CrossRef]

41. Wilcoxon, F. Individual Comparisons by Ranking Methods. *Biom. Bull.* **1945**, *1*, 80. [CrossRef]
42. Cerrai, D.; Wanik, D.W.; Bhuiyan, M.A.E.; Zhang, X.; Yang, J.; Frediani, M.E.; Anagnostou, E.N. Predicting storm outages through new representations of weather and vegetation. *IEEE Access* **2019**, *7*, 29639–29654. [CrossRef]
43. Yang, F.; Watson, P.; Koukoula, M.; Anagnostou, E.N. Enhancing Weather-Related Power Outage Prediction by Event Severity Classification. *IEEE Access* **2020**, *8*, 60029–60042. [CrossRef]
44. Yang, F.; Wanik, D.W.; Cerrai, D.; Bhuiyan, M.A.E.; Anagnostou, E.N. Quantifying Uncertainty in Machine Learning-Based Power Outage Prediction Model Training: A Tool for Sustainable Storm Restoration. *Sustainability* **2020**, *12*, 1525. [CrossRef]
45. Wanik, D.W.; Anagnostou, E.N.; Hartman, B.M.; Frediani, M.E.B.; Astitha, M. Storm outage modeling for an electric distribution network in Northeastern USA. *Nat. Hazards* **2015**, *79*, 1359–1384. [CrossRef]
46. Guikema, S.D.; Quiring, S.M.; Han, S.R. Prestorm Estimation of Hurricane Damage to Electric Power Distribution Systems: Prestorm Estimation of Hurricane Damage. *Risk Anal.* **2010**, *30*, 1744–1752. [CrossRef] [PubMed]

sustainability

MDPI

Article

The Relationship between Distance and Risk Perception in Multi-Tier Supply Chain: The Psychological Typhoon Eye Effect

Ming-Xing Xu [1,2], Shu Li [3,4,*], Li-Lin Rao [2,3] and Lei Zheng [5]

1 School of Transportation, Fujian University of Technology, Fuzhou 350108, China; xumx@fjut.edu.cn
2 CAS Key Laboratory of Behavioral Science, Institute of Psychology, Chinese Academy of Sciences, Beijing 100101, China; raoll@psych.ac.cn
3 Department of Psychology, University of Chinese Academy of Sciences, Beijing 100049, China
4 Department of Psychology, School of Humanities and Social Sciences, Fuzhou University, Fuzhou 350108, China
5 School of Economics and Management, Fuzhou University, Fuzhou 350108, China; leizh@fzu.edu.cn
* Correspondence: lishu@psych.ac.cn; Tel.: +86-10-64841536

Abstract: Previous research has shown that an individual's proximity to the epicenter can influence their perception and response to risk. However, this aspect has been largely overlooked in the supply chain risk literature. This paper aims to fill this gap by investigating the impact of distance on the perception and response to supply chain disruption risk. An online survey was conducted with 1055 managers working within the supply chain of ZTE, a Chinese multinational company providing integrated communications and information solutions. The survey aimed to examine how their distance from the disruption epicenter (i.e., ZTE) affected their risk perception and subsequent managerial responses. The findings indicate that those closer to the epicenter perceive a lower risk of disruption compared to those farther away, resulting in a reduced likelihood of taking management action. This phenomenon is referred to as the "psychological typhoon eye" (PTE) effect in supply chain disruption risk. Further analysis revealed that risk information quality mediated the relationship between distance and risk perception, while an individual's job position level moderated the relationship between risk information quality and disruption risk perception. To mitigate the PTE effect in the multi-tier supply chain, the focal firm must prioritize high-quality information synchronization, extending beyond single-company initiatives.

Keywords: disruption risk perception; psychological typhoon eye effect; multi-tier supply chain; risk information quality

Citation: Xu, M.-X.; Li, S.; Rao, L.-L.; Zheng, L. The Relationship between Distance and Risk Perception in Multi-Tier Supply Chain: The Psychological Typhoon Eye Effect. *Sustainability* **2023**, *15*, 7507. https://doi.org/10.3390/su15097507

Academic Editors: Esmaeil Zarei, Samuel Yousefi and Mohsen Omidvar

Received: 31 March 2023
Revised: 26 April 2023
Accepted: 28 April 2023
Published: 4 May 2023

1. Introduction

Today's supply chains are increasingly vulnerable to disruption risk due to operational shutdowns directly or indirectly caused by a wide array of events, including climatological disasters, epidemics, political conflicts, terrorism, and financial scams [1,2]. To minimize the high costs of supply chain disruptions, firms need to actively access and manage the direct or indirect disruption risks from upstream or downstream of the supply chain [3]. A significant number of research studies on supply chain disruption risk management have been carried out to date [4], and most of the literature has assumed that supply chain managers can make optimal decisions based on objective risk assessment [5]. However, research has found that intuitive or emotional responses play a key role in human decision-making, leading people to make biased decisions that systematically deviate from rational judgment [6,7]. In addition, people usually rely on heuristic strategies (such as availability heuristics, anchoring heuristics, and representativeness heuristics) rather than a rational model to make judgments and decisions under uncertainty because risks cannot be accurately assessed [8]. Managerial responses to supply chain disruptions are triggered by managers' risk perceptions rather than by the disruptions themselves [9,10]. Moreover, in decision-making, managerial risk perceptions are more influential than purely objective risk assessments alone [11]. Disruption risk perception refers to an individual manager's subjective assessment of the risk inherent in a disruption [12]. In recent years, there has been a surge in academic research on supply chain disruption risk perception [10], with these studies underscoring the importance of risk perception in supply chain risk decision-making. However, a thorough review reveals that most of the existing literature focuses on buyers' perceptions of supply disruption risk within two-tier supply chains (i.e., a dyadic relationship between supplier and buyer), while a supply chain typically consists of a focal firm and numerous upstream and downstream members in a multi-tiered structure.

When the focal firm of a supply chain experiences a disruption event or faces disruption risks (as illustrated in Figure 1), its upstream and downstream members may be directly (tier-1 suppliers/customers) or indirectly (suppliers/customers at tier-2 and beyond) impacted [13]. For instance, when the Hynix memory maker in China experienced a fire, computer manufacturers and parts suppliers perceived the risk and quickly purchased as much inventory as possible to secure better prices, which pushed up prices and created shortages [13]. Consequently, it is essential to understand not only how managers of the focal firm perceive its disruption risk but also how managers of upstream and downstream firms in the supply chain perceive the disruption risk.

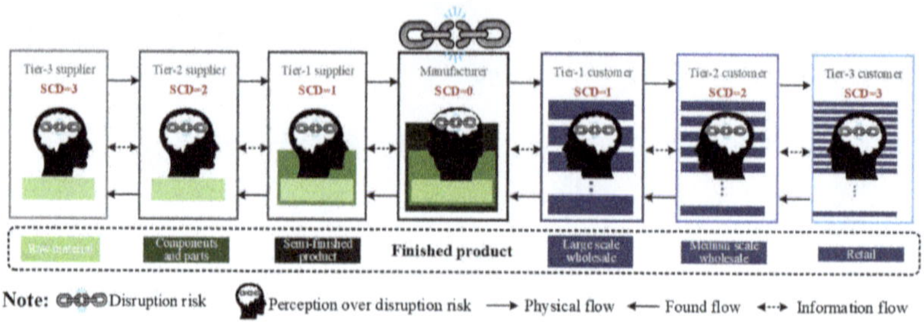

Figure 1. Managers' risk perception over manufacturer disruption in a multi-tier supply chain.

Subjective risk perception is not a direct reflection of objective danger or threat (such as supply chain disruption), and individuals perceive the same danger differently [14,15]. Intuitively, closer proximity to the epicenter correlates with a higher perceived risk of the threat. However, research has demonstrated that this is not always the case; field evidence indicates that individuals closer to the epicenter are often less worried or fearful than those farther away. This phenomenon is known as the "psychological typhoon eye" (PTE) effect, as the epicenter of a typhoon is relatively calm. The phenomenon of the PTE effect highlights the intricate interplay between risk perception and distance, which encompasses various aspects such as the geographical distance from the epicenter [16,17], interpersonal relationship distance from the affected party [18], and the level of involvement in the threat of danger [19]. Hence, the primary objective of the present study is to gain insights into the impact of distance on managers' perception of supply chain disruption risk. We build on the literature to investigate whether another version of the PTE effect (i.e., "supply chain distance" version) exists in the context of supply chain disruption risk perception. In this study, supply chain distance is conceptualized as the number of tiers between the disrupted firm and its upstream or downstream members. For instance, the supply chain distance between the disrupted firm and its first-tier suppliers/customers is 1, the distance between the disrupted firm and its second-tier suppliers/customers is 2, and so on (as depicted in Figure 1).

Numerous studies have demonstrated that risk information significantly influences risk perception (e.g., [20,21]). People's perceptions of risks are shaped by the information available to them [22]. Firms facing greater environmental uncertainty must gather and process more information. Managerial perceptions of disruption risk depend on comprehensive and high-quality information about the disruption, and managers utilize relevant information to assess disruption risk and make decisions [23]. As a result, disseminating relevant information about the disruption to supply chain partners has become a critical risk mitigation strategy [24]. Accordingly, research on the role of available risk-related information in supply chain managers' disruption risk perception is anticipated to raise intriguing questions about risk perception mechanisms that warrant further investigation. Therefore, the second aim of the present study is to empirically test whether the risk information can account for the supply chain distance version of the PTE effect.

Job position level can be another important factor affecting risk perception in the context of supply chain disruptions. Higher-level managers may possess a more expansive view of the supply chain and greater experience with potential disruptions, enabling them to better identify risks and threats [25]. In contrast, lower-level managers may have a more limited perspective and less experience with supply chain disruptions, leading to a decreased sensitivity to disruption risks. Additionally, higher-level managers may be responsible for making strategic decisions that have a greater impact on the organization, leading to more cautious decision-making in the face of disruption risks [26]. However, it is also possible that top managers perceive the risks as less threatening because they have more resources and support. Based on the above analysis, it is worth studying whether job position level will affect the relationship between risk information and managers' risk perception, which is also the third research objective of the present study.

Given the above, this study will investigate the following research questions:

RQ1:*If a focal firm in a multi-tier supply chain is at risk of disruption, does a PTE effect exist in the perception of this risk among managers in different tiers of the supply chain?*

RQ2:*What role does risk information play in the perception of disruption risk among supply chain managers?*

RQ3:*Does job position level moderate the relationship between risk information and disruption risk perception?*

The answers to these questions will not only extend the existing literature by facilitating the understanding of managers' risk perception mechanisms in the context of multi-tier supply chains but also help broaden the application of psychological theory in operations and supply chain management.

The remainder of this study is structured as follows: We first review the extant research and develop our theoretical model and hypotheses. We then overview the methodology, statistical analyses, and findings. Finally, we highlight the academic and managerial implications of our finds, the limitations of the study, and opportunities for future research.

2. Theoretical Constructs and Hypotheses

2.1. Effect of Distance upon Risk Perception

Risk refers to the uncertainty and potential severity of consequences associated with an activity that is valued by humans. Risk perception refers to the subjective judgment of individuals regarding such risks. Traditionally, risk perception is measured by multiplying the probabilities of risk occurrence by the magnitude of the risk impact [27]. This method is considered rational, yet it has limitations. Sociologists and psychologists have shown that laypeople tend to perceive risk based on emotions, intuition, and direct judgment, whereas a rational risk assessment is typically processed by experts [28]. These emotional and intuitive perceptions of risk can be irrational and influenced by factors such as risk descriptions, previous experiences, effect, imagination, trust, values, and worldviews [29].

Recent behavioral supply chain research has only touched on the linkages between various factors and how they impact people's perception of and response to disruption risk. For example, Sarafan et al. demonstrate how individualism–collectivism negatively affects how individuals perceive risk and supplier-switching intention in the face of a supply disruption [12]. Other researchers have emphasized the importance of attributions and emotions in explaining differences in managerial decisions following the occurrence of a disruption [30]. Vanpoucke and Ellis examine the relationship between buyers' perceptions of disruption risk and their adoption of buffer- and process-oriented risk mitigation tactics [31]. These limited studies have contributed to our understanding of the psychological and social factors affecting people's perception of disruption risk from the perspective of a one-tier supply chain, but little attention has been paid to the effect of distance on disruption risk perception. Therefore, our present research aims to explore how distance affects an individual's perception of disruption risk in the context of a multi-tier supply chain.

There is an extensive body of literature demonstrating how distance affects individuals' risk perception and behavior [32]. According to the PTE effect theory, people who are closer to the center of an adverse event are less concerned or fearful about the event. For example, Maderthaner et al. found that in a local attitude survey about a nuclear reactor in Vienna, residents living farther from the reactor perceived it to be riskier than those living closer [33]. Tilt discovered that industrial workers who labored under highly polluted conditions provided lower risk ratings than farmers and commercial/service sector workers who were farther from the polluting sources [34]. In a study by Li et al., a convenience sample of 2262 adults was surveyed about their post-earthquake concerns regarding safety and health after the Wenchuan earthquake, and their findings suggested that people who were farther from the earthquake area (i.e., more remote) were more likely to have a higher estimation of their post-earthquake concern [16]. During the SARS epidemic, it was reported that the level of exposure to SARS was not a primary determinant of experienced anxiety, and nearness to the center of the epidemic was negatively related to anxiety levels [35]. Similarly, studies conducted during the COVID-19 pandemic have come to similar conclusions (e.g., [17,36]).

Within the context of the supply chain disruption risk, the disrupted firm is the epicenter of the risk (as the manufacturer in Figure 1), and the upstream and downstream supply chain members' reactions and responses to the disruption risk are enhanced with the increment of supply chain distance, which increases the negative effect of the disruption risk [13]. Previous studies have shown that individual managers' disruption risk perception has a positive correlation with their reactions and responses to the risk [12,37,38]. This implies that the supply chain members' level of risk perception may also increase with the increment of supply chain distance.

Therefore, we hypothesize that:

Hypothesis 1 (H1): *The distance between managers' firms from the disrupted firm and their levels of disruption risk perception are positively related.*

H1 posits that as the upstream and downstream firms in the supply chain move closer to the disrupted firm, their concern regarding the risk associated with the disruption of the focal firm decreases. In other words, when the focal enterprise in the supply chain faces a disruption risk, the perception of managers at different tiers of the supply chain towards this risk is influenced by the proximity of their firms to the focal firm, demonstrating the presence of the PTE effect.

2.2. The Mediating Role of Available Risk Information

Previous studies have shown that people's perceptions about any risk are shaped by the information available to them [22]. In most cases, individuals tend to perceive a greater risk when they have more knowledge about the adverse event, as they are aware that its consequences could be severe [39]. However, in many cases, people perceived less risk when they had sufficient information about the adverse event, which was likely driven by familiarity bias [40]. That is, people who are more familiar with the risks are likely to perceive them as less frightening. As commonly acknowledged, individuals who are located at a considerable distance from the epicenter typically receive second-hand information regarding risks, while those who are in close proximity not only have access to first-hand risk-related information but also have their own direct experiences to draw upon. Not surprisingly, this leads to a difference in their perception of risk. Based on this, it is reasonable to speculate that the available risk-related information may be the underlying mechanism for the PTE effect.

Information quantity and information quality are two important aspects of available risk-related information [41]. More recently, Yang and colleagues discovered that the proportion of risk information (RIP) played a significant role in explaining the PTE effect concerning COVID-19 risk perception in Wuhan. Specifically, the RIP acts as a mediator between the respondents' distance from Wuhan and their level of concern and perception of risk regarding the epidemic that took place in the city [42]. In their study, RIP was defined as the ratio of "the amount of information related to the occurrence of risk events in a certain area" and "the total amount of information about all events in a certain area". Yang et al.'s research [42] can be seen as an explanation of the PTE effect mechanism in terms of the quantity dimension of the available information, while the present study attempts to examine whether another dimension of the available information, i.e., information quality, can explain the PTE effect in the supply chain disruption risk.

In the initial stages of a supply chain disruption risk, the available information is often uncertain and ambiguous. As individuals move closer to the epicenter of the risk, the information becomes more certain and less ambiguous [12,24]. This means that the disrupted firm's tier-1 suppliers/customers can relatively easily obtain disruption risk-related information that is of high quality, but their distant (tier-2 and above) suppliers/customers cannot [43]; they only have indirect access to the disrupted firm's second-hand information that is filtered, altered, and likely to be inaccurate via their partners, mass media, social media, etc. Additionally, firms typically postpone public announcements of disruptions [44], underreport, or hide disruption information [45], which reduces the level of quality of disruption information available to managers. Low information quality is characterized as delayed, incomplete, and ambiguous, which prevents supply chain managers from having a clear picture of what is actually happening in the disrupted firm [46], leading to supply chain managers far from the epicenter overestimating the disruption risk [47]. In other words, the PTE effect in the supply chain disruption risk may be caused by differences in the quality of risk information. That is, with the increment of managers' distance to the disrupted firm, the quality of disruption risk information declines, which leads managers to overestimate the disruption risk.

Therefore, we formulate the following research hypothesis:

Hypothesis 2 (H2): *Risk information quality played a mediating role between distance and disruption risk perception.*

2.3. The Moderating Role of Job Position Level

When people make judgments or decisions, they may be influenced by their prior beliefs, attitudes, and values [48]. Many studies have shown that people are susceptible to the "belief bias" effect and tend to accept or reject conclusions based on their consistency with everyday knowledge, regardless of whether these conclusions validly deviate from their premises [49–51]. Pre-existing beliefs can cause bias for people's perception of risk, leading them to over- or underestimate the likelihood or severity of a risk based on their existing beliefs [52,53]. For example, a person who strongly believes in the safety of a particular technology may underestimate the risks associated with that technology, while a person who strongly opposes that technology may overestimate the risks.

Most non-experts lack professional expertise and enough experience to assess risk [29]. As a result, they often rely on various cues available to them to aid in their judgment and decision-making [21]. Experts, because of their training and experience, are more likely to have knowledge (i.e., expertise and experience) about a certain hazard or adverse event unavailable to the average citizen. Therefore, experts do not need too much information about the adverse event in order to make their risk assessments [54]. The role of a supply chain manager is highly analytical and typically involves tasks such as planning, scheduling, and coordinating supplies [38]. High-level supply chain managers have a large amount of knowledge and experience in the field of supply chain risk management [55]. The more knowledge and expertise they have about disruption risks, the more they feel certain about their risk assessments, and the less disruption risk information from external sources they use in the assessment process [23]. Based on this, it is reasonable to speculate that top managers' perception of disruption risk is less affected by risk information quality than lower-level managers. Therefore, we hypothesize that:

Hypothesis 3 (H3): *Job position level moderates the relationship between risk information quality and disruption risk perception. Specifically, the effect of risk information quality on disruption risk perception gets stronger among low-level managers, but it is attenuated among high-level managers.*

2.4. Perceived Risk Influences Individuals' Response

The risky decision-making theory provides an explanation for the relationship between managerial risk perceptions and their subsequent behavioral response in the face of a supply chain disruption [56]. This response includes the actions taken by managers to minimize the impact of disruption. Zsidisin and Wagner demonstrated that managers who perceive the extended supply chain as a potential risk source are more likely to take action to mitigate such risks [57]. Meanwhile, Ellis et al. found that buyers who perceive high levels of overall supply disruption risk tend to seek alternative sources of supply to mitigate such risks [37]. In addition, Kull et al. revealed that cognitive and behavioral factors, inducing risk perceptions in uncertain supplier selection situations, can lead to a higher preference for suppliers with more certain outcomes [38]. Sarafan et al. conducted a scenario-based experiment to investigate the effect of cultural value orientations on individuals' perception of risk and supplier-switching intention in the face of a supply disruption. They found that higher levels of disruption risk perception led to significantly higher supplier switching intention [12].

Therefore, we follow previous studies by offering the following hypothesis:

Hypothesis 4 (H4): *Higher perceived disruption risk is associated with a higher propensity to take action in the face of supply chain disruption.*

Based on the above analysis, we depict the conceptual research model in Figure 2.

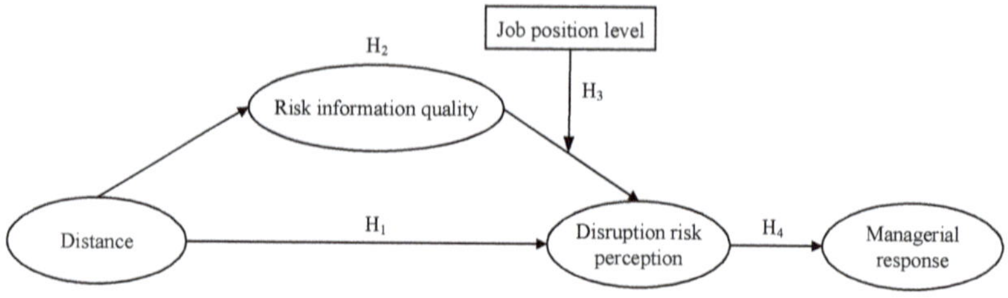

Figure 2. Theoretical model.

3. Methodology

3.1. Empirical Study Setting

To test the above hypothesis, we selected an information and communication technology supply chain with ZTE as the focal firm and conducted a scenario-based empirical investigation according to the similarity study [37,58]. Specifically, we focused on managers of ZTE and its upstream and downstream firms in the supply chain and how to perceive ZTE's disruption risk. This study setting is suitable for the hypotheses test for three reasons. First, more and more companies, such as telecom equipment makers, are at a high level of disruption risk caused by adverse events such as geostrategic conflict or COVID-19.

Second, the company had been barred by the U.S. Commerce Department from purchasing components from American companies in April 2018; therefore, professionals from the ZTE supply chain have a better understanding of ZTE's disruption risk and risky decision-making. Third, ZTE is a prominent global telecom equipment manufacturer that strives to deliver cutting-edge technologies and comprehensive solutions to a diverse clientele comprising governments, enterprises, and consumers in over 160 countries. Due to ZTE's vast network of upstream and downstream companies, it is convenient for us to recruit professionals affiliated with ZTE and its partners as research respondents.

3.2. Respondents and Data Collection

The survey was conducted on a computerized response system to facilitate the completion and collection of data throughout China from March to November 2020. We invited professionals from upstream and downstream firms of the ZTE supply chain through the authors' personal social networks, CFLP (China Federation of Logistics & Purchasing), and a web-based surveys platform (www.wjx.cn, accessed on 4 March 2020). A total of 1735 respondents engaged in our survey. Respondents were first asked to confirm their network position in the ZTE supply chain (i.e., ZTE's first-tier supplier/customer is 1, second-tier supplier/customer is 2) and then answer a 24-item questionnaire on their perceived psychological distances of ZTE's disruption, subjective perception of ZTE's disruption risk, disruption risk information quality, managerial response, risk propensity, and demographic characteristics.

Finally, we received 1055 usable responses, and all questionnaires were valid because we set the online survey system so it did not allow missing data. Among the sampled firms, there were 295 from ZTE Corporation, 359 from ZTE's upstream suppliers (tier-1 and tier-2 suppliers), and 401 from ZTE's downstream customers (tier-1 wholesalers and tier-2 wholesalers/retailers). More than half of sampled firms are medium and large enterprises. Characteristics of the sampled firms are presented in Table 1.

Table 1. Characteristics of sampled firms (N = 1055).

		Frequency	Percentage (%)
Supply chain position	ZTE corporation	295	28.0
	Tier-1 suppliers	193	18.3
	Tier-2 suppliers	166	15.7
	Tier-1 wholesalers	207	19.6
	Tier-2 wholesalers/retailers	194	18.4
Number of employees	≤10	38	3.6
	11–50	166	15.7
	51–100	250	23.7
	101–500	335	31.8
	501–1000	129	12.2
	≥1001	137	13.0
Annual sales revenue (CNY)	<1 million	137	13.0
	1–4.99 million	246	23.3
	5–9.99 million	233	22.1
	10–49.99 million	188	17.8
	≥50 million	251	23.8

It is noteworthy that the participants from ZTE Corporation not only serve as its employees but also function as organizers and managers of the supply chain. As a result, they hold a central position in relation to disruption risk and experience it directly, thereby being the most directly impacted stakeholders. ZTE employees acquire pertinent information regarding disruption directly, and function as disseminators of disruption risk information. They communicate information pertaining to disruption risk through various channels, such as news media, partners, and social networks.

To ensure the external validity of results [38], the overwhelming majority of participants had more than three years of experience in related operations or supply chain management areas, approximately 52.5% of the respondents were men, and over 80% of respondents held a bachelor's degree or above. The detailed demographic information on the respondents is presented in Table 2.

Table 2. Demographic data in the surveys (N = 1055).

		Frequency	Percentage (%)
Gender	Male	554	52.5
	Female	501	47.5
Age	≤30	404	38.3
	31–40	423	40.1
	41–50	160	15.2
	51–60	64	6.1
	>60	4	0.4
Education level	Secondary education certificate	5	0.5
	Senior school diploma	49	4.6
	Three-year college diploma	178	16.9
	Bachelor's degree	691	65.5
	Graduate degree	132	12.5
Work experience	<3 years	155	14.7
	3–5 years	222	21.0
	6–10 years	295	28.0
	11–20 years	219	20.8
	>20 years	164	15.5
Job Function	Planning and purchasing	313	29.7
	Operations and production	233	22.1
	Warehousing and logistics	179	17.0
	Research and development	25	2.4
	Sales and marketing	305	28.9
Job position level	Executive-level manager	148	14.0
	Middle-level manager	247	23.4
	Low-level manager	326	30.9
	Ordinary employee	334	31.7

It was found that the average age of ordinary employees was 32.14 years old, with more than 80% having over 3 years of work experience. Junior managers had an average age of 34.35 years, with 51.3% having more than 5 years of work experience. Middle managers had an average age of 38.6 years, with 22.3% having more than 10 years of work experience. Top managers had an average age of 47.26 years, with 29.1% having more than 20 years of work experience.

3.3. Measure Development

We conducted a comprehensive review of the literature on supply chain disruption, risk perception, and behavioral decision-making to establish operational definitions and survey measurement items. To ensure content validity, we adapted items from previous studies to our research setting when applicable. The participants were asked to rate their level of agreement on a 10-point scale (1 = Strongly disagree, to 10 = Strongly agree) in response to a series of statements concerning their work experience and principles.

3.3.1. Dependent Variable

The dependent variable includes the focal dependent variable and the ultimate dependent variable. The focal dependent variable is disruption risk perception, and the ultimate dependent variable is the managerial response to ZTE's supply chain disruption.

Disruption risk perception (DRP). We employed the psychometric paradigm as a research framework to measure the risk perception of ZTE's supply chain disruption. We adopted three items from Xie et al. [35] and Zheng et al. [19] to measure the disruption risk. The participants were asked if they thought the negative performance impact caused by ZTE's supply chain disruption was serious, dreadful, and uncontrollable. The statements of the three items are as follows: serious, "We will face a severe threat caused by ZTE's supply chain disruption" (DRP1); dreadful, "We extremely concern about the threat to my company caused by ZTE's supply chain disruption" (DRP2); uncontrollable, "We are unable to avoid the threat caused by ZTE's supply chain disruption" (DRP3). The higher the values, the higher level of risk they perceived from ZTE's supply chain disruption.

Managerial response to ZTE's supply chain disruption (MR). We developed a three-item scale to measure respondents' managerial response regarding ZTE's supply chain disruption based on previous studies [37,59]. The items are: "We will take action immediately to ZTE's supply chain disruption" (MR1); "We will take effective measures to ZTE's supply chain disruption" (MR2); "We will undertake an adequate response to ZTE's supply chain disruption" (MR3). Respondents were asked to rate each item based on the degree to which they agreed with the statement. The higher the values, the more propensity to take action regarding ZTE's supply chain disruption.

3.3.2. Independent Variable

Supply chain distance. Supply chain distance was determined by the network position of the respondent's company in ZTE's supply chain and was an objective distance. At the beginning of the questionnaire, respondents were asked to confirm their company's role in the ZTE supply chain: the supply chain distance of ZTE's first-tier supplier/customer is 1, the second-tier supplier/customer's supply chain distance is 2, and so on.

In addition to the objective distance (supply chain distance), we also measured the subjective distance between the respondents and ZTE's disruption risk, which was operationalized as psychological distance. In a range of risk domains—from climate change to nuclear energy, from food safety to health—the association of psychological distance on an individual's perception and response to risk has been proven to be robust [60–62].

Psychological distance (PD). We created a set of four items to assess the participants' perception of the psychological distance of ZTE's disruption. This includes spatial distance, temporal distance, social distance, and hypotheticality [63]. In this present study, spatial distance is not the geographical distance between the supply chain actor and the location where a disruption triggers but the supply chain distance instead. Supply chain distance was defined as the "distance between actor and the disruptive incidents, or 'position' of an actor in a supply chain network" according to Birkie and Trucco [64] and Ozkul and Barut [65]. The distance was estimated subjectively by the participants, so it also can be called subjective supply chain distance. Therefore, spatial distance (i.e., supply chain distance) was measured by asking "My company's 'position' in the ZTE's supply chain network determines we are far from ZTE's disruption" (PD1). The items used to measure temporal and social distance were based on Spaccatini et al.'s work [32], and the item used to measure the hypotheticality of ZTE's supply chain disruption was based on Ellis et al.'s work [37]. Social distance was measured with "ZTE's supply chain disruption will have a little impact on my company" (PD2). Hypotheticality was measured by asking "There is a low probability that ZTE will experience a supply chain disruption" (PD3). The temporal distance was measured with "If ZTE's supply chain would be disrupted by adverse events such as COVID-19 and U.S.-China conflict, that will be something for a long time to come" [66] (PD4). Each of these four items measured a distinct dimension of psychological distance, with high values indicating greater psychological distance and low values indicating less psychological distance.

3.3.3. Mediate Variable

Perceived risk information quality (PRIQ). Timeliness, credibility, and being easily understandable are important dimensions of information quality, and they will affect the individual's judgment and decision [67]. Therefore, higher risk information quality would make participants suffer from less risk information illusion. PRIQ was operationalized from four dimensions using four items, which refers to information that is readily accessible, timely, credible, and understandable [68]. They are: "Information about ZTE's supply chain disruption is easy access" (PRIQ1); "Information about ZTE's supply chain disruption is timely" (PRIQ2); "Information about ZTE's supply chain disruption is credible" (PRIQ3); "Information about ZTE's supply chain disruption is understandable" (PRIQ4). The higher values represented a lower degree of information illusion.

3.3.4. Moderate Variable

Job position level (JPL) includes four levels: executive-level manager, middle-level manager, low-level manager, and ordinary employee.

3.3.5. Control Variables

In addition to the above-mentioned key variables, respondents' risk propensity (RPr) was measured for validation and control variables [12]. Four items for measuring participants' risk propensity were chosen from the risk propensity scale developed by Hung and Tangpong [69]: "I like to take chances, although I may fail" (RPr1); "I like to try new things, knowing well that some of them will disappoint me" (RPr2); "To earn greater rewards, I am willing to take higher risks" (RPr3); "I seek new experiences even if their outcomes may be risky" (RPr4). The higher the values, the more likely individuals have a greater risk propensity. We also controlled a series of factors to maximize internal validity and rule out other explanations, such as age, gender, educational background, and work experience. Other than these individual-level control variables, we also controlled for the firm size (measured by the number of employees and annual sales revenue). All of the above control variables were kept as categorical variables.

3.4. Construct Validity and Reliability

Since some measurement items are used for the first time in the context of operational and supply chain management, we examined the reliability and validity of scales through exploratory factor analysis (EFA) and confirmatory factor analysis (CFA), as suggested by Anderson and Gerbing [70]. First, EFA is used to evaluate whether the measurement items are consistent with theoretical expectations, and then the internal structure of the scale is further confirmed by CFA.

We used SPSS 22.0 to perform the EFA. Results indicate the presence of five factors based on the criteria of eigenvalues greater than 1 (72.72% of total variance explained). The criterion we followed to determine whether to keep an item on a factor was that the item should have a loading of at least 0.40 on the primary factor and not have significant dual loadings (i.e., >0.30 on more than one factor) [71]. One item was problematic: PD4 had a loading of 0.542 on factor 2 and 0.559 on factor 4, and was excluded. Five factors and 16 items were prepared for the subsequent CFA, as shown in Table 3.

To further assess the reliability and validity of the constructs, we performed CFA and employed four tests to evaluate the convergent validity and internal consistency of the reflective constructs: Cronbach's alpha, average variance extracted (AVE), composite reliability (CR), and item loading of the measures. The goodness of fit of the measurement model was evaluated through five common indices, including the ratio of Chi-square to the degree of freedom (χ^2/df), comparative fit index (CFI), the goodness of fit index (GFI), Tucker–Lewis index (TLI), and root mean square error of approximation (RMSEA). The CFA was conducted using AMOS 22.0 software [72]. The fit statistics indicated a satisfactory fit between the predicted and observed model with χ^2/df = 4.80 [χ^2(113) = 542.14 and p = 0.00], CFI = 0.96, GFI = 0.94, TLI = 0.95, and RMSEA = 0.06 [73,74]. As shown in Table 4, all Cronbach's alpha and CR statistics exceeded the 0.7 cut-off recognized in the literature, suggesting good construct reliability. The convergent validity of our multi-item scales is adequate since the AVE was larger than 0.5 and most items had factor loading exceeding 0.7 on their construct [37,75]. We used Harman's one-factor test to assess the presence of common method variance (CMV). This EFA result indicated that CMV was not a potential issue in this study [76]. Moreover, discriminant validity was assessed by comparing the correlation coefficient of each construct with other constructs to the square root of its AVE. Our findings indicated that the square root of AVE for each construct was greater than its correlation coefficient with other constructs, which supported the discriminant validity of our measures [75].

Table 3. The items and factor loadings of the five-factor model.

Measurement Items	Factor 1 (PRIQ)	Factor 2 (MR)	Factor 3 (PD)	Factor 4 (DRP)	Factor 5 (RPr)
PRIQ 1	**0.854**	−0.029	−0.072	−0.252	0.095
PRIQ 2	**0.858**	0.089	−0.034	−0.202	0.047
PRIQ 3	**0.898**	0.013	−0.019	−0.118	0.073
PRIQ 4	**0.719**	−0.256	−0.014	−0.095	0.164
MR 1	−0.056	**0.867**	0.106	0.276	0.007
MR 2	−0.048	**0.897**	0.124	0.142	0.006
MR 3	−0.044	**0.884**	0.103	0.179	0.040
PD 1	−0.039	0.134	**0.897**	0.075	0.003
PD 2	−0.019	0.082	**0.890**	0.101	0.002
PD 3	−0.058	0.118	**0.864**	0.101	−0.012
PD 4	0.035	**0.542**	0.171	**0.559**	0.059
DRP 1	−0.260	0.158	0.048	**0.731**	−0.198
DRP 2	−0.268	0.239	0.131	**0.774**	−0.090
DRP 3	−0.251	0.229	0.108	**0.785**	−0.103
RPr 1	0.044	0.079	0.004	−0.212	**0.709**
RPr 2	0.084	0.038	−0.019	−0.060	**0.787**
RPr 3	0.068	0.129	0.024	−0.106	**0.752**
RPr4	0.142	−0.281	−0.019	0.115	**0.637**
Proportion variance (%)	29.246	16.525	10.832	10.122	5.993
Cumulative (%) of variance explained	29.246	45.771	56.603	66.725	72.718

Table 4. Measures used in proposed constructs.

Construct	Item	Cronbach's α	AVE	CR	Loading	t-Value	SE
Psychological Distance (PD)		0.88	0.71	0.88			
	PD1				0.88	-	-
	PD2				0.84	31.35	0.03
	PD3				0.80	29.74	0.03
Perceived Risk Information Quality (PRIQ)		0.88	0.66	0.88			
	PRIQ1				0.86	-	-
	PRIQ2				0.85	34.55	0.03
	PRIQ3				0.87	35.93	0.03
	PRIQ4				0.64	22.88	0.03
Risk Propensity (RPr)		0.70	0.40	0.72			
	RPr1				0.61	-	-
	RPr2				0.74	15.12	0.06
	RPr3				0.69	15.05	0.06
	RPr4				0.43	10.97	0.06
Disruption Risk Perception (DPR)		0.83	0.63	0.83			
	DRP1				0.70	-	-
	DRP2				0.84	23.47	0.05
	DRP3				0.83	23.36	0.05
Managerial Response (MR)		0.92	0.81	0.93			
	MR1				0.89	-	-
	MR2				0.90	41.05	0.03
	MR3				0.91	41.76	0.03

4. Results

Data analysis consisted of three steps. In the first step, we performed descriptive statistics and examined Pearson's product–moment correlations among sociodemographic characteristics, firm size, and the five factors in Table 4, which include psychological distance, disruption risk information quality, disruption risk perception, risk propensity, and managerial response. The second step was a mediation analysis: to investigate the indirect effect of distance on disruption risk perception via disruption risk information quality. The third step was a multivariate analysis: we conducted hierarchical moderated regression analyses in order to test the moderating effect of risk propensity and job position level.

4.1. Descriptive Analyses

Table 5 displays the variable means, standard deviations, and correlation coefficients between variables for this study. The demographic profile of the respondents was shown in the first five rows of the table.

Table 5. Means, SDs, and pairwise correlations of the measures.

	M	SD	1	2	3	4	5	6	7	8	9	10	11
1 Gender	0.47	0.50											
2 Age	1.90	0.90	0.12 ***										
3 Educational background	3.85	0.71	0.07 *	−0.10 **									
4 Work experience	3.01	1.28	0.17 ***	0.82 ***	−0.15 ***								
5 Job position level	2.20	1.04	−0.04	0.00	0.03	0.09 **							
6 Number of employees	3.72	1.33	−0.02	−0.04	0.25 ***	−0.05	0.01						
7 Annual sales revenue	3.16	1.36	0.03	−0.10 **	0.23 ***	−0.07 *	0.02	0.67 **					
8 Risk propensity	5.02	1.37	0.05	−0.02	0.04	−0.03	0.04	−0.02	-0.05				
9 Psychological distance	5.94	1.92	−0.01	−0.19 ***	0.08 **	−0.19 ***	0.04	0.05	0.09 **	−0.02			
10 Perceived risk information quality	4.88	1.68	0.01	0.09 **	0.04	0.11 ***	−0.02	−0.02	−0.07 *	0.24 ***	−0.11***		
11 Disruption risk perception	6.39	1.98	0.03	−0.18 ***	0.11 ***	−0.18 ***	−0.04	0.11 ***	0.18 ***	−0.27 ***	0.24 ***	−0.48 ***	
12 Managerial response	5.45	1.93	−0.00	−0.18 ***	0.08 **	−0.17 ***	0.04	0.08 *	0.14 ***	−0.02	0.26 ***	−0.14 ***	0.44 ***

Note: M = mean; SD = standard deviation. Variables were coded as follows—Gender: 1 = female, 0 = male; Age: 1 = below 30 years old, 2 = 31–40 years old, 3 = 41–50 years old, 4 = 51–60 years old, 5 = above 61 years old; Education: 1 = secondary education certificate, 2 = senior school diploma, 3 = three-year college diploma, 4 = bachelor's degree, 5 = graduate degree; Work experience: 1 = less than 3 years, 2 = 3–5 years, 3 = 6–10 years, 4 = 11–20 years, 5 = more than 20 years; Job position level: 1 = ordinary employee, 2 = low-level manager, 3 = middle-level manager, 4 = executive-level manager; Number of employees: 1 = fewer than 10 employees, 2 = 11–50 employees, 3 = 51–100 employees, 4 = 101–500 employees, 5 = 501–1000 employees, 6 = more than 1001 employees; Annual sales revenue: 1 = fewer than CNY 1 million, 2 = CNY 1–4.99 million, 3 = CNY 5–9.99 million, 4 = CNY 10–49.99 million, 5 = more than CNY 50 million. * $p < 0.05$, ** $p < 0.01$, *** $p < 0.001$.

4.2. Direct Effect of Distance on Disruption Risk Perception

4.2.1. Supply Chain Distance Affects Disruption Risk Perception

The risk perception ratings of the participants varied significantly depending on their supply chain distance (F (4, 1055) = 41.16, $p < 0.001$, and $\eta^2 = 0.04$, by ANOVA). The scores for their risk perception from lowest to highest in the ZTE supply chain were: ZTE, ZTE's tier-1 supplier/customer, and ZTE's tier-2 supplier/customer (see Figure 3). Fisher's least significant difference (LSD) post hoc test further revealed that the ZTE group reported the lowest risk perception (M = 5.87, SD = 2.03), significantly lower than the ratings given by tier-1 suppliers (M = 6.54, SD = 1.93) and tier-2 suppliers (M = 7.00, SD = 1.68) (F (2, 693) = 9.45, p <0.001, η^2 = 0.03), which was also significantly lower than the ratings given by tier-1 customers (M = 6.26, SD = 2.13) and tier-2 customers (M = 6.67, SD = 1.82) (F (2, 651) = 19.68, p <0.01, η^2 = 0.06). In the upstream, tier-2 suppliers perceived more risk than tier-1 suppliers ($p < 0.05$). In the downstream, similarly, tier-2 customers perceived more risk than tier-1 customers ($p < 0.05$). However, the rated risk perception between tier-1 suppliers and tier-1 customers had no significant difference ($p > 0.05$); likewise, no significant difference was found between tier-2 suppliers and tier-2 customers ($p > 0.05$).

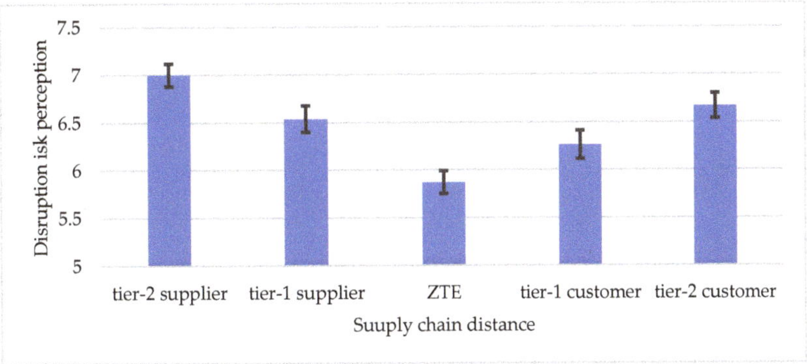

Figure 3. Supply chain distance and disruption risk perception, ***Note:*** Bar heights indicate mean values, and error bars indicate standard error.

The above results indicated that the greater the supply chain distance, the higher the risk perception toward ZTE's disruption risk. That is, in the supply chain with ZTE as the focal firm, the farther away the upstream or downstream members were from ZTE, the higher the risk they perceived from ZTE's supply chain disruption. The hypothesized "supply chain distance" version of the PTE effect was thus observed in the present study. Therefore, the results supported H1.

4.2.2. Psychological Distance Affects Disruption Risk Perception

We respectively conducted a hierarchical regression analysis to reveal the impact of psychological distance on disruption risk perception upstream and downstream of the ZTE supply chain. Participants' age, gender, educational background, work experience, and firm size were entered as control variables in this analysis. Upstream supply chain members consist of ZTE, tier-1 suppliers, and tier-2 suppliers; downstream supply chain members consist of ZTE, tier-1 customers, and tier-2 customers.

The results of the hierarchical regression analysis for the ***upstream*** of ZTE's supply chain are presented in Table 6. When all variables were included in the model, it accounted for 21.9% of the variance in the perception of a higher risk for ZTE's disruption. In model 1, demographical variables were entered as controls; the overall model was significant, and $R^2 = 0.170$, F (7, 646) = 18.879, $p < 0.001$. In model 2, the psychological distance was entered as a predictor; the overall model remained significant, and $R^2 = 0.219$, F (8, 645) = 22.659, $p < 0.001$. The psychological distance was found to be a significant predictor of disruption risk perception (B = 0.233, $p < 0.001$).

We conducted the same regression analysis to analyze the effect in the ***downstream*** of ZTE supply chain, and the results are shown in Table 7. Overall, with all variables entered, the model explained 15.8% of the variance in having a higher risk perception of ZTE's disruption. In model 1, demographical variables were entered as controls; the overall model was significant, and $R^2 = 0.065$, F (7, 688) = 14.579, $p < 0.001$. In model 2, the psychological distance was entered as a predictor; the overall model remained significant, and $R^2 = 0.094$, F (8, 687) = 16.123, $p < 0.001$. The psychological distance was a significant predictor of risk perception ($B = 0.182$, $p < 0.001$).

Table 6. Hierarchical regression analysis of demographical variables and psychological distance on disruption risk perception in the upstream of ZTE's supply chain (N = 654).

Variable	Model 1			Model 2		
	B	*SE*	β	B	*SE*	β
Step 1						
(Constant)	7.257 ***	0.534		5.643 ***	0.576	
Gender	0.009	0.145	0.002	−0.008	0.141	−0.002
Age	−0.084	0.142	−0.038	−0.036	0.138	−0.016
Educational background	0.243 *	0.107	0.086	0.240 *	0.104	0.085
Work experience	−0.239 *	0.100	−0.155	−0.200 *	0.097	−0.129
Risk propensity	−0.392 ***	0.051	−0.278	−0.377 ***	0.049	−0.268
Number of employees	0.061	0.075	0.040	0.063	0.073	0.041
Annual sales revenue	0.232 ***	0.072	0.160	0.213 ***	0.070	0.147
Step 2						
Psychological distance				0.233 ***	0.036	0.228
F value			18.879			22.659
R^2			0.170			0.219
Adj. R^2			0.161			0.210
ΔAdj R^2			0.170			0.050

Note: * $p < 0.05$, *** $p < 0.001$.

Table 7. Hierarchical regression analysis of demographical variables and psychological distance on disruption risk perception in the downstream of ZTE's supply chain (N = 696).

Variable	Model 1			Model 2		
	B	*SE*	β	B	*SE*	β
Step 1						
(Constant)	8.233 ***	0.542		7.139 ***	0.579	
Gender	0.438 **	0.148	0.108	0.440 **	0.146	0.108
Age	−0.223	0.139	−0.099	−0.179	0.137	−0.080
Educational background	0.127	0.105	0.045	0.099	0.103	0.035
Work experience	−0.199 *	0.101	−0.124	−0.172 *	0.099	−0.107
Risk propensity	−0.383 ***	0.054	−0.256	−0.382 ***	0.053	−0.255
Number of employees	−0.080	0.071	−0.054	−0.069	0.070	−0.046
Annual sales revenue	0.184 **	0.069	0.126	0.168 **	0.068	0.115
Step 2						
Psychological distance				0.182 ***	0.038	0.174
F value			14.579			16.123
R^2			0.129			0.158
Adj. R^2			0.120			0.148
ΔAdj R^2			0.129			0.029

Note: * $p < 0.05$, ** $p < 0.01$, *** $p < 0.001$.

The aforementioned analysis results indicate that psychological distance has a significant positive impact on disruption risk perception in both the upstream and downstream of ZTE's supply chain. Figure 4 depicts the diagram of the results. To better observe and compare the results, the psychological distance scores of the upstream were converted to negative values, and the left and right half of the diagram represent the results of the upstream and downstream, respectively.

An interaction effect test was conducted in order to further test whether there is a symmetrical relationship between the results of the upstream and downstream. The results indicate that both the overall model (F (9, 1045) = 14.331, $p > 0.05$) and the interaction coefficient ($B = -0.016$, $p > 0.05$) were not significant, which indicates that there is no interaction effect between the upstream and downstream. In other words, the impact of psychological distance on disruption risk perception is symmetric (consistent) between the upstream and downstream. Due to this symmetry, the following analysis no longer distinguishes between the upstream or downstream. Therefore, the tier-1 supplier and tier-1 customer are merged (hereinafter referred to as tier 1), and the tier-2 supplier and tier-2 customer are also merged (hereinafter referred to as tier 2).

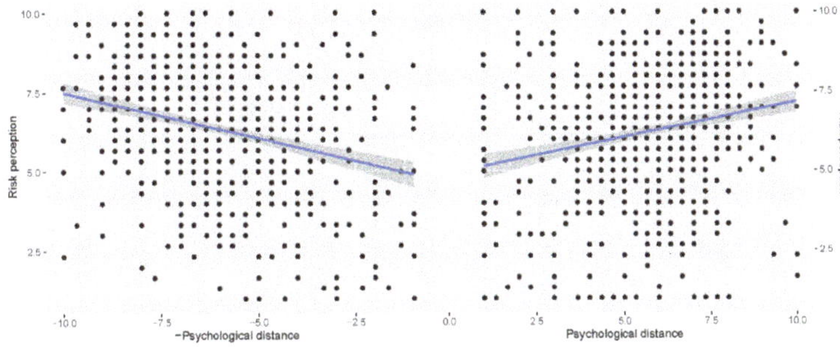

Figure 4. Scatterplot of the relationship between psychological distance and risk perception of supply chain disruption. The best-fitting regression line is depicted in the center.

4.3. Indirect Effect of Distance on Disruption Risk Perception Via Perceived Risk Information Quality

H2 predicted that objective distance and subjective distance would have a direct impact on the perceived quality of risk information and disruption risk perception, as well as an indirect impact on risk perception through perceived risk information quality. We conducted two separate mediation analyses for objective distance (i.e., supply chain distance) and subjective distance (i.e., psychological distance) on disruption risk perception. In order to investigate the mediated impacts of the perceived risk information quality on the relationship between distance and disruption risk perception, we utilized the PROCESS V3.3 tool to test multiple mediators and analyze the overall effects.

4.3.1. The Mediated Effects of Objective (Tier) Distance on Disruption Risk Perception via Perceived Risk Information Quality

The respondents' ratings of perceived disruption risk information quality differed significantly depending on their supply chain distance (F (2, 1052) = 89.078, $p < 0.001$, and $\eta^2 = 0.145$, by ANOVA). That is, their perceived risk information quality from highest to lowest in the supply chain were: ZTE (*PRIQ score* = 5.876), ZTE's tier-1 supplier/customer (*PRIQ score* = 4.678), and ZTE's tier-2 supplier/customer (*PRIQ score* = 4.883). Fisher's Least Significant Difference (LSD) post hoc test indicated that the ZTE group's perceived risk information quality was significantly higher than those of the tier-1 and tier-2 suppliers/customers. ($p < 0.001$).

Supply chain distance (multi-categorical variable containing three groups: ZTE, tier 1, and tier 2) was entered as the predictor and was encoded as a dummy variable (ZTE was set as the control group), and perceived risk information quality was considered a mediator, with ZTE's disruption risk perception serving as the outcome or dependent variable. The results of our analysis were assessed using a bootstrap estimation approach, which involved 5000 samples and is presented in Table 8. Specifically, we examined the total, direct, and mediated effects of the supply chain distance on disruption risk perception.

Following Hayes [77], we began by examining the total effect of supply chain distance on the risk perception of ZTE's disruption (i.e., the effect of supply chain distance on disruption risk perception without the presence of any mediating effects), and found a positive and significant effect (the control group (ZTE) was used as the reference, group of tier 1: $B = 0.476$, $SE = 0.146$, $p < 0.001$, 95% CI = [0.190, 0.762]; group of tier 2: $B = 0.908$, $SE = 0.150$, $p < 0.001$, 95% CI = [0.614, 1.201]). Next, we examined the complete model, which includes both the direct and mediated effects of supply chain distance on disruption risk perception. As shown in Table 8, the relative direct effect of supply chain distance on disruption risk perception in the presence of mediators becomes nonsignificant (the control group (ZTE) was used as the reference, group of tier 1: $B = -0.167$, $SE = 0.136$, $p > 0.05$, 95% $CI = [-0.434, 0.099]$; group of tier 2: $B = 0.057$, $SE = 0.143$, $p > 0.05$, 95% $CI = [-0.224, 0.338]$), indicating full mediation through perceived risk information quality. In addition, supply chain distance showed a relative indirect effect on disruption risk perception through perceived risk information quality (the control group (ZTE) was used as the reference, group of tier 1: $B = 0.643$, $SE = 0.085$, 95% $CI = [0.482, 0.813]$; group of tier 2: $B = 0.850$, $SE = 0.093$, 95% $CI = [0.677, 1.033]$). The bias-corrected bootstrap confidence intervals for the indirect effects were entirely above zero, indicating significant mediation effects.

Table 8. Mediating effects of perceived risk information quality on disruption risk perception (ZTE was set as the control group).

Path of Mediating Effect	Point Estimate	SE	95% CI	
			Low	High
Group of Tier 1:				
Relative total effect (Tier 1 → disruption risk perception)	0.476	0.146	0.190	0.762
Relative direct effect (Tier 1 → disruption risk perception)	−0.167	0.136	−0.434	0.099
Relative mediating effect (Tier 1 → Perceived risk information quality → disruption risk perception)	0.643 [a]	0.085	0.482	0.813
Group of Tier 2:				
Relative total effect (Tier 2 → disruption risk perception)	0.908	0.150	0.614	1.201
Relative direct effect (Tier 2 → disruption risk perception)	0.057	0.143	−0.224	0.338
Relative mediating effect (Tier 2 → Perceived risk information quality → disruption risk perception)	0.850 [a]	0.093	0.677	1.033

Note: "a" indicates mediating effect is significant.

4.3.2. The Mediated Effects of Subjective (Psychological) Distance on Disruption Risk Perception via Perceived Risk Information Quality

We conducted the same analysis of the subjective (psychological) distance. In this model, objective (tier) distance was entered as control. We first examined the total effect of psychological distance on the risk perception of ZTE's disruption (i.e., the effect of psychological distance on disruption risk perception without the mediated effects) and found a positive and significant effect ($B = 0.208$, $SE = 0.031$, $p < 0.001$, 95% $CI = [0.147, 0.269]$). We then examined the direct and mediated effects. As shown in Figure 5, psychological distance had a positive effect on perceived risk information quality ($B = -0.080$, $SE = 0.027$, $p < 0$ 0.01, 95% $CI = [0.112, 0.219]$), which was positively related to willingness to use (B = −0.535, SE = 0.031, $p < 0.001$, 95% $CI = [-0.595, -0.474]$; indirect effect of psychological distance: $B = 0.043$, $SE = 0.017$, 95% $CI = [0.012, 0.077]$), indicating a mediation effect. The direct effect of psychological distance on disruption risk perception in the presence of mediators was significant ($B = 0.165$, $SE = 0.027$, $p < 0.001$, 95% $CI = [0.112, 0.219]$). As shown in Figure 5, the results suggest that perceived risk information quality partially mediates the effect of psychological distance on disruption risk perception.

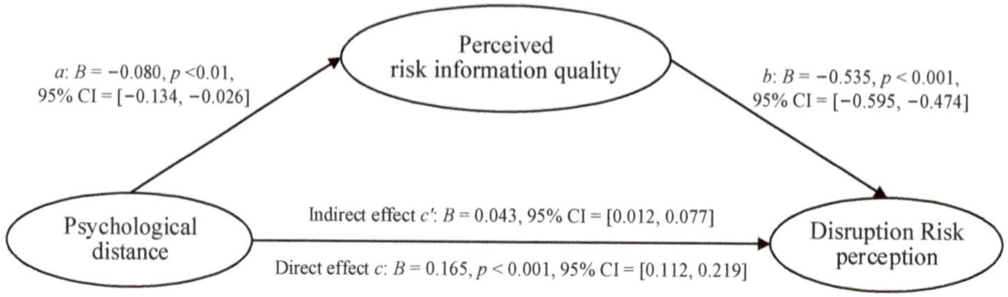

Figure 5. The mediation model illustrates the effects of subjective (psychological) distance and perceived risk information quality on the risk perception of ZTE's supply chain disruption.

Altogether, mediation analyses confirmed our H2, i.e., the effect of distance on disruption risk perception is mediated through perceived risk information quality.

4.4. The Moderating Effect of Job Position Level

We used the hierarchical linear regression model to examine the moderating effect of job position level. Before performing the regression analysis, we centered all non-nominal variables to alleviate the threats of multi-collinearity between the component measures [78]. Models 1–3 and 4–5, respectively, employed supply chain disruption risk perception to take management response as dependent variables. For models 1–3, sociodemographic and firm size were entered into the first layer as control variables (i.e., model 1), and then the perceived risk information quality and job position level were entered as the second layer (i.e., model 2). It should be noted that job position level is a categorical variable and was encoded as a dichotomous variable (0 = low-level positions, 1= high-level positions). In the final layer of the regression model, the interaction term between job position level and perceived risk information quality was included (i.e., model 3). For models 4–5, the first layer is the control variables (i.e., model 4), and the second layer is the supply chain disruption risk perception (i.e., model 5).

Model 1: Supply chain disruption risk perception = B_0 + Controls + e

Model 2: Supply chain disruption risk perception = B_0 + B_1(Perceived risk information quality) + B_2(Job position level) + Controls + e

Model 3: Supply chain disruption risk perception = B_0 + B_1(Perceived risk information quality) + B_2(Job position level) + B_3(Job position level × Perceived risk information quality) + Controls + e

Model 4: Managerial response = B_0 + Controls + e

Model 5: Managerial response = B_0 + B_1(Supply chain disruption risk perception) + Controls + e

When assessing the moderating effect, it is common practice to use the regression coefficient and significance of interaction terms to determine the presence of such an effect. According to the findings presented in Table 9, job position level serves as a significant moderator in the association between perceived risk information quality and disruption risk perception ($B_3 = 0.153$, $p < 0.05$).

Table 9. Hierarchical linear regressions (for "Disruption risk perception" and "Managerial response").

	Dependent Variable: DRP						Dependent Variable: MR			
	Model 1		Model 2		Model 3		Model 4		Model 5	
	Estimate	SE	Estimate	SE	Estimate	SE	Estimate	SE	Estimate	SE
Layer 1: Control Variables										
(Constant)	−0.543	0.373	−0.887 **	0.447	−0.847 **	0.336	−0.097	0.381	0.131	0.348
Gender	0.265 *	0.116	0.197 †	0.105	0.197 †	0.104	0.060	0.119	−0.051	0.109
Age	−0.183	0.112	−0.235 *	0.101	−0.243 *	0.101	−0.286 **	0.114	−0.209 *	0.104
Education	0.184 *	0.085	0.262 ***	0.076	0.259 ***	0.076	0.100	0.086	0.022	0.079
Experience	−0.168 *	0.080	−0.055	0.072	−0.049	0.072	−0.069	0.081	0.002	0.074
Rpr	−0.393 ***	0.042	−0.240 ***	0.039	−0.234 ***	0.039	−0.033	0.043	0.132 ***	0.040
Number of employees	−0.030	0.058	−0.009	0.052	−0.015	0.052	−0.034	0.059	−0.021	0.054
Annual sales revenue	0.213 ***	0.056	0.161 **	0.051	0.160 **	0.051	0.179 **	0.058	0.089	0.053
Layer 2: Main effect										
PRIQ			−0.506 ***	0.032	−0.558 ***	0.038				
JPL			−0.211 *	0.107	−0.207 *	0.106				
DRP									0.421 ***	0.029
Layer 3: Interaction effect										
JPL × PRIQ					0.158 *	0.065				
R^2	0.140		0.311		0.315		0.051		0.212	
Adjusted R^2	0.134		0.305		0.308		0.044		0.206	
R^2 change	0.140		0.171		0.004		0.051		0.161	
F-statistic	24.376		52.356		47.926		8.006		35.230	

Note: † $p < 0.1$, * $p < 0.05$, ** $p < 0.01$, *** $p < 0.001$.

To simplify the interpretation of the interaction terms, we utilized the "pick-a-point" technique to identify the conditional impact of perceived risk information quality on disruption risk perception, based on low and high levels of job position level (where 0 = low level, 1 = high level), in accordance with the findings of the moderating effect analysis. Subsequent simple slope tests revealed that the association between perceived risk information quality and disruption risk perception remains significant in both low and high levels of job position level (slope $_{\text{low job position level}} = -0.558$, $p < 0.001$; slope $_{\text{high job position level}} = -0.400$, $p < 0.001$). As Figure 6 presents, in the case of a high job position level, perceived risk information quality has a weaker negative effect on disruption risk perception. Thus, H3 was supported.

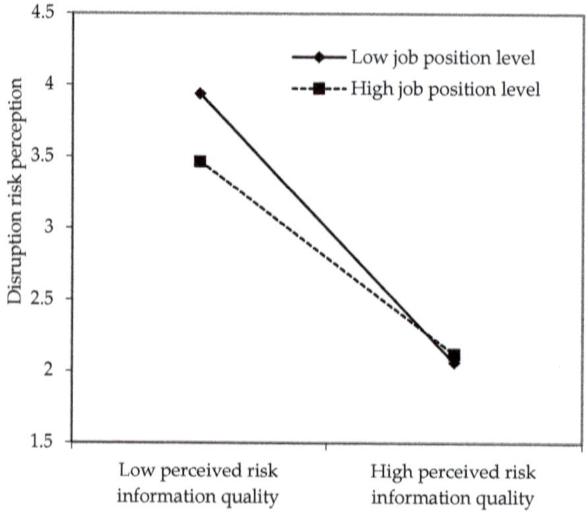

Figure 6. Spotlight analyses of moderating effect.

Finally, with regard to the influence of disruption risk perception on managerial response, the outcomes displayed in Table 9 provide evidence in favor of Hypothesis 4, indicating that heightened disruption risk perception results in a markedly stronger managerial response ($B = 0.421$, $p < 0.001$).

5. Discussion

5.1. Theoretical Contributions

The PTE effect has been observed in different risk areas such as earthquakes (e.g., [16]), terrorist attacks (e.g., [79]), epidemic outbreaks (e.g., [35,36]), environmental pollution (e.g., [19]), etc. Unlike these studies, where the risk perception was a function of geographical distance, the PTE effect in the present study is a function of a new type of distance, i.e., supply chain distance. As far as we are aware, this is the initial empirical investigation that examines the PTE effect in the realm of operations and supply chain management. Our study addresses the need for research that incorporates psychological risk theories into operations and supply chain management by extending upon previous research on supply chain disruption risk [12,80].

This study investigates the risk perception of supply chain upstream and downstream toward the focal firm's disruption. We moved beyond the single-tier buyer–seller relationships towards a multi-tier supply chain context to provide empirical insights into managers' perception of disruption risk. Although there has been considerable research on supply chain disruption risk perception, there is predominantly a focus on managers' perception of supply disruption risk within the purchasing domain [10,12,37]. This dyadic buyer–supplier perspective considers only the existence of direct relationships within a supply chain where most studies have drawn their research boundaries. However, such investigations fail to fully encompass the numerous, distinct, and interdependent interactions that coexist throughout the supply chain [58,81,82]. A supply chain network is vulnerable to disruptions not only because of the direct impacts of those disruptions but also because of the risk propagation [1,83]. Therefore, if the lead firm is at risk of supply chain disruption, the company itself and its first-tier and low-tier suppliers/customers would assess and respond to the direct or indirect risk [84,85]. This requires us to understand not only the focal firm's disruption risk perception but also how upstream and downstream firms perceive the disruption risk. The present study makes a preliminary exploration and provides some inspiration for future studies in this area.

This study also contributes to our understanding of the mechanism of the PTE effect. Risk perception is all about thoughts, beliefs, and constructs [86], and people's perception of risk is based on experience and available information [87]. Available information has a significant impact on the assessments and judgments that are made, thereby influencing the emotions and actions of supply chain participants towards the disrupted firm [88]. Yang et al.'s research [42] has explored the PTE effect mechanism in terms of the quantity dimension of the available information in the context of the COVID-19 pandemic, while the present study validated the PTE effect through the quality dimension of the available information in the context of supply chain disruption risk. Specifically, supply chain members who are close to the disrupted echelon have easier and more timely access to first-hand disruption risk information, i.e., they have access to high-quality information about the disruption risk. High-quality available information helps managers reasonably assess the actual level of disruption risk so that they do not over- or underestimate the risk. Supply chain members who are far away from the disrupted echelon can only receive disruption risk information from media reports or their partners. However, the disruption risk information has problems with distortion, delay, and untrustworthiness, i.e., supply chain managers receive lower quality information about the disruption risk. Low-quality information has the potential to lead to confusion and limit an individual's capacity to adequately process and react to the information presented [89], which makes it impossible for supply chain managers to learn the truth of the disrupted firm, ultimately leading them to overestimate the disruption risk [46,47].

We also found that job position level moderated the relationship between risk information quality and risk perception. Senior supply chain managers, who are experts with extensive risk management experience and expertise, are more inclined to use intuition to assess risk and require less risk-related information [23,90,91], so they are less influenced by external information and, accordingly, the level of risk information quality has less impact on their disruption risk perception. As with the general public, supply chain managers at lower levels conduct risk assessments based primarily on the risk information obtained, or rather on the basis of information or evidence-based assessments [86,92], therefore, the level of information quality has a greater impact on their perception of supply chain disruption risk.

In uncertain environments, people rely on heuristic strategies to make judgments and decisions [93]. According to the theory, supply chain managers are more likely to believe information that they are exposed to. Managers of focal companies and their direct trading partners observe or personally experience the disruption risk, and their assessment of disruption risk will be closer to reality. However, upstream and downstream members far from the focal company primarily learn about its disruption risk information through media or other channels. They tend to use this distorted and amplified information to assess risk, leading to a significant increase in their perception of disruption risk. Additionally, senior managers tend to assess risk based on their experience rather than the information they receive. In a word, supply chain managers often use heuristic strategies to assess disruption risk. However, differences in information quality, experience, etc., result in a perception bias of disruption risk known as the PTE effect.

5.2. Managerial Implications

The theory of risky decision-making offers an explanation for the association between managerial risk perceptions and their utilization of mitigation strategies [37]. A higher perception of disruption risk would lead supply chain managers to take action on behalf of their home organization in order to minimize the likelihood of being affected by a supply-side or demand-side disruption. For instance, when the focal firm faces supply chain disruption risk, its upstream partners will cut the exclusive supply capacity, and downstream partners will implement alternative sourcing [57] or switch suppliers [12]. This may exacerbate the focal firm's overall operational risk. Therefore, how to alleviate the PTE effect upstream and downstream is the key to supply chain disruption risk management for the focal firm.

Supply chain disruption risk is characterized by information uncertainty and the gap between the information available and the information needed to estimate and respond to the risk [94]. The present study shows that the high-quality risk information perceived by supply chain managers can reduce the proportion of disruption risk information so as to reduce their focus illusion and would finally reduce the level of perception of the focal firm's disruption risk. As such, effective management of disruption risk within and between firms necessitates a collective commitment to high-quality information synchronization, which ensures that disruption risk information is readily accessible, timely, credible, and comprehensible. This effort must extend beyond a single company initiative and involve all firms in the supply chain [68,82,95].

Specifically, first, the focal firm must ensure that upstream and downstream enterprises have easy access to real-time information related to disruption risk. It is widely acknowledged that many companies lack adequate information about their lower-tier partners in multi-tier supply chains [96]. This is compounded by narrow information sharing and communication channels, which restrict the efficiency of supply chain risk management efforts [82]. Therefore, the focal firm must take proactive measures to improve supply chain risk visibility and communication, empowering lower-tier suppliers to easily obtain reliable disruption risk information [97,98], making it difficult for them to gain misinformation or illusory information about the disruption risk. Secondly, the focal firm should timely release disruption risk-related information, disallowing a window of time for disruption risk misinformation to spread. The timely release and updating of disruption risk-related information is an effective measure to stop the spread of misinformation and avoid partners being overly concerned about disruptions [99]. The focal firm can release information and announcements through the official website, email, social media, and other channels so that supply chain members can obtain timely disruption risk information. Thirdly, the focal firm should enhance the credibility of the disruption risk information. Accurate and reliable information helps to eliminate supply chain professionals' information illusions and reduce their concerns and misevaluation of the disruption risk [100,101]. Lastly, the disruption risk information provided by the disrupted firm needs to be easy to understand. Compared with unclear and ambiguous information, specific and easy-to-understand information can reduce an individual's risk perception [102]. This requires the disrupted firm to release information relevant to the disruption risk in an easy-to-understand manner (e.g., text combined with pictures or videos) to facilitate supply chain members to assess the disruption risk properly.

Furthermore, it is important to note that definitions and interpretations of disruption risk terms may vary across organizations due to differences in business contexts and cultures [11]. This variation can lead to misunderstandings in shared disruption risk information and can ultimately impede effective supply chain risk management efforts within and between firms [98]. To address this challenge, it is recommended that a unified risk information language be established within and between supply chain firms to ensure consistent and clear communication about disruption risk. This approach will support objective disruption risk assessment and effective disruption risk communication along the entire supply chain of the focal firm.

Compared to high-level managers, lower-level supply chain managers' disruption risk perception towards the focal firm is more likely to be affected by their perceived risk information quality. So, it is more necessary to provide them with more timely, credible, and understandable disruption risk information and make it easier for them to acquire this information.

It is worth noting that different representations of the same information can lead to different judgments and decisions [103]. Therefore, in addition to the disruption risk information quality, the representation of disruption risk information can also affect managers' risk perception and decision-making. This inspires us that when focal companies release disruption risk information, they can flexibly design the framework to reasonably weaken the threat of disruption information, strengthen the information of disruption mitigation and control, mitigate excessive concerns of supply chain members far from the focal company about disruption risk, and reduce their irrational operational decisions.

6. Limitations and Future Work

Risk perception among managers in a multi-tier supply chain has received limited attention. This current study provides initial insights into this area, but it still has several potential limitations that must be highlighted to motivate future research. First, the supply chain management literature identifies several factors that may also impact perceptions of supply chain disruption risk [38]. The present study empirically examines the influence of distance on risk perception within the context of a multi-tier supply chain. Other contingency factors such as cultural value orientations [12,104], uncertainty [10,38], or trust [54] would affect an individual's disruption risk perception; future research could examine whether there is a PTE effect, which may contribute to providing a richer understanding of risk perception in the multi-tier supply chain. Second, the understanding of the underlying mechanism of the PTE effect in the multi-tier supply chain raises a challenging question: how can a disrupted focal firm enable its supply chain partners to receive high-quality disruption risk information to mitigate the PTE effect? While the present study does not focus on specific tactics, it does provide a rich avenue for future scientific research to design and examine the PTE effect mitigation strategies. Third, it is worth noting that our data was gathered through a cross-sectional survey, which is susceptible to respondents' subjective judgment and may involve some level of arbitrariness or variability. To further support and validate our results, future research could utilize alternative data collection methods and research designs, such as longitudinal studies (e.g., [14]) or laboratory experiments (e.g., [38,105]). Additionally, it is important to acknowledge that this study was conducted in mainland China, with its unique social norms and economic system. Thus, in order to enhance the external validity of our findings, it is recommended that further research be conducted in other countries/regions with different social and cultural norms.

7. Conclusions

The perception of disruption risks by supply chain managers can significantly impact their subsequent management responses, particularly in an era where supply chain disruption risks have become increasingly common due to events such as the 2011 Great Tohoku Earthquake, the COVID-19 pandemic, and the Russia–Ukraine conflict. This study investigates the risk perception of supply chain disruption in ZTE and its upstream and downstream members. The results indicate that as supply chain members are farther from the epicenter (i.e., ZTE), their risk perception of the disruption at the epicenter increases, a phenomenon we refer to as the PTE effect in supply chain disruption risk. Further research reveals that both supply chain distance and psychological distance influence disruption risk perception through risk information quality, and job position level moderates the relationship between risk information quality and disruption risk perception. These findings suggest that the focal firm must go beyond single-company initiatives and prioritize high-quality information synchronization to mitigate the PTE effect within the supply chain.

Author Contributions: Conceptualization: S.L., M.-X.X. and L.-L.R.; Methodology: M.-X.X. and L.Z.; Formal analysis and investigation: S.L. and M.-X.X.; Writing—original draft preparation: M.-X.X. and S.L.; Writing—review and editing: M.-X.X., S.L., L.-L.R. and L.Z. All authors have read and agreed to the published version of the manuscript.

Funding: This study was supported by the National Natural Science Foundation of China (Grant No. 71761167001), the MOE (Ministry of Education of China) Youth Foundation Project of Humanities and Social Sciences (Grant No. 19YJC630194), the Natural Science Foundation of Fujian Province (Grant No. 2020J01902), and the Major Projects of Fujian Social Science Research Base (Grant No. FJ2020JDZ068).

Institutional Review Board Statement: All procedures performed in studies involving human participants were in accordance with the ethical standards of the Institutional Review Board of the Institute of Psychology of the Chinese Academy of Sciences and with the 1964 Helsinki Declaration and its later amendments or comparable ethical standards.

Informed Consent Statement: Informed consent was obtained from all individual participants included in the study.

Data Availability Statement: The datasets generated during and/or analyzed during the current study are available from the corresponding author upon reasonable request.

Acknowledgments: We thank our participants for their time and for responding to our survey, and the four anonymous reviewers for providing valuable feedback.

Conflicts of Interest: All authors declare no competing interests.

References

1. Li, Y.; Zobel, C.W. Exploring supply chain network resilience in the presence of the ripple effect. *Int. J. Prod. Econ.* **2020**, *228*, 107693. [CrossRef]
2. Moosavi, J.; Fathollahi-Fard, A.M.; Dulebenets, M.A. Supply chain disruption during the COVID-19 pandemic: Recognizing potential disruption management strategies. *Int. J. Disast. Risk Res.* **2022**, *75*, 102983. [CrossRef] [PubMed]
3. Kamalahmadi, M.; Shekarian, M.; Mellat Parast, M. The impact of flexibility and redundancy on improving supply chain resilience to disruptions. *Int. J. Prod. Res.* **2021**, *60*, 1992–2020. [CrossRef]
4. Pournader, M.; Kach, A.; Talluri, S. A review of the existing and emerging topics in the supply chain risk management literature. *Decis. Sci.* **2020**, *51*, 867–919. [CrossRef] [PubMed]
5. Heckmann, I.; Comes, T.; Nickel, S. A critical review on supply chain risk–Definition, measure and modeling. *Omega* **2015**, *52*, 119–132. [CrossRef]
6. De Martino, B.; Kumaran, D.; Seymour, B.; Dolan, R.J. Frames, biases, and rational decision-making in the human brain. *Science* **2006**, *313*, 684–687. [CrossRef]
7. Tversky, A.; Kahneman, D. Judgment under Uncertainty: Heuristics and Biases: Biases in judgments reveal some heuristics of thinking under uncertainty. *Science* **1974**, *185*, 1124–1131. [CrossRef]
8. Gigerenzer, G.; Gaissmaier, W. Heuristic decision making. *Annu. Rev. Psychol.* **2011**, *62*, 451–482. [CrossRef]
9. Chadist, P. Factors Underlying Companies Response to Supply Chain Disruption: A Grounded Theory Approach. Ph.D. Dissertation, City University London, London, UK, 2012.
10. Sato, Y.; Tse, Y.K.; Tan, K.H. Managers' risk perception of supply chain uncertainties. *Ind. Manag. Data Syst.* **2020**, *120*, 1617–1634. [CrossRef]
11. March, J.G.; Shapira, Z. Managerial perspectives on risk and risk taking. *Manag. Sci.* **1987**, *33*, 1404–1418. [CrossRef]
12. Sarafan, M.; Squire, B.; Brandon-Jones, E. The effect of cultural value orientations on responses to supply-side disruption. *Int. J. Oper. Prod. Manag.* **2020**, *40*, 1723–1747. [CrossRef]
13. Scheibe, K.P.; Blackhurst, J. Supply chain disruption propagation: A systemic risk and normal accident theory perspective. *Int. J. Prod. Res.* **2018**, *56*, 43–59. [CrossRef]
14. Burns, W.J.; Peters, E.; Slovic, P. Risk perception and the economic crisis: A longitudinal study of the trajectory of perceived risk. *Risk Anal.* **2012**, *32*, 659–677. [CrossRef] [PubMed]
15. Mitchell, V.W. Organizational risk perception and reduction: A literature review. *Br. J. Manag.* **1995**, *6*, 115–133. [CrossRef]
16. Li, S.; Rao, L.L.; Ren, X.P.; Bai, X.W.; Zheng, R.; Li, J.Z.; Liu, H. Psychological typhoon eye in the 2008 Wenchuan earthquake. *PLoS ONE* **2009**, *4*, e4964. [CrossRef]
17. Chen, Y.; Feng, J.; Chen, A.; Lee, J.E.; An, L. Risk perception of COVID-19: A comparative analysis of China and South Korea. *Int. J. Disast. Risk Res.* **2021**, *61*, 102373. [CrossRef]
18. Li, S.; Rao, L.L.; Bai, X.W.; Zheng, R.; Ren, X.P.; Li, J.Z.; Zhang, K. Progression of the "psychological typhoon eye" and variations since the Wenchuan earthquake. *PLoS ONE* **2010**, *5*, e9727. [CrossRef]

19. Zheng, R.; Rao, L.L.; Zheng, X.L.; Cai, C.; Wei, Z.H.; Xuan, Y.H.; Li, S. The more involved in lead-zinc mining risk the less frightened: A psychological typhoon eye perspective. *J. Environ. Psychol.* **2015**, *44*, 126–134. [CrossRef]
20. Ha, T.M.; Shakur, S.; Do, K.H.P. Linkages among food safety risk perception, trust and information: Evidence from Hanoi consumers. *Food Control* **2020**, *110*, 106965. [CrossRef]
21. Guo, Y.; Li, Y. Online amplification of air pollution risk perception: The moderating role of affect in information. *Inform. Commun. Soc.* **2018**, *21*, 80–93. [CrossRef]
22. Chisty, M.A.; Islam, M.A.; Munia, A.T.; Rahman, M.M.; Rahman, N.N.; Mohima, M. Risk perception and information-seeking behavior during emergency: An exploratory study on COVID-19 pandemic in Bangladesh. *Int. J. Disaster Risk Res.* **2021**, *65*, 102580. [CrossRef]
23. Tazelaar, F.; Snijders, C. Operational risk assessments by supply chain professionals: Process and performance. *J. Oper. Manag.* **2013**, *31*, 37–51. [CrossRef]
24. Hult, G.T.M.; Craighead, C.W.; Ketchen, D.J., Jr. Risk uncertainty and supply chain decisions: A real options perspective. *Decis. Sci.* **2010**, *41*, 435–458. [CrossRef]
25. Tse, Y.K.; Chung, S.H.; Pawar, K.S. Risk perception and decision making in the supply chain: Theory and practice. *Ind. Manag. Data Syst.* **2018**, *118*, 1322–1326. [CrossRef]
26. Singh, N.P.; Hong, P.C. Impact of strategic and operational risk management practices on firm performance: An empirical investigation. *Eur. Manag. J.* **2020**, *38*, 723–735. [CrossRef]
27. Aven, T.; Renn, O. On risk defined as an event where the outcome is uncertain. *J. Risk Res.* **2009**, *12*, 1–11. [CrossRef]
28. Rundmo, T. Associations between affect and risk perception. *J. Risk Res.* **2002**, *5*, 119–135. [CrossRef]
29. Slovic, P.; Fischhoff, B.; Lichtenstein, S. Facts and fears: Understanding perceived risk. In *Societal Risk Assessment*; Springer: Boston, MA, USA, 1980; pp. 181–216.
30. Polyviou, M.; Rungtusanatham, M.J.; Reczek, R.W.; Knemeyer, A.M. Supplier non-retention post disruption: What role does anger play? *J. Oper. Manag.* **2018**, *61*, 1–14. [CrossRef]
31. Vanpoucke, E.; Ellis, S.C. Building supply-side resilience–a behavioural view. *Int. J. Oper. Prod. Manag.* **2019**, *40*, 11–33. [CrossRef]
32. Spaccatini, F.; Pancani, L.; Richetin, J.; Riva, P.; Sacchi, S. Individual cognitive style affects flood-risk perception and mitigation intentions. *Appl. Soc. Psychol.* **2021**, *51*, 208–218. [CrossRef]
33. Maderthaner, R.; Guttmann, G.; Swaton, E.; Otway, H.J. Effect of distance upon risk perception. *J. Appl. Psychol.* **1978**, *6*, 380–382. [CrossRef]
34. Tilt, B. Perceptions of risk from industrial pollution in China: A comparison of occupational groups. *Hum. Org.* **2006**, *65*, 115–127. [CrossRef]
35. Xie, X.F.; Stone, E.; Zheng, R.; Zhang, R.G. The 'Typhoon Eye Effect': Determinants of distress during the SARS epidemic. *J. Risk Res.* **2011**, *14*, 1091–1107. [CrossRef]
36. Lateef, T.; Chen, J.; Tahir, M.; Lateef, T.A.; Chen, B.Z.; Li, J.; Zhang, S.X. Typhoon eye effect versus ripple effect: The role of family size on mental health during the COVID-19 pandemic in Pakistan. *Glob. Health* **2021**, *17*. [CrossRef] [PubMed]
37. Ellis, S.C.; Henry, R.M.; Shockley, J. Buyer perceptions of supply disruption risk: A behavioral view and empirical assessment. *J. Oper. Manag.* **2010**, *28*, 34–46. [CrossRef]
38. Kull, T.J.; Oke, A.; Dooley, K.J. Supplier selection behavior under uncertainty: Contextual and cognitive effects on risk perception and choice. *Decis. Sci.* **2014**, *45*, 467–505. [CrossRef]
39. Wachinger, G.; Renn, O.; Begg, C.; Kuhlicke, C. The risk perception paradox—Implications for governance and communication of natural hazards. *Risk Anal.* **2013**, *33*, 1049–1065. [CrossRef]
40. Wang, M.; Keller, C.; Siegrist, M. The less you know, the more you are afraid of—A survey on risk perceptions of investment products. *J. Behav. Financ.* **2011**, *12*, 9–19. [CrossRef]
41. Letzring, T.D.; Wells, S.M.; Funder, D.C. Information quantity and quality affect the realistic accuracy of personality judgment. *J. Personal. Soc. Psychol.* **2006**, *91*, 111–123. [CrossRef]
42. Yang, S.W.; Xu, M.X.; Kuang, Y.; Ding, Y.; Lin, Y.X.; Wang, F.; Li, S. An Agenda-Setting Account for Psychological Typhoon Eye Effect on Responses to the Outbreak of COVID-19 in Wuhan. *Int. J. Environ. Res. Public Health* **2023**, *20*, 4350. [CrossRef]
43. Yoon, J.; Talluri, S.; Rosales, C. Procurement decisions and information sharing under multi-tier disruption risk in a supply chain. *Int. J. Prod. Res.* **2020**, *58*, 1362–1383. [CrossRef]
44. Hendricks, K.B.; Singhal, V.R. An empirical analysis of the effect of supply chain disruptions on long-run stock price performance and equity risk of the firm. *Prod. Oper. Manag.* **2005**, *14*, 35–52. [CrossRef]
45. Schmidt, W.; Raman, A. *When Supply-Chain Disruptions Matter. Working Paper No. 13-006 (July 20, 2012)*; Harvard Business School: Cambridge, MA, USA, 2012; Available online: www.hbs.edu/research/pdf/13-006.pdf (accessed on 13 November 2012).
46. Wang, L.; Foerstl, K.; Zimmermann, F. Supply chain risk management in the automotive industry: Cross-functional and multi-tier perspectives. In *Dynamic and Seamless Integration of Production, Logistics and Traffic*; Springer: Cham, Switzerland, 2017; pp. 119–144.
47. Chatterjee, S.; Chaudhuri, R.; Vrontis, D. Role of fake news and misinformation in supply chain disruption: Impact of technology competency as moderator. *Ann. Oper. Res.* **2022**, 1–24. [CrossRef] [PubMed]
48. Albarracin, D.; Wyer, R.S., Jr. The cognitive impact of past behavior: Influences on beliefs, attitudes, and future behavioral decisions. *J. Personal. Soc. Psychol.* **2000**, *79*, 5. [CrossRef]

49. Evans, J.S.B.; Barston, J.L.; Pollard, P. On the conflict between logic and belief in syllogistic reasoning. *Mem. Cogn.* **1983**, *11*, 295–306. [CrossRef] [PubMed]
50. Thompson, V.A. Reasoning from false premises: The role of soundness in making logical deductions. *Can. J. Exp. Psychol.* **1996**, *50*, 315–319. [CrossRef]
51. Torrens, D. Individual differences and the belief bias effect: Mental models, logical necessity, and abstract reasoning. *Think. Reason* **1999**, *5*, 1–28. [CrossRef]
52. Thalmann, A.T.; Wiedemann, P.M. Beliefs and emotionality in risk appraisals. *J. Risk Res.* **2006**, 453–466. [CrossRef]
53. Vaughan, E.; Nordenstam, B. The perception of environmental risks among ethnically diverse groups. *J. Cross-Cult. Psychol.* **1991**, *22*, 29–60. [CrossRef]
54. Siegrist, M.; Keller, C.; Kastenholz, H.; Frey, S.; Wiek, A. Laypeople's and experts' perception of nanotechnology hazards. *Risk Anal.* **2007**, *27*, 59–69. [CrossRef]
55. Nikookar, E.; Yanadori, Y. Preparing supply chain for the next disruption beyond COVID-19: Managerial antecedents of supply chain resilience. *Int. J. Oper. Prod. Manag.* **2022**, *42*, 59–90. [CrossRef]
56. Sitkin, S.B.; Pablo, A.L. Reconceptualizing the determinants of risk behavior. *Acad. Manag. Rev.* **1992**, *17*, 9–38. [CrossRef]
57. Zsidisin, G.A.; Wagner, S.M. Do perceptions become reality? The moderating role of supply chain resiliency on disruption occurrence. *J. Bus. Logist.* **2010**, *36*, 70–90. [CrossRef]
58. Chae, S.; Lawson, B.; Kull, T.J.; Choi, T. To insource or outsource the sourcing? A behavioral investigation of the multi-tier sourcing decision. *Int. J. Oper. Prod. Manag.* **2019**, *39*, 385–405. [CrossRef]
59. Durach, C.F.; Wiengarten, F.; Choi, T.Y. Supplier-supplier coopetition and supply chain disruption: First-tier supplier resilience in the tetradic context. *Int. J. Oper. Prod. Manag.* **2020**, *40*, 1041–1065. [CrossRef]
60. Carmi, N.; Kimhi, S. Further than the eye can see: Psychological distance and perception of environmental threats. *Hum. Ecol. Risk Assess.* **2015**, *21*, 2239–2257. [CrossRef]
61. Chung, M.; Lim, Y.S. When health organization answers the question: Differential effects of dialogic messages in website and Twitter through social presence and psychological distance. *Health Commun.* **2020**, *37*, 685–695. [CrossRef]
62. Geng, L.; Liu, T.; Zhou, K.; Yang, G. Can power affect environmental risk attitude toward nuclear energy? *Energy Policy* **2018**, *113*, 87–93. [CrossRef]
63. Liberman, N.; Trope, Y. The psychology of transcending the here and now. *Science* **2008**, *322*, 1201–1205. [CrossRef]
64. Birkie, S.E.; Trucco, P. Do not expect others do what you should! Supply chain complexity and mitigation of the ripple effect of disruptions. *Int. J. Logist. Manag.* **2020**, *31*, 123–144. [CrossRef]
65. Ozkul, A.; Barut, M. Measuring supply chain relationships: A social network approach. *Int. J. Integr. Supply Manag.* **2009**, *5*, 38–61. [CrossRef]
66. Chu, H.; Yang, J.Z. Risk or efficacy? How psychological distance influences climate change engagement. *Risk Anal.* **2020**, *40*, 758–770. [CrossRef]
67. Keller, K.L.; Staelin, R. Effects of quality and quantity of information on decision effectiveness. *J. Consum. Res.* **1987**, *14*, 200–213. [CrossRef]
68. Janjua, N.K.; Nawaz, F.; Prior, D.D. A fuzzy supply chain risk assessment approach using real-time disruption event data from Twitter. *Enterp. Inf. Syst.* **2021**, *17*, 1959652. [CrossRef]
69. Hung, K.T.; Tangpong, C. General risk propensity in multifaceted business decisions: Scale development. *J. Manag. Iss.* **2010**, *22*, 88–106.
70. Anderson, J.C.; Gerbing, D.W. Structural equation modeling in practice: A review and recommended two-step approach. *Psychol. Bull.* **1988**, *103*, 411–423. [CrossRef]
71. Myers, M.G.; Stein, M.B.; Aarons, G.A. Cross validation of the Social Anxiety Scale for Adolescents in a high school sample. *J. Anxiety Disord.* **2002**, *16*, 221–232. [CrossRef] [PubMed]
72. Hair, J.F.; Black, W.C.; Babin, B.J. *Multivariate Data Analysis: A Global Perspective: International Version*, 7th ed.; Pearson Education: London, UK, 2013.
73. Hu, L.T.; Bentler, P.M. Cutoff criteria for fit indexes in covariance structure analysis: Conventional criteria versus new alternatives. *Struct. Equ. Model. A Multidiscip. J.* **1999**, *6*, 1–55. [CrossRef]
74. Iacobucci, D. Structural equations modeling: Fit indices, sample size, and advanced topics. *J. Consum. Psychol.* **2010**, *20*, 90–98. [CrossRef]
75. Fornell, C.; Larcker, D.F. Evaluating structural equation models with unobservable variables and measurement error. *J. Mark. Res.* **1981**, *18*, 39–50. [CrossRef]
76. Podsakoff, P.M.; MacKenzie, S.B.; Lee, J.Y.; Podsakoff, N.P. Common method biases in behavioral research: A critical review of the literature and recommended remedies. *J. Appl. Psychol.* **2003**, *88*, 879–903. [CrossRef] [PubMed]
77. Hayes, A.F. *Methodology in the Social Sciences. Introduction to Mediation, Moderation, and Conditional Process Analysis: A Regression-Based Approach*; Guilford Press: New York, NY, USA, 2013.
78. Aiken, L.S.; West, S.G. *Multiple Regression: Testing and Interpreting Interactions*; Sage: Thousand Oaks, CA, USA, 1991.
79. Li, S.; Li, J.L.; Yang, S.W.; Wu, X.J.; Chen, J.F.; Ding, Y.; Zheng, R. The psychological typhoon eye effect in responses to terrorism. *J. Pac. Rim Psychol.* **2020**, *14*, e21. [CrossRef]

80. Cantor, D.E.; Blackhurst, J.V.; Cortes, J.D. The clock is ticking: The role of uncertainty, regulatory focus, and level of risk on supply chain disruption decision making behavior. *Transp. Res. Part E Logist.* **2014**, *72*, 159–172. [CrossRef]
81. Mena, C.; Humphries, A.; Choi, T.Y. Toward a theory of multi-tier supply chain management. *J. Supply Chain Manag.* **2013**, *49*, 58–77. [CrossRef]
82. Wang-Mlynek, L.; Foerstl, K. Barriers to multi-tier supply chain risk management. *Int. J. Logist. Manag.* **2020**, *31*, 465–487. [CrossRef]
83. Ivanov, D.; Sokolov, B.; Dolgui, A. The Ripple effect in supply chains: Trade-off 'efficiency-flexibility-resilience' in disruption management. *Int. J. Prod. Res.* **2014**, *52*, 2154–2172. [CrossRef]
84. Marchese, K.; Paramasivam, S. The Ripple Effect How manufacturing and retail executives view the growing challenge of supply chain risk. *Deloitte DEV LLC.* **2013**.
85. Park, Y.W.; Blackhurst, J.; Paul, C.; Scheibe, K.P. An analysis of the ripple effect for disruptions occurring in circular flows of a supply chain network. *Int. J. Prod.* **2022**, *60*, 4693–4711. [CrossRef]
86. Sjöberg, L. Worry and risk perception. *Risk Anal.* **1998**, *18*, 85–93. [CrossRef]
87. Plapp, T.; Werner, U. Understanding risk perception from natural hazards: Examples from Germany. In *RISK21-Coping with Risks Due to Natural Hazards in the 21st Century*; CRC Press: Boca Raton, FL, USA, 2006; pp. 111–118.
88. Sarkar, S.; Kumar, S. A behavioral experiment on inventory management with supply chain disruption. *Int. J. Prod. Econ.* **2015**, *169*, 169–178. [CrossRef]
89. Gao, J.; Zhang, C.; Wang, K.; Ba, S.L. Understanding online purchase decision making: The effects of unconscious thought, information quality, and information quantity. *Decis. Support Syst.* **2012**, *53*, 772–781. [CrossRef]
90. Dane, E.; Pratt, M.G. Exploring intuition and its role in managerial decision making. *Acad. Manag. Rev.* **2007**, *32*, 33–54. [CrossRef]
91. Messina, D.; Barros, A.C.; Soares, A.L.; Matopoulos, A. An information management approach for supply chain disruption recovery. *Int. J. Logist. Manag.* **2020**, *31*, 489–519. [CrossRef]
92. Kumar, S.; Himes, K.J.; Kritzer, C.P. Risk assessment and operational approaches to managing risk in global supply chains. *J. M. Technol. Manag.* **2014**, *25*, 873–890. [CrossRef]
93. Mousavi, S.; Gigerenzer, G. Risk, uncertainty, and heuristics. *J. Bus. Res.* **2014**, *67*, 1671–1678. [CrossRef]
94. Bode, C.; Wagner, S.M.; Petersen, K.J.; Ellram, L.M. Understanding responses to supply chain disruptions: Insights from information processing and resource dependence perspectives. *Acad. Manag. J.* **2011**, *54*, 833–856. [CrossRef]
95. Ellinger, A.E.; Chen, H.; Tian, Y.; Armstrong, C. Learning orientation, integration, and supply chain risk management in Chinese manufacturing firms. *Int. J. Logist-Res. App.* **2015**, *18*, 476–493. [CrossRef]
96. Chopra, S.; Sodhi, M.; Lücker, F. Achieving supply chain efficiency and resilience by using multi-level commons. *Decis. Sci.* **2021**, *52*, 817–832. [CrossRef]
97. Basole, R.C.; Bellamy, M.A. Supply network structure, visibility, and risk diffusion: A computational approach. *Decis. Sci.* **2014**, *45*, 753–789. [CrossRef]
98. Cao, S.F.; Bryceson, K.; Hine, D. Improving supply chain risk visibility and communication with a multi-view risk ontology. *Supply Chain Forum* **2020**, *21*, 1–15. [CrossRef]
99. Li, Y.; Zobel, C.W.; Russell, R.S. Value of supply disruption information and information accuracy. *J. Purch. Supply Manag.* **2017**, *23*, 191–201. [CrossRef]
100. Liu, M.; Zhang, H.; Huang, H. Media exposure to COVID-19 information, risk perception, social and geographical proximity, and self-rated anxiety in China. *BMC Public Health* **2020**, *20*, 1649. [CrossRef] [PubMed]
101. Lu, P.; Kong, D.; Shelley, M. Risk perception, preventive behavior, and medical care avoidance among American older adults during the COVID-19 pandemic. *J. Aging Health* **2021**, *33*, 577–584. [CrossRef] [PubMed]
102. Lee, S.; Lee, N.; Dockter, C.E. Effects of message presentation type on gm food risk perception, similarity judgment, and attitude. *Health Commun.* **2021**, *36*, 1666–1676. [CrossRef]
103. Tversky, A.; Kahneman, D. The framing of decisions and the psychology of choice. *Science* **1981**, *211*, 453–458. [CrossRef]
104. Gupta, M.; Gupta, S. Influence of national cultures on operations management and supply chain management practices-a research agenda. *Prod. Oper. Manag.* **2019**, *28*, 2681–2698. [CrossRef]
105. DuHadway, S.; Carnovale, S.; Kannan, V.R. Organizational communication and individual behavior: Implications for supply chain risk management. *J. Supply Chain Manag.* **2018**, *54*, 3–19. [CrossRef]

MDPI
St. Alban-Anlage 66
4052 Basel
Switzerland
www.mdpi.com

Sustainability Editorial Office
E-mail: sustainability@mdpi.com
www.mdpi.com/journal/sustainability

www.ingramcontent.com/pod-product-compliance
Lightning Source LLC
Chambersburg PA
CBHW040808250926
51850CB00007BA/1111
* 9 7 8 3 0 3 6 5 8 7 5 8 5 *